America's Environmental Legacies

Franklin Kalinowski

America's Environmental Legacies

Shaping Policy through Institutions and Culture

Franklin Kalinowski
Warren Wilson College
Morganton, North Carolina, USA

ISBN 978-1-349-95670-8   ISBN 978-1-349-94898-7 (eBook)
DOI 10.1057/978-1-349-94898-7

© The Editor(s) (if applicable) and The Author(s) 2016
Softcover reprint of the hardcover 1st edition 2016
This work is subject to copyright. All rights are solely and exclusively licensed by the Publisher, whether the whole or part of the material is concerned, specifically the rights of translation, reprinting, reuse of illustrations, recitation, broadcasting, reproduction on microfilms or in any other physical way, and transmission or information storage and retrieval, electronic adaptation, computer software, or by similar or dissimilar methodology now known or hereafter developed.
The use of general descriptive names, registered names, trademarks, service marks, etc. in this publication does not imply, even in the absence of a specific statement, that such names are exempt from the relevant protective laws and regulations and therefore free for general use.
The publisher, the authors and the editors are safe to assume that the advice and information in this book are believed to be true and accurate at the date of publication. Neither the publisher nor the authors or the editors give a warranty, express or implied, with respect to the material contained herein or for any errors or omissions that may have been made.

Cover image © Premium Stock Photography GmbH / Alamy

Printed on acid-free paper

This Palgrave Macmillan imprint is published by Springer Nature
The registered company is Nature America Inc. New York

*This book is dedicated to
Pomatomus saltatrix, Branta canadensis,
and especially to my old friend,
Odocoileus virginianus*

Acknowledgments

For an author, one of the pleasures of finishing a book is the opportunity to reflect back over their life, think of all those who helped and encouraged the writing effort, and acknowledge the gratitude owed to these mentors, colleagues, friends, and partners. In my case, the list must start with those teachers who shaped my thinking and pointed me toward a life of scholarship. Paul Dolan, Paul Shepard, and especially the late John Rodman made imprints on my character that can be found on every page of this book. During the years of actual writing, John Casey and Phil Otterness provided the stimulating conversations and friendly collegiality that helped keep me focused on the project. My deepest appreciation goes to Sandra Hinchman and William Vitek, whose guidance and encouragement went far beyond the usual assistance of colleagues. Sandy and Bill read chapters, made suggestions, sat on conference panels, and comforted my too-easily-bruised ego as the effort slowly matured to fruition. They would be the first to point out that the thinking in this book does not always conform to their opinions, and I hereby accept that the arguments and mistakes contained within are exclusively my own, but their support was invaluable. I cannot imagine this project would have been completed without their efforts. There is one person who does not fit into any clear category, for during the more than 40 years of our relationship David Ingersoll has been my professor, my graduate thesis advisor, my department chairman, and my enduring friend. Acknowledging my many unpaid debts to him is a task I assume with respect and humility. Other friends who have accompanied me on my life's journey include Jerry Conti, Eric Zencey, and Randy Murphy, all of whom have left impressions on my thinking and writing. Through all of this, the questioning, debate, and support provided by my many, many former students have made my life as a college professor a joyous calling. More than any other, however, it is the love and support of my dear wife that most needs acknowledgment. While I sat at my computer and wrote, Judy went out and worked every day to protect and preserve the Natural surroundings we both love so much. My deepest hope is that my small effort can somehow combine with her real accomplishments to move our land in a better direction toward long-lasting harmony.

Franklin Kalinowski
Morganton, NC, USA

Contents

1 Introduction: Interpreting America's Two
   Constitutions—Looking Through an Environmental Lens   1

Part I   Environmental Politics and Our
         Written Constitution   25

2 The Scope and Limits of Mainstream
   Environmentalism   27

3 Radical Environmentalism: Challenging Our
   Institutions and Beliefs   61

4 The Constitution and the Environment: Ecological
   Principles and Liberal Policy Making   105

Part II Cultural Legacies: America's Unwritten
         Constitution   139

5 Political Theory and the American Founding:
   The Tension Between Logic and History   141

6   The Environmental Legacy of Thomas Jefferson:
    Cultivating the Rooted Citizen   177

7   The Environmental Legacy of Alexander Hamilton:
    Manufacturing Power from Delusion   215

8   The Environmental Legacy of James Madison:
    Pursuing Stability in a World of Limits   253

Part III   The Heritage: Environmentalism and
           the Evolving Constitution   285

9   The Constitution After 100 Years: Environmental
    Theory in the Gilded Age   287

10 Living with the Legacies: Our Culture Confronts Our
    Environment   323

Index   357

CHAPTER 1

Introduction: Interpreting America's Two Constitutions—Looking Through an Environmental Lens

From time to time, usually during periods of social upheaval, Americans are drawn to reinvestigate and reinterpret their Constitution. This periodic rethinking of our origins is a good thing, for when we take on this examination, citizens of the United States learn something about their institutions and their culture: we come to see them through original perspectives and in a new light. No matter what the causal apprehensions or the resulting scholarship, the reinterpretation of the Constitution breathes fresh life into our understanding of that Founding Document and the culture that supports it, and for those reasons alone, justifies the occasional renewal of analysis. At the same time that we learn something about the early creation of our republic, we also learn something about the present condition of the American psyche; for whatever trends or concerns precipitate the reinterpretation, these will likely be the issues upon which the innovative exploration will concentrate. During this *ridurre ai principii* (as Niccolo Machiavelli would call it), Americans can gain a clearer appreciation of their contemporary anxieties over the direction in which their Founding Document leads them. The Constitution in this sense becomes a cultural mirror where we can view ourselves more clearly and investigate the social blemishes that may be signs of deeper, more serious, infection. Previous interpretations arose out of economic crises in this nation's past and explored the impact of the Constitution on America's search for economic justice. Other historical writers have perceived the country was in a political emergency and reinterpreted the Constitution in an attempt to flesh out the democratic, or anti-democratic, tendencies within it.[1]

There is evidence in the form of many people arguing the existence of an environmental or ecological crisis that we are presently living during a period that calls for another such deep, critical assessment. To the extent this is the case, we should expect to see reinterpretations of the American Constitution with a specific focus on the relationship between politics, institutions, and culture on the one hand, and the nonhuman environment on the other. Such an interpretation would need to be broadly interdisciplinary and synthesizing. It would take the many analyses of our current environmental situation and unite them with the several prominent interpretations of American political structures and national values. An interpretation is an explanation from a particular perspective. It is a way of viewing certain events, facts, or descriptions and placing them within some larger context in order to flesh out meaning and consequences.

Of course, there are a great number of people who deny the existence of serious environmental problems, and they certainly reject the notion of a widespread ecological crisis. For these individuals and their supporting groups, the entire environmental debate is a well-orchestrated hoax perpetuated by left-leaning alarmists who are out to undermine our economic system and further a political agenda of increased governmental interference in what these conservatives see as our perfectly operating free market. At some level, these environmental deniers might grasp the connections between ecological dangers, economic structures, and political responses, but their apprehensions regarding those responses lead them to build an interpretation of environmental politics upon the rejection of any evidence that ecological problems confront us. But all interpretations are not equally supportable: they are more than mere opinions. Explanations of social events that rest upon the rejection of the best, most current scientific evidence assume a large, blind, and rationally unsupportable faith in the status quo that becomes weaker and less tolerable with every new ecological disaster or scientific report.

The result, as we shall see, is a multitude of interpretations regarding humanity's relationship with its environment stretching from this cornucopian belief that no significant environmental problems exist and that abundant, clean, never-ending exponential growth is our future to deep ecology's dread that an ecological apocalypse is imminent unless humanity fundamentally reorders its outlook toward Nature. There are so many environmental interpretations, in fact, that one begins to wonder if all these cacophonous antagonists are really talking about the same planet. A review of the voluminous literature on humanity's relationship to the environment reveals at least ten significant interpretations, perspectives, narratives, or "discourses." These environmental interpretations lead either to the conviction that no problems exist, or to the assurance that difficulties exist but are manageable within our current set of problem-solving institutions, or to the belief that a crisis is unfolding which will cause us to fundamentally reconsider our social arrangements. Sorting out and understanding all of this can be made a bit less complicated if we focus on the various explanations, the proponents and groups associated with each, and if we direct our attention to some basic questions each interpretation must address. For each perspective or discourse, we need to specify and make explicit first, the severity (or lack thereof) of environmental dangers as each sees it; second, the interconnection both among various environmental dangers and among these threats and our economic and political policies; and third, our nation's ability to resolve these issues within our prevailing institutions and the social values that comprise the ethical fabric of American culture. Dangers, interconnections, and responses comprise the heart of an environmental interpretation. Looking at the American Constitution through the lens of these interpretations will permit us to see our Founding Document in a new way. It will also permit us to judge the adequacy of this eighteenth-century document and its supporting set of values, beliefs, and habits for the times in which current, and future, Americans live and will live. To this extent, our present and future cannot be fully understood without also considering our past. How we as a nation interpret and respond to various threats becomes, over time, woven into the fabric of our shared communal record. It becomes an integral part of our history, and the legacy of those responses shapes and conditions our reactions to contemporary challenges. There is, therefore, an interconnected pattern that unites science, economics, politics, and history, and although this pattern may be extremely complicated, it is, nevertheless, comprehensible. The key, at first, is to analyze dangers and look for connections.

A contemporary reader, scouring the literature on environmental topics, will find a host of excellent studies on the various dangers we face, the scientific evidence underpinning the issues, and (in some cases) the connections among differing ecological threats. Rather than a detailed study of risks and connections, the book you are currently reading will focus on one big issue in the environmental debate: the capacity of the American political system to respond to these numerous ecological challenges. Before inquiring into this capacity to respond, however, some appreciation of the dangers and connections seems called for. Species extinction, human economic intrusion into the habitats of native plants and animals, and the unending increase of the human population are one set of connections. Another, not unrelated, set of connections we might look at is climate change, the contribution the burning of fossil fuels makes to warming the planet, the seemingly unquenchable need for energy consumption by our economy, and the links between energy, economic growth, global economic instability, and political unrest. These circumstances pose threats that are significant in their own right, but become compounded when tied together in a pattern of effects, causes, and feedbacks. Among the various researchers who have addressed environmental concerns, no one has done a better job summarizing the dangers and interconnections than Oberlin College Professor David Orr. Widely regarded as America's foremost environmental educator, Orr has authored a number of books that spell out the threats this nation faces. His 2009 offering, <u>Down to the Wire: Confronting Climate Collapse</u>, is particularly helpful in pointing out the environmental, economic, and political dangers we face, demonstrating how they fundamentally interconnect with each other, and critically analyzing the negative consequences of either societal inaction or shallow responses that fail to probe to the sources of the risks. "[T]he hardest tests for our Constitution and democracy are just ahead," he asserts, "and have to do with the relationship between governance, politics, and the dramatic changes in Earth systems now under way."[2]

The first, and arguably the most widely discussed, environmental challenge is the threat to ecosystem stability and human civilization posed by the alteration of climate patterns currently being brought on by the burning of fossil fuels, the release of carbon dioxide (CO_2) into the atmosphere, and the resulting greenhouse effect that is warming the planet to levels not experienced for thousands of years. With each new scientific report, the evidence becomes stronger and more undeniable that humanity is tampering with the Earth's climate in unprecedented and dangerous ways. For hundreds of thousands (perhaps millions) of years, the amount of CO_2 in the atmosphere remained fairly constant at approximately 280 parts per million (ppm). Since the onset of the Industrial Revolution and the subsequent burning of coal, oil, and natural gas, that level has steadily, and exponentially, risen to over 400 ppm of atmospheric CO_2. These human-caused changes are scientifically verifiable. From these empirical facts, respected experts, such as James Hansen and his colleagues, have concluded that the "safe" concentration of CO_2 is somewhere around 350 ppm.[3] In other words, we have already overshot the long-term CO_2 levels consistent with a stable climate. Responsible governance demands action to halt the increase in greenhouse gas emissions and move toward the reduction of CO_2 in the atmosphere. The impacts on the economy of reducing current CO_2 to this safe level are subject to controversy, and the political policies needed to make this reduction are even more contentious, but the actualities of rising greenhouse gas emissions and the subsequent climate change being produced are no longer matters subject to differing interpretations. These are scientific facts and those who deny them deny reality.

The second ecological challenge to governance is the breakdown of ecosystems and the ecological services they provide. In addition to climate change, the destruction of wetlands, pollution, and human population encroachment into previously wild areas are impairing, or causing us to lose, such vital aids to civilization as flood control, water purification and transport, pollination services, waste absorption, pest control, an incalculable storehouse of genetic information, and a host of other benefits that society could neither replace nor do without.[4] There are interconnections and feedbacks here: climate change and other factors are degrading ecosystems, degraded ecosystems make our ecological and social systems less adaptive and resilient, and these damaged systems make it more difficult for us to deal with challenges such as climate change.

The third challenge we face is the peaking of global oil extraction with the subsequent collapse of the fossil fuel–based economy. America's withdrawal of oil from traditional land and offshore reserves reached its yearly maximum in 1972. As predicted by many petroleum geologists using the model developed by M. King Hubbert, world petroleum extraction hit its highest point in 2008, but in the last few years, events have reworked the debate over peak oil and world energy supplies. Spokesmen for the oil and gas industry have declared that a bonanza in "unconventional energy resources" such as shale gas, tight oil (shale oil), and tar sands will usher in an era of energy abundance capable of fueling exponential economic growth for another century and making the USA an energy exporter comparable to the Arab Gulf States. There are, however, reasons to be skeptical regarding these declarations. Extraction techniques associated with shale gas, notably hydraulic fracturing (or "fracking") of deep underground reserves, are fraught with environmental risks; prospective sites, or "plays," for both shale gas and tight oil have very high decline rates as production drops off rapidly after the initial drilling; and the extreme costs (both economic and environmental) of energy from tar sands make that resource very low in "net-energy" or "energy return on energy invested." Noted Canadian geoscientist J. David Hughes has studied these claims of forthcoming energy abundance and declared, "the projections by pundits and some government agencies that these technologies can provide endless growth heralding a new era of 'energy independence', in which the U.S. will become a substantial net exporter of energy are entirely unwarranted based on the fundamentals. At the end of the day, fossil fuels are finite and these exuberant forecasts will prove to be extremely difficult or impossible to achieve."[5]

There are other important considerations involved in this debate over conventional and unconventional energy sources, however, which a focus on environmental, economic, and political connections would bring to the surface. Bill McKibben, environmental activist and resident scholar at Middlebury College, has performed calculations that expose dilemmas surrounding climate change, energy extraction, and vulnerabilities in our economic system. Given the amount of CO_2 humans have already pumped into the atmosphere, the planet is currently committed to a 1.8 degree Fahrenheit (1° Celsius) rise in average global temperature. This amount of climate change will have significant and dangerous impacts. From these facts, a large and increasing number of scientists and political leaders have concluded that a 3.6° F (2° C) increase represents the upper limit beyond which "we're entering the guaranteed-catastrophe zone."[6] Those who are warning we must keep global temperature increase below this 3.6° F limit include the Major Economies Forum (a group comprising all countries that burn vast amounts of carbon), the participants at the Copenhagen Conference on Climate Change, German Chancellor Angela Merkel, and President Barack Obama. The concern is the amount of additional CO_2 it would take to generate this unacceptable increase. We have reached the point where we must ask, what is the maximum quantity of additional CO_2 humans can add to the atmosphere before we cross this "guaranteed-catastrophe" line? That figure, according to reputable geophysicists, is 565 gigatons of CO_2 over the next 40 years. These, then, are the critical numbers that measure the dangers of climate change: 350 ppm of CO_2 in the atmosphere is the "safe" level (we have already crossed that line, which is why we are going to have the 1.8° F increase); a 3.6° F rise in average global temperature would seriously threaten civilization; and 565 gigatons more CO_2 in the atmosphere (over the next 40 years) would generate that 3.6° F increase. To give this some perspective: a gigaton is one billion tons. Each year, Americans account for the release of 22 tons of CO_2 per person. With a population of around 320 million, that means Americans release about 7 gigatons of CO_2 per year. Even if we could adopt policies to hold our population and our CO_2 emissions at their current levels (two immense undertakings for our political system), Americans would still release 280 gigatons of CO_2, or almost half of the 565 gigaton target for the entire world, into the atmosphere over the next 40 years.[7] Put simply, our institutions of governance and economics face an enormous challenge in keeping CO_2 releases below that perilous 565 gigaton level.

This is the point where McKibben draws the connection between climate change and energy extraction, for according to sources within the world's fossil fuel companies, they already have located enough coal, oil, natural gas, tight oil, and tar sands to generate, if extracted and burned, 2795 gigatons of CO_2.[8] That is five times the amount of CO_2 scientists say it is sensible to release. McKibben states the conclusion with simple arithmetical clarity: in order to avoid climate catastrophe, 80 % of the world's currently known fossil fuel reserves must be kept in the ground. Transitioning to a post-carbon society based upon alternative energy fuels is essential if we are to restrict and lower CO_2 emissions into the atmosphere. From this perspective, it is almost irrelevant whether or not vast reserves of fossil fuels remain to be extracted. If the withdrawal of fossil fuels has peaked, we need to transition away from a reliance on these diminishing sources. If reserves have not peaked and large quantities of unconventional fossil fuels remain, we must leave the vast majority of them in the ground in order to hold CO_2 at, or below, the upper limit of 565 gigatons. In either case, the era of cheap, abundant coal, oil, and natural gas has come to an end.

But, once again, if we are looking for connections, we need to ask: what would be the economic and social consequences of keeping 80 % of fossil fuel reserves in the ground? And here, the truths of climate change meet the requisites of corporate finance, for it is a fact of capitalist investment that the value of stocks in these fossil fuel companies have already priced in the extraction of these carbon reserves. Underground reserves are considered corporate assets, and if they were taken off the books, the value of those corporations would be devastated. The current prices of the stock in coal, oil, and gas companies as listed on the New York Stock Exchange, and the other equities exchanges around the world, are based on the assumption that these reserves will be extracted and sold.[9] If, as is ecologically prudent, we were to choose not to extract and burn those reserves, the fossil fuel industry would suffer such an economic loss that they would, no doubt, financially collapse.

McKibben's analysis reveals multiple connections and paradoxes. If fossil fuel extraction has peaked, then our carbon-based economy and social system is headed for an extended period of decline and crisis. If fossil fuel extraction has not peaked (as the optimistic proponents of nonconventional fuels contend), then we are headed for a climate change catastrophe if we choose to "drill, baby, drill," since given our currently dominant cultural values, it is a virtual certainty that if extracted, these fossil fuels will be burned. On the other hand, if we implement policies designed to keep 80 % of known reserves in the ground and dramatically reduce CO_2 emissions, then the fossil fuel industry (and probably a sizable portion of the rest of the economy) will suffer disastrous economic loss. Scylla is climate change calamity, while Charybdis is a financial and economic crisis.

Converting American society from fossil fuels to some mix of alternative energy and dramatically lower energy consumption will place a strain on our political and economic systems that they have probably never faced before, for the sad, but undeniable, truth is that since the end of World War II, the United States has designed housing patterns, transportation infrastructure, and commercial networks based on the illusionary belief that fossil fuels would be safe, cheap, and plentiful for any time into the foreseeable future. As a nation, we built suburban, single-family housing subdivisions that are entirely dependent on the automobile to connect residents to the larger society. Without cars, people living within these single-use zoning patterns are separated from jobs, businesses, stores, and schools and are helpless to carry out the simplest of daily routines. An expansive network of interstate highways furthered our addiction to oil as suppliers and distributors of almost every product became more and more separated from each other, and the trucking industry became the life blood of our economy. Social commentator James Howard Kunstler, in <u>The Long Emergency</u>, has accurately labeled this pattern of suburban houses, freeways, malls, office parks, and fast food eateries "the greatest misallocation of resources in world history."[10] The obstacles to the creation of a sustainable and stable economic system go beyond routine monetary and fiscal policy and lead us to reconsider some of the most fundamental assumptions underpinning our economy.

The continued dependence on debt-funded consumer spending as the engine driving the economy needs also to be questioned. Consumer spending now accounts for 71 % of our gross domestic product (GDP) and should it stagnate or decline, the repercussions on the overall economy would be dire. Recent measurements of consumer confidence indicate, however, that average Americans may be tiring of the "shop til you drop" lifestyle. Coupled with family incomes that have been stagnant, or falling, for a number of years, Americans may be growing weary of filling up their basements, garages, and rented storage spaces with even more stuff. But the economic truth is that without exponentially rising consumer spending, the American economy, as presently constituted, would be in deep trouble. This all points to the central incongruity between our environment and our economic system: that is the absolute requirement within capitalism for a continuous, ever expanding, increase in the output of goods and services. Economic growth is the *sine qua non* of our free market capitalist economy. It is economic growth that allows us to discount the future and receive compound interest on investments. The possibility and desirability of economic growth is the core value that unites all members of America's upper-class elites. Democrats and Republicans may disagree on many issues, but they share a solid consensus that economic growth is an absolute requirement. Beyond the elites, economic growth is the major inducement held out to average Americans that they too can have a higher standard of living in a world where everyone gets more and that our economic system is essentially just. This exponential growth demanded by our economic system is stretching our finite ecosystem to compound breaking points. Our economy, in other words, is being conducted under the premise that commercial activity is disconnected and independent from the ecosystem, when the fact is they are inextricably linked through energy, resources, sinks, and services. More needs to be said about the issue of economic growth and the environment, but for now, suffice it to observe that ignoring, denying, or managing these structural flaws within free market capitalism has been an important part of the mental portfolio of many economists and financiers over the past several decades.

Even the most transparent, honest, and competent financial system would find it difficult to manage these ecological and economic challenges, but recent events may lead us to doubt both the foresight and the integrity of those who direct the activities of our banks, hedge-funds, and brokerage houses. The 2008 financial crisis introduced average Americans to a multitude of financial instruments such as collateralized debt obligations, credit/default swaps, mortgage-backed securities, and derivatives, along with suspect accounting practices between Wall Street and rating agencies ostensibly charged with determining the risk worthiness of financial instruments but functioning instead to confuse and dupe would-be investors. Expressions such as "skin in the game," "too big to fail," and "front-running" became linked to policies with clever acronyms such as "TARP" that left glassy-eyed citizens with the uncomfortable suspicion that all of the complexity was not so much an inevitable consequence of an intricate and innovative business, but rather an immense nefarious scheme to create an illusion of wealth and opportunity which bore little resemblance to the much more modest performance of the actual economy.[11] Even Federal Reserve Chairman Alan Greenspan was forced to admit his assumption that investment institutions operated in the best interest of their clients may not be true. At the base of it all, Wall Street had become disconnected from the economy in a similar way that the economy had become disconnected from the ecosystem.

American political institutions face the daunting task of simultaneously reforming an economic system with deep structural flaws built upon questionable premises and a financial system plagued with corruption. Yet five years after the recent financial meltdown, Massachusetts Senator Elizabeth Warren reports America's four largest banks are 30 % bigger than they were in 2008.[12] The law that was passed, commonly called the Dodd-Frank Act, did pathetically little to change either the structure or mindset within which financial managers operate. In spite of its official title, the "Wall Street Reform and Consumer Protection Act" more accurately demonstrates the power of Washington lobbyists to constrain and manipulate any attempt to seriously address underlying assumptions, practices, and institutions.

These environmental, economic, and financial challenges to governance point to an irony that surrounds all of America's institutions. In spite of our great size and apparently tremendous strength, there is an underlying weakness and fragility to our basic social structure. Protestations to the contrary notwithstanding our financial system, economy, and ecological foundations are so interdependent that anything that would seriously disrupt any one of these would have a negative rippling effect throughout the entire society. An oil refinery fire in Venezuela, a devastating typhoon in Asia, or civil wars in Bosnia or the Middle East can have damaging impacts on the global economy. Without a coherent, timely, and effective response, these various threats expose a lack of flexibility, resilience, and adaptability that makes the overall social system vulnerable, and as almost every edition of the evening news demonstrates, there are numerous points, both domestically and internationally, where that vulnerability can be tested. There is a price to be paid for being the world's largest consumer of resources.

The growth of our economy depends on access to resources and markets throughout the world, which depends on international trade and finance, which, in turn, depends on a stable and predictable global regime. Over the years, these connections and America's vital interest in global stability, coupled with our pre-eminent military power, have driven us to an assorted mix of alliances and interventions. For most of the post–World War II era, the U.S. was quite willing to turn a blind eye to domestic repression committed by despotic regimes as long as those governments maintained friendly relations with us (i.e., they did not tilt toward leftist revolutionaries) and adhered to the openness and predictableness of international economics. The consequence of this policy, which should have been seen by almost everyone, was an enormous amount of resentment felt by the populations of these unfortunate nations. As a result, the urge for liberation became linked with a powerful anti-American animosity. As the demand for freedom exploded into political expression, the resentment was also released. The perceived need to contain this "blowback" has resulted in huge defense budgets, the positioning of 737 US military bases around the world, and wars in Iraq and Afghanistan costing an estimated $1 trillion dollars annually.[13] There are, however, additional factors complicating American foreign policy beyond our apparent national interest in perpetuating economic growth and political stability.

While there is, no doubt, a sizable portion of America's leadership and public who view events through this prism of cold, calculating *Realpolitik*, there is also a powerful strain in our culture that is sensitive to concerns involving universal rights. At least since the time of Woodrow Wilson, this global idealism reacts with revulsion at the sight of the genocidal slaughtering of civilians, the aggressive invasion of weaker neighbors, "ethnic cleansing" of hated minorities, terrorism conducted by religious fanatics, or the use of torture, poison gas, and police brutality. Among these global idealists, the feeling is strong that if the United States can use its moral, economic, or (hopefully as a last resort) military might to prevent or stop this carnage, our national principles demand that we act. Our culture values both self-interest and human rights, and although a concern for rights is usually insufficient in and of itself, when wedded to a real or alleged national interest, leaders can often count on the support of the public. The ecological, economic, and political costs of this intervention might only become revealed later and then a third element in our culture, a long established American parochialism and a wish not to get involved in the affairs of others, will express itself and demands will be voiced against America becoming "the world's policeman." In America, the politics of interest, the politics of rights, and the politics of isolationism coexist in interwoven contradiction. Reconciling these incongruous impulses and policies would challenge even the most rational, calculating, or sensitive system of governance.

These, then, are five challenges to governance as outlined by one recognized authority. It is certainly possible to modify this list or add other threats. Orr's failure to discuss the important issue of human population growth, for example, is clearly a notable and regrettable omission. Still, these five dangers represent a good lens through which to begin an interpretation of the American Constitution. Our capacity to respond to climate change, denigrated ecological services, the peaking of energy supplies, structural and operational flaws in our economic and financial systems, and the interconnected challenges of international politics will severely test our governance system. Although it is unquestionable that more scientific research needs to be conducted and transmitted to the public in a comprehensible manner, once the dangers and interconnections are understood, we must also inquire into the possible public policies that could be formulated and implemented within the decision-making process institutionalized in the American Constitution.

Responding to these multiple dangers will necessitate policies that are based on the best scientific evidence and formulated with a clear eye on the interests of the American people and the ecological setting in which they live. There is certainly no shortage of recommendations. From deregulation and a trusting faith in the free market, to "structured incentives" to tax carbon and other pollutants, to strict command-and-control regulations to proscribe certain behaviors, to substantial commitments to alternative technologies, to a complete redesign of our economic system, to a national population policy, to restrictions on trade and a radical decentralization of our society, to a rejection of the fundamental principles of progressive civilization, each environmental discourse will contain strategies for responding to the dangers and interconnections consistent with their perspective. Are there criteria we could use to choose among these competing and often contradictory policy proposals? At the most fundamental level, policy analysis can inquire whether the proposed response would work; and whether this recommended policy can actually be passed into law and implemented. In other words, we can begin to evaluate recommended policies according to the extent to which they would be effective and whether or not they are feasible.

Of course, the potential effectiveness of a law or policy will vary according to who is doing the analysis, the perspectives and biases they bring to the subject, and the narrowness or breadth of the recommended solutions. Given the seriousness of the challenges America faces, and the interconnections that tie issues together, this last question of policy scope is particularly important. A policy may be ineffective because it does not resolve the problem it was designed to correct, or it may be unsuccessful because in the process of correcting one issue, it creates a new problem at least as serious as the targeted challenge. Finally, a policy may be a failure if it merely shifts the composition and point of impact of a problem without reducing either its severity or its destructiveness. This is sometimes referred to as "tinkering," and recommendations that merely tinker with a challenge need to be exposed for what they are: "solutions" that give the illusion that a challenge has been addressed when, in actuality, little or nothing has actually been accomplished. An example within the context of the current discussion might be helpful.

Recently, the suggestion has been circulated that America might combat climate change and seriously reduce its CO_2 emissions if we switched from coal to natural gas as the primary fuel for generating electricity. Natural gas, as abundant as coal according to some, burns much cleaner than coal and releases considerably less CO_2 into the atmosphere for any given quantity of electricity produced. Public policy—so the argument goes—should aim to increase the extraction of natural gas (e.g., reduce or eliminate restrictions on fracking) and convert existing coal-fired plants to burning natural gas. The argument is sometimes extended to include using natural gas as the fuel source for automobiles and other forms of ground travel. The thesis of this policy proposal is that natural gas is the "clean," "environmentally safe" fossil fuel of the future.

But is it? Additional facts need to be appreciated and the full consequences of the pro-natural gas policy taken into account before the proposal can be fairly evaluated. Most reservoirs of natural gas found around the world are comprised of 80–90 % methane (CH_4) per volume. Raw natural gas must be extensively processed to remove almost all other materials (such as ethane, propane, butane, and other hydrocarbons), leaving pure CH_4 as the final commercial product. Now, here is the significant fact that must be considered. Although it is true that CH_4, when burned, releases far less CO_2 than coal or oil, CH_4 itself—unburned—is 72 times more powerful as a greenhouse gas than CO_2. If only a very small percentage of CH_4 is leaked into the atmosphere from the production, refining, distribution, or dispensing operations (picture tens of thousands of CH_4 filling stations for automobiles), then the ostensible gains from switching to natural gas would be offset or completely disappear. Since it would take less than a 1 % leakage of natural gas to eliminate the alleged benefits of this fossil fuel, natural gas production would have to be virtually faultless—not 90 % or even 98 % faultless—in order to avoid absolutely ruinous consequences. The natural gas policy option assumes we could build and operate a system that was at least 99.5 % perfect. There could be no margin of error; no human or mechanical mistakes could ever occur; no unforeseen or uncontrolled events could ever take place. Otherwise, changing electric production from coal to natural gas would be nothing more than a shift from CO_2 to CH_4 as the driving cause of global climate change. The overall ecological impacts may not be reduced; in fact, climate change could considerably worsen. This proposed policy of converting from coal and oil to natural gas represents a clear example of tinkering, and as David Orr has so nicely stated, "It makes no sense whatsoever to choose policies that switch from potentially catastrophic problems to those that are merely ruinous."[14]

Workable, effective environmental policies necessitate laws and regulations that are coherent, coordinated, and comprehensive. They would appreciate and counter the interconnected threats as well as responding to the narrowly defined immediate danger. Effective policy builds on a clear, long range vision of what a sustainable society might look like; it matches policy means with societal ends and spells out logical, consistent steps to move us from where we are to the intended goal or outcome. In other words, effective environmental policy would have to be based on rational decision-making. Can such policies be formulated, enacted, and implemented within this country? This is the question that opens up an ecological interpretation of the Constitution of the United States. With an eye to the environmental dangers we face, our attention can turn to the Constitution, we can examine how it functions, and we can analyze the probability of effective environmental policy being produced by the American government.

Any written constitution performs several key tasks for the nation. First, constitutions structure the government. They create institutions and stipulate the relations among them. In America, our Constitution establishes a federal system of government with state governments assigned certain broad powers (historically, these have included the authority to regulate property and the police power to declare certain acts illegal), as well as a Federal, or national government, given specific, enumerated powers (taxation, spending, regulating interstate commerce, and formulating foreign policy). Where environmental issues fit in this federal scheme is subject to dispute. The Constitution creates three branches of the Federal government and two branches of the Legislature. The duties and responsibilities of the institutions are spelled out with more, or less, specificity. The second function constitutions perform is designating the manner in which these institutions will be staffed. For example, the office of President is filled by elections structured through the Electoral College. US Senators were first chosen by state legislatures until this system was altered by the Seventeenth Amendment. Members of the House of Representatives have always been chosen by direct popular vote. The Supreme Court is nominated by the President and approved by the Senate with terms lasting "during good behavior" (i.e., lifetime terms). In addition to these, the Constitution also provides for a number of officers and functionaries to man the Executive Branch. Senior appointments are normally made by the President, with some requiring Senatorial "advise and consent" while others (the President's personal White House staff) do not. Since 1883, a professional civil service, comprised of tenured experts, has filled a large number of positions within the departments, offices, and agencies of the Federal government.

The third function of constitutions is of most interest to our present discussion. Written constitutions establish the procedure by which proposed policies are enacted into law. Bicameralism, separation of powers, checks and balances, the executive veto, the absence of a Federal line-item veto, guaranteed procedural and substantive rights, the rule of law, Congressional oversight, and judicial review are some of the factors determining the course of action in the policy-making process. They also have a large impact on shaping and constraining the product of this process. Either explicitly, or by strong implication, these factors can be found in the 1787 document written in Philadelphia. Other elements of the policy process have evolved over time and become so much a part of the American political system that the government could not operate without them.

Political parties, the President's cabinet, and the Congressional committee system are certainly indispensable. Other innovations are less obviously imperative. The filibuster in the Senate, a vast network of paid Washington lobbyists, and recently, computerized redistricting of seats in the House of Representatives guaranteeing invulnerable, one-party rule have generated controversy and calls for reform. Which of these elements can be considered part of our Constitution and which cannot depends on how narrow or expansive a definition we use. Yale's Sterling Professor of Political Science, Robert Dahl, argues that critical inquiry needs to be built upon the recognition that a thorough investigation of the Constitution must recognize that "the Constitution of the United States" should be taken to mean not only the document written in Philadelphia and the formally adopted Amendments that have been added but also what he calls "the Constitutional system" of structures and practices that have developed over the years.[15]

Another question that must be addressed is whether or not workable policies are feasible within our Constitutional setting. Even if effective environmental policy could be designed, could such proposals work their way through the legislative labyrinth and be enacted into law? Are they practical? One person who is convinced America is capable of responding to our environmental challenges within the framework of our Constitution is Al Gore. He is certainly familiar with Washington politics. Prior to his unsuccessful presidential bid in 2000, Gore served four terms in the House of Representatives, eight years in the Senate, and two terms as Vice President. This experience, coupled with his well-documented passion for environmental topics, should give him insight into the essential ingredients for effective and practical policy. Gore has confidence that our Constitution and Constitutional system pose no serious obstacles to the creation of sound environmental policy. He considers our Constitution a "daring and wonderfully effective invention," and has faith that "a document written more than two hundred years ago is still universally recognized as the world's most forward-looking charter for self-government."[16] From his perspective, the American Constitution provides the framework within which effective, workable policy that is also practical and feasible can be designed, enacted, and implemented.

In 2009, Gore published a book titled *Our Choice: A Plan to Solve the Climate Crisis*. In the beginning, the former Vice President declares his optimistic belief that "we have at our fingertips all of the tools we need to solve three or four climate crises—and we only need to solve one."

He then makes a truly remarkable statement. "The only missing ingredient is collective will."[17] Two hundred and sixty pages later, after discussing numerous technologies for shifting to a low carbon society (such as building a "super grid" for electrical transmission), he reiterates the same claim: "The technologies necessary to build a super grid are all fully developed and available now. *The only missing ingredient is political will.*"[18] With some reflection, the astonishing implications of Gore's pronouncements become clear. More than technology, more even than comprehensive and practical policy, it is the collective, political will of a nation that is the essential missing ingredient standing between America and the solution to our multiple challenges to governance. Attentive readers might be led to expect that if this is the case, the bulk of his book would be an analysis of "collective will" and "political will." Unfortunately, these attentive readers would be disappointed. The overwhelming content of Gore's writings is committed to promoting technologies and practices he believes would be helpful. There is nothing explaining or analyzing collective, political will: the "missing ingredient" he says is critical. The reader is left wondering; what is the "collective will" of a nation? What gives a country the "political will" to act and in which direction does this will lead us? Can a nation as large and complex as the United States have multiple, contradictory wills tugging it in different directions? Where can we search to find an articulation of America's political will?

Volition, determination, or the inclination to act is contained in a nation's values and historical practices. A country's collective will is defined by its political culture. It is, as Alexis de Tocqueville wrote, our "habits of the heart." It is our national character as Americans; our ethics in the original sense of our *ethos* or way of life. It may be only a bit melodramatic to say our collective will is our national soul. This opens a second definition for the word "constitution," for if we consult the dictionary, the word refers both to the fundamental laws and principles of a government and also to "the way in which a thing is made up; the physical or mental make-up of a person." There are, in other words, two somewhat distinct meanings for the word. The term "constitution" can refer either to the fundamental law of the land—in America, this is our written Constitution—or it can refer to our cultural values, habits, and traditions. This is the second American constitution and understanding it is indispensable if we are to fully interpret and comprehend our ecological challenges.

Most nations have a written constitution, but all nations have unwritten constitutions. Getting Americans to understand and appreciate this is difficult. When we hear the words "the American constitution," our minds immediately focus on the written document. This lack of attention does nothing, however, to diminish the importance of our cultural, unwritten constitution. In fact, the deeply embedded quality of our values, perspectives, and habits may make them all the more meaningful. Robert Dahl understands this. He astutely observes,

"I'm aware that the distinction between our formal or written Constitution and our informal or unwritten constitution may be puzzling to some of my American readers. Unlike the British, who have lived for centuries with an unwritten constitution that is nowhere laid out in a single document that one could call the British constitution, Americans may find it hard to realize that we are accustomed to certain traditional political practices, institutions, and procedures that we tend to take for granted as essential aspects of our American system of government, even though they are not prescribed by a written constitution." Elsewhere, Dahl acknowledges the tremendous importance of "national histories, political cultures, and perceptions of internal and strategic threats to survival" and asserts, "If a country is to maintain its democratic institutions through its inevitable crises, it will need a body of norms, beliefs, and habits that provide support for the institutions in good times and bad—a democratic culture transmitted from one generation to the next."[19] The Constitution as a formal document is the Founding Charter we look *at*. As a set of traditions and principles, the constitution is a cultural lens we look *through*. David Orr is correct; taking action on our environmental, economic, and political challenges will severely test our Constitution, but Robert Dahl's observations are equally true; responding to our current crises will also test our cultural values, our unwritten constitution, our national character as a people. In order to explain and understand these challenges, we must analyze how our written Constitution constrains policy, and we must go looking for our unwritten constitution; the cultural values and legacies that shape our collective will. An interpretation of environmental politics that does not delve into both of America's constitutions would be incomplete, shallow, and defective.

The first and best place to look for the unwritten constitution of the United States is in the writings of our Founding Fathers. These men were trying to justify a revolution and form a new system of government. In order to accomplish their task, they needed to make appeals to what they perceived as the core values of American leaders and citizens. The Founders did not create those values, but they articulated principles, perspectives, goals, and visions of the future that became what we today refer to as American political thought. Two hundred years ago, they were listened to because they accurately understood and gave public expression to the national character of the American people. As the history of this nation unfolded, this unwritten constitution underwent changes; various elements of American political culture became shuffled and rearranged; but the legacies of these core values have stayed with us. The most prominent of those Founding Fathers—Alexander Hamilton, James Madison, and Thomas Jefferson—each had distinct ideas concerning the values, anxieties, possibilities, and goals of this nation. Within the political thought of each man can be found lines of reasoning built upon assumptions regarding human nature, the place of reason and passion in society, the importance of private interest and public ethics in human concerns, the threats this nation faces, the role and proper structure of government, and the direction in which the political economy of America should head.

And there is another aspect in the writings of each of these men. Too often overlooked and underappreciated by later scholars, the philosophies of America's Founding Fathers contain assumptions, logic, and visions of Americans' relationship with the nonhuman environment within which the politics of this country takes place. The role that Nature might play in the future, whether it was an adversary to be conquered, a force that would set bounds on our potential, or a testing ground upon which we could display our moral improvement was embedded in and underpinned the more expressly political components of their thought. Just as the social, economic, and political elements of American political thought have rearranged and evolved, so too our environmental legacies, bequeathed to us from our Founding, have become part of our heritage that sculpts our collective will. If this collective, political will is the essential missing ingredient in determining how we today respond to our ecological challenges, then recognizing the origins and impacts of this unwritten environmental constitution is the key to interpreting where we are, how we got here, and where we might go.

The political theories of Hamilton, Madison, and Jefferson are quite different and very often contradictory. Nowhere are these contradictions more evident than in the environmental components of each man's thought. It should come as no surprise, therefore, if we find in present-day America, various environmental perspectives that dramatically conflict with each other and implicitly rely on the environmental legacies articulated by the Founders, passed down through our history, and simultaneously pulling our cultural will in diverse directions. It would be naïve not to acknowledge that there is something in our national character that leads us to believe that technology, money, and power can solve all problems and drives us to ransack our environment. It would be equally foolish not to admit that many, many of our fellow countrymen approach these issues with a passive acquiescence, focus their attention on the daily routines of their private lives and optimistically hope that the regular workings of our economic and political system will make everything turn out for the best. Still, as the ongoing struggles of a vibrant environmental movement demonstrate, there is also something in our political culture that makes us care about Nature, react with pain and anger at the destruction of something we hold dear, and causes us to summon the will to resist injury and promote harmony. Each of these attitudes is with us. We need to only understand their origins, consider the implications of each, and choose the ones we want to honor and promote.

How we as a nation decide to act and the form that actions takes will, therefore, be decided by three broad categories of concern: the perspectives, narratives, or discourses we bring to the environmental challenges and the policies contained within these discourses; the constraints and shaping of outcomes built upon the policy-making process of our written Constitution; and our unwritten constitution and the determination we muster to address these tests of our national character. These concerns will be the focus of this study. Put simply, this is a book about politics, policy, and political will. Each of these factors deserves careful analysis, but it is this last component that will occupy the largest part of this study. As Americans, we need to recognize the power of our unwritten constitution and the evolving values and habits of our culture to affect the choices we make, for it is those choices—whether to ransack, ignore, or care for the environment—that will be the controlling feature of our response.

During recent years in academic circles, a great deal has been written about "paradigms." Anxiety over the relationship between American society and America's environment has led several scholars to call for a "paradigm shift" in our political, ecological, and economic "world views." The argument—completely rational on its surface—is that our perceptions of politics and our view of ecological relationships are so intertwined that if our environmental problems (or crises) are being at least partially caused by our perspective on humans and the environment, then changing that perspective would seem to be called for.[20] From a cursory glance, this view seems to make sense. The weakness of the "paradigm shift" position, however, is the fact that it is cursory and superficial. Adherence to a program of paradigm shift rests upon several premises that need to be spelled out. First, it is assumed that Americans have a clear understanding of the political and social perspectives that shape their culture. Second, it is assumed that the connection between our political thought and our environmental outlooks is established and understood. Third, it is generally postulated that the relationship between our political culture and environment is almost exclusively negative, that is, little analysis is extended to those elements in American political thought that might actually contribute to the protection and defense of our ecosystem. And fourth, the premise is made that America's culture, as articulated at least in part by the history of our political thought, consists of a monolithic, consistent, broad consensus that must change completely and utterly. None of these assumptions is intuitively obvious or self-evident. In fact, a careful analysis might reveal each is false. We do not have to completely change our political culture, as the "paradigm shift" discussion seems to imply. Even if it were possible, we do not have to become a nation of Buddhists. Still less do we need to become a nation of Marxists. It is true that the sources of our ecological malaise are rooted deep in American values, habits, and perspectives, but it is equally valid to say that the resolution of that crisis can be found in an American political and social culture that is richly diverse, and potentially at least, environmentally sustainable.

For those Americans who deeply love their country but who are also worried about the direction in which we are headed, this should come as good news. There is no need for an overly cerebral discussion of "paradigms" and "paradigm shifts." Instead, we need to seriously and yet passionately understand which elements in our diverse culture are leading us in environmentally dangerous directions and which elements in the thought of our forbearers we can draw upon to help us through what looks like difficult days ahead. America's future is not determined. There are various policies we can employ in response to our situation, and the choices we make will influence what we, as a nation, will become. Neither, however, are our options unlimited. The American constitution, in the sense of both the formal document and in the meaning of our culture shapes and constrains the alternatives available to us. Not every policy option is possible; some alternative worlds belong in the realm of fiction. One further observation needs to be made: the social choices facing us are not without enormous moral implications. Our future is not determined, it is not unconstrained, and it is not ethically relativistic. Some choices we might make, such as believing we can continue down our current path of exponential economic growth, a politics based on domination, or the belief that the entire world was created for exclusive exploitation by our species, will invariably lead to results that are ecologically destructive, politically repressive, and ethically unconscionable. Other choices might just as likely lead to an America that is ecologically healthy, economically and politically sustainable, and just for all members of our expansive human and nonhuman community. To expect an ecological and political interpretation to be neutral, objective, or value free, therefore, is to expect something that is neither possible nor desirable.[21]

America's environmental predicament would be difficult enough if it were only a matter of deciding which of the many environmental perspectives to apply to a stable and given set of institutions built upon a shared and acknowledged collection of values, beliefs, and priorities. The situation is made even more complex, however, by the fact that the current research on the American Constitution and the American Founding is also in a state of flux. What we are presented with is an assortment of incompatible interpretations of our environmental situation being imposed upon an equally contentious and incompatible set of perspectives on America's fundamental principles. Sifting through these environmental narratives, recognizing how they have blended with differing components of our political culture, and tracing the evolution of our responses to national challenges will constitute an ecological interpretation of America's two constitutions.

NOTES

1. Early critiques include Charles A. Beard, <u>An Economic Interpretation of the Constitution of the United States</u>, originally published in 1913 with a new author's introduction in 1935 (New York: The Fress Press, 1965). Challenging the undemocratic tendencies of the Constitution was done by J. Allen Smith, <u>The Spirit of American Government</u> (Cambridge, MA: The Belknap Press of Harvard University Press, 1965), originally published in 1907. Contemporary appraisals are Daniel Lazare, <u>The Frozen Republic: How the Constitution is Paralyzing Democracy</u> (New York: Harcourt Brace & Company, 1996); Sanford Levinson, <u>Our Undemocratic Constitution: Where the Constitution Goes Wrong</u> (New York and Oxford: Oxford University Press, 2006); and Robert A. Dahl, <u>How Democratic is the</u>

American Constitution?, Second Edition (New Haven and London: Yale University Press, 2003).
2. David W. Orr, Down to the Wire: Confronting Climate Collapse (Oxford and New York: Oxford University Press, 2009).
3. Ibid., 19. Good introductions to the science and politics behind climate change are Spencer R. Weart, The Discovery of Global Warming (Cambridge, MA: Harvard University Press, 2003), and Richard E. Mann, The Hockey Stick and the Climate Wars (New York: Columbia University Press, 2012).
4. Ibid., 22. On species extinction, see Elizabeth Kolbert, The Sixth Extinction: An Unnatural History (New York: Henry Holt and Company: 2014). A good review of ecological services is found in Eric Zencey, The Other Road to Serfdom (Hanover and London: University Press of New England, 2012).
5. J. David Hughes, Drill, Baby, Drill: Can Unconventional Fuels Usher in a New Era of Energy Abundance? (Santa Rosa, CA: Post Carbon Institute, 2013).
6. Bill McKibben, "A Matter of Degrees: The Arithmetic Of A Warming Climate," Orion, Vol. 31, No. 4 (July/August 2012), 10–11.
7. Orr, Down to the Wire, 71. Some further considerations; according to the US Energy Information Agency, world CO_2 releases for the year 2011 totaled 31.6 gigatons. Once again assuming no increase in yearly emissions, at this rate, atmospheric CO_2 levels would hit the 565 gigaton catastrophe line in a little under 18 years or around the year 2029, long before the 40-year destructive target.
8. Ibid., 11.
9. Ibid.
10. James Howard Kunstler, The Long Emergency (New York: Grove Press, 2005), 233.
11. Justin Fox, The Myth of the Rational Market (New York: Harper Business, 2009); Joseph E. Stiglitz, Freefall: America, Free Markets, and the Sinking of the World Economy (New York: W.W. Norton & Co., 2010); Michael Lewis, Flash Boys (New York: W.W. Norton & Co., 2014).
12. "In a fiery speech, Warren calls for limiting the size of banks," Noah Bierman, The Boston Globe (online edition), September 12, 2013.
13. Orr, Down to the Wire, op.cit., 24–245. Orr attributes his use of the term "blowback" to Chalmers Johnson, Blowback: The Costs and Consequences of American Empire (New York: Henry Holt, 2000).
14. Orr, Down to the Wire, op. cit., 29.
15. Robert A. Dahl, How Democratic is the American Constitution?, 2nd Edition (New Haven: Yale University Press, 2003), 41.

16. Senator Al Gore, Earth in the Balance: Ecology and the Human Spirit (Boston and New York: Houghton Mifflin Company, 1992), 171.
17. Al Gore, Our Choice: A Plan to Solve the Climate Crisis (Emmaus, PA: Rodale Inc., 2009), 15.
18. Ibid., 275. Emphasis added.
19. Robert A. Dahl, How Democratic is the American Constitution?, 159–160 and 138.
20. For example, William Ophuls and Stephen Boyan's Ecology and the Politics of Scarcity (New York: W.H. Freeman and Company, 1992), Lester Milbrath, Environmentalists: Vanguard for a New Society (Albany: State University of New York Press, 1984), Charles Bednar, Transforming the Dream (Albany: State University of New York Press, 2003). See also John Rodman, "Paradigm Change in Political Science: An Ecological Perspective," American Behavioral Scientist, Vol. 24, No.1, September/October 1980, 49–78.
21. For a good discussion of anti-determinism and anti-relativism, see Herman Daly and Joshua Farley, Ecological Economics (Covelo, CA: Island Press, 2004), 44–47.

PART I

Environmental Politics and Our Written Constitution

CHAPTER 2

The Scope and Limits of Mainstream Environmentalism

Americans are, or like to think of themselves as, "practical" people. If a particular policy gains the reputation of being "practical," then many people in this society assume it should be adopted or at least given serious attention. Conversely, to label an action or thought as "impractical" or "unrealistic" is usually considered tantamount to condemning it. Given that we as a culture place such a high value on the idea of "practical solutions," it would be helpful to understand precisely what is meant by this term. This is particularly true in the realm of environmental politics, where policy options have been advocated that range from a cornucopian faith in free markets, self-interest, and unbridled growth to deep ecology's call for a radical transformation away from anthropocentric ethics and industrial society. Which of these policies are the most "practical," and which should be rejected as "unrealistic?"

If one explores the literature on environmental politics and policy, it becomes clear that authors use the term "practical solutions" in two different, and not always compatible, senses. On the one hand, "practical solutions" can be defined as those policies that are capable of being put into practice. They are feasible.[1] They are adoptable because they fit within the present institutions and rules of the game that define "the art of the possible." Alternatively, "practical solutions" can be defined as those policies that will accomplish the task set for them. They are workable. Practical policies in this second sense are those policies that will be successful in isolating and addressing the root causes of the problem or crisis. They are practical because they are workable, and they are workable because they will successfully resolve the issue. The problem will no longer be a problem. Obviously, the ideal situation would be to construct policies that are practical in both senses of the term: they would solve the problem, and we could get them put into practice. The dilemma occurs when society becomes confronted with policy issues that require us to move outside mainstream politics in order to discover and formulate workable policies. Unfortunately, environmental issues often seem to confront us with examples of the incompatibility between the two notions of practicality. The continued persistence of environmental degradation, despite over four decades of concerted policy effort, suggests that environmental policies that are "practical" in the first sense of feasibility may not be "practical" in the second sense of goal achievement. What we can get implemented may not work, and what would work may not be capable of implementation.

These nuanced meanings of "practical" allow us to organize environmental agendas into two broad categories. One group of environmental commentators—those we can label "mainstream environmentalists"—believe it is possible to successfully address environmental concerns within the existing institutions of our society. They see no essential contradiction between a healthy, sustainable environment and our political system of representative democracy, our economic system of market-based capitalism, and our ethical beliefs, which we might characterize as individualistic humanism. Most mainstream environmentalists recognize ecological challenges exist, but they see them as threats *to* our system rather than risks posed *by* our system. A second set of environmental narratives takes a very different view regarding American institutions. In recent years, a vocal group of writers and activists—those we might label foundational or radical environmentalists—have contended that it is the constant growth in goods and services required by our economic system; the fragmented and dysfunctional operations of our political system; and the narrow anthropocentrism of our ethics that are leading us to disregard and devastate the ecological systems of our planet. Mainstream environmentalists may be considered "practical" in the sense that their policies are feasible; while radical environmentalists would argue their policies are the only ones that are truly workable.

Since the intensified focus on environmental issues began with "the environmental decade" of the 1970s, a variety of agendas have been suggested defining the issue that needs to be resolved, nailing down a set of concepts and procedures that can move the process forward, and putting forward some idea—however vague—regarding what a "solution" might look like.² Whether implicit or explicit (and the arguments that follow will try to make each as explicit as possible), the choice of a policy agenda is a choice in theoretical perspective. It is a selection of values, but it is also a determination of what will be considered facts for that group's followers. It is the definitions, support, and perspective that a group is willing to struggle to see adopted, and in this sense, there is truth in E.E. Schattschneider's observation; "the definition of alternatives is the supreme instrument of power."³ This chapter will investigate the legacies of five narratives that fit within the category of mainstream environmentalism.

The Anti-Environmental Cornucopians

If by "practical" we mean that set of policies that could most easily be put into practice, then obviously nothing is more practical than maintaining the status quo. The main current of American economic and political thought has long been built upon a faith in individualism, competition, scientific and technological progress, and economic growth. Children of a humanistic Enlightenment, Americans have faith that one person's increased wealth need not be purchased by decreasing the wealth of another. In an expanding economy, it is assumed that all humans can benefit. In addition to believing we are a practical people, Americans also think of themselves as optimists.

In environmental politics, one group of writers has taken this basic optimism to the point of denying the very existence of problems between humans and the rest of Nature. These are the cornucopians. When forced on occasion to admit there may be some "local problems" involving the environment, cornucopians assert their faith that the solutions will be easy, painless, and virtually automatic.⁴ Characterizing them as "cornucopians" places emphasis on their view that Nature is, or can be made, an endless supply of resources and wealth for humans. Since most members of this group also exhibit an uncritical faith in technology to solve any and all social problems, John Dryzek, in his widely respected study, *The Politics of the Earth*, adopts the label "Prometheans."⁵ Whether one prefers the image of the ever-abundant cornucopia or the view of Prometheus teaching humans how to use fire (and being severely punished for his technological theft), the supporters of this political agenda dedicate significant effort in attacking nearly all forms of environmentalism, arguing for relaxing or abandoning most of the environmental protections society has enacted over the past century, and picturing an upbeat future for America and the world as economic production increases, human populations grow, new energy sources are found, produced, or invented, and the world is made increasingly humanized.

Over the years, there has been no shortage of recruits willing to join the campaign against pro-environmental activism. Members of the group include the late director of the RAND Institute, Herman Kahn, climate change nay-sayers Fred Singer, Patrick Michaels, Frederick Seitz, and Richard Lindzen; economists Wilfred Beckerman, Michael C. Lynch, and Jagdish Bhagwati and political scientist Bjorn Lomborg. Popular attacks on environmentalists have come from Gregg Easterbrook, Martin Lewis, and Peter Huber. Propagandists, with no particular background or expertise, include the impeccably quaffed T.V. celebrity John Stossel and fiction author Michael Crichton.[6] Their work is well-financed by right-wing billionaires such as Sheldon Adelson, the Koch brothers, and Art Pope; promoted by conservative think tanks such as the Heritage Foundation, the American Enterprise Institute, and FreedomWorks; given voice in Congress by spokesmen such as Oklahoma Senator James Inhofe and Texas Representative Joe Barton, and disseminated through outlets such as Fox News, *The Wall Street Journal*, and CNBC. The battle between cornucopians and contemporary environmentalists dates back to the early 1980s with the emergence of a voice that would set the standard for virtually all subsequent cornucopian arguments. At this time, professor of economics and business administration at the University of Maryland, Julian Simon, established himself as the pre-eminent representative of the anti-environmental movement. The elements of Simon's argument consist of the use of trend data, a kind of logic, an extreme faith in technology, and the rejection of any reasoning or facts that contravene his conclusions.

Regarding trend data, in a 1994 debate with environmentalist Norman Myers, Simon proclaimed his belief in "the common wisdom" "that in the absence of additional information, the best first approximation for a variable tomorrow is its value today, and the best second approximation is that the variable will change at the same rate in the future that is has in the past."[7] In other words, absent conflicting data, we should assume linear extrapolations of the status quo into the future. Simon's analysis begins with the *assumption* things will not substantially change and then uses this premise to reach the conclusion that things will not substantially change. Examples of this supposition continue today. In 2012, the North Carolina

General Assembly, under the control of conservative Republicans, passed legislation specifically rejecting scientific projections of a rise in sea levels caused by climate change and requiring state coastal policy to be guided exclusively by "historical data."[8] Since sea levels have not dramatically risen in the past, North Carolina law is now based on the assumption that they will not rise appreciably in the future. The debate over the future of humanity's relationship with its environment hinges, however, on whether or not events are about to take an abrupt, nonlinear turn for the worse. Basing arguments on linear trends ignores both scientific data and a critical methodology for predicting forthcoming events.

Another example of the puzzling logic employed by cornucopians can be found in Simon's response to the population issue. In 1980, in response to a directive from President Jimmy Carter, the Council on Environmental Quality and the Department of State produced a study, The Global 2000 Report to the President, on the condition of the environment and on the prospects for the future relationship between humans and the environment.[9] The report's findings were not rosy. Among other issues, Global 2000 used commonly recognized demographic data on fertility rates, population momentum, and the possibility of a demographic transition leveling off world population figures before the middle of the twenty-first century, and, similar to other reputable studies, concluded the human population will approach or exceed ten billion sometime in the first half of the twenty-first century, and this number of humans will put extreme, negative pressure on the carrying capacity of the Earth. Specifically, Global 2000 said that the world will become "more crowded." In 1984, Julian Simon and Herman Kahn published The Resourceful Earth: A Response to Global 2000, which denied and claimed to refute these findings.[10]

Simon and Kahn focus on the phrase "more crowded," assume that a significant difference exists between the fact of "more populated" and the concept of "more crowded," and then proceed to construct an argument to demonstrate that although the Earth may have more humans occupying space and consuming resources in the next century, it will actually be *less* "crowded." In other words, the fact of increased population is converted into the abstraction of "crowded." For the rest of their argument, the fact is ignored and only the abstraction is addressed. The issue of population is further abstracted by transforming it from a physical issue concerning the number of humans to an economic issue regarding the incomes of this population. They assume that only human thought and human labor produce wealth (thereby begging the environmental question), contend that more people can and will produce more wealth, and conclude that with more wealth humans will be able to purchase those amenities necessary to make the world "less crowded." In two sentences, the issue of human population and resource use is denied, not by facts, but by economic assumptions dating back to John Locke and Karl Marx, and by logic that does not date back to anyone. "[A] growing population does not imply that humans living on the globe will be more 'crowded' in any meaningful fashion. As the world's people have increasingly higher incomes, they purchase better housing and mobility."[11]

Their reasoning is this: more people equals more labor equals more wealth equals more house space and travel equals less crowded. In other words, the more people the world has, the less crowded it becomes. Simon admits his position "defies common sense," but, undaunted, he and other cornucopians have pressed on with their rhetoric.[12]

The lynchpin of cornucopian thought is an absolute faith in technology, a belief that human economic growth and control over Nature can continue forever, and a denial of any limits on human behavior. Julian Simon provides the starkest example of this way of thinking. "We now have in our hands—in our libraries, really—the technology to feed, clothe, and supply energy to an ever-expanding population for the next 7 billion years." To reasonable minds, this is moving beyond simple optimism to preposterous Pollyanna, but cornucopians have a response for those who deny their environmental denials: "Hearers of the messages in this book are often incredulous and ask, 'But what about the others side's data?' There is no other data."[13] An arrogance toward those who question their perspective combines with an unquestioning devotion to human ingenuity to reject all assertions that public policies are needed to regulate environmental challenges. For cornucopians, these policies are unnecessary because the problems do not exist.

The denial of ecological limits is extended to the rejection of the very notion of carrying capacity—that most fundamental principle of ecology that states an organism's population is limited by the availability of the resources it needs to exist. For Simon, principles of ecology have lost all relevance. "Because of increases in knowledge, the earth's 'carrying capacity' has been increasing throughout the decades and centuries and millennia to such an extent that the term 'carrying capacity' has by now no useful meaning." John Dryzek, in his analysis of the corncopian/Promethean discourse, has drawn the shocking implication, "For Prometheans, natural resources, ecosystems, and indeed nature itself, do not exist."[14] This may seem a bit harsh, but in actuality, Dryzek is correct. If Nature exists, then there is something outside of human control that may limit human economic and social activity, but if these limits do not exist, then Nature, in a real, meaningful, scientific sense, is not there. The first question environmental politics must ask is "does Nature exist, and, if it does, what are the implications of that fact?"

This points to an interesting incongruity in the cornucopian argument: while expressing a boundless faith in technology, this narrative rejects both the principles and evidence of the best science. The unsustainability of exponential growth, the buildup of CO_2, greenhouse warming, climate change, the finite characteristic of fossil fuels, the limited (and decreasing) availability of clean water, pollution of water and air, overharvesting, endangered species, and the loss of wetlands (to name but a few) are all dismissed as "meaningless." Science and technology, wedded together in the western Enlightenment since at least the time of Francis Bacon, have somehow become disjointed in their minds, with science becoming denied and only the worship of technology remaining.

There will certainly be a place for technology in America's response to our environmental challenges, but contrary to what cornucopians may believe, there are reasons to temper our trust in purely technological solutions. First, technological innovations are almost always accompanied by unintended consequences. Coal-fired electrical generating plants and the internal combustion engine were not designed for the purpose of emitting CO_2 into the atmosphere. That negative outcome was a byproduct of the technology's primary aim, but more and more, risks being faced by our environment can have their sources traced to inventions designed for quite different purposes. Faith in technology must be accompanied by humility and the recognition we may be causing new crises as we correct old problems. The second reason for technological skepticism is the historical fact that as technology has made us more efficient in the use of some resources, the consumption of those resources has actually increased. Sometimes referred to as Jevons Paradox, the argument builds on an important distinction between the efficiency with which a resource is used as juxtaposed to the scale of its use.[15] For example, a car that delivers 60 miles per gallon is much more efficient than one that only gets 15 miles per gallon, but if the technologically more sophisticated vehicle is driven 10 times as far, it will consume more gasoline in spite of its superior efficiency. Efficiency and scale are different, and it is a dangerous illusion to think that because we are becoming more efficient, we are also lowering the scale of our impact.

The third reason to moderate technological enthusiasm is the rather obvious fact that technology cannot, by itself, make the choice between competing technologies. Consider, once again, the issue of carbon buildup in the atmosphere in this light of technological alternatives. America can possibly respond by shifting electrical generation to appropriate technologies that work with the forces of Nature such as wind and solar; or we can choose hard technologies that seek to dominate Nature, such as the continued use of fossil fuels, but move toward complex (and very expensive) techniques such as CO_2 capture, compression, and underground storage; or we can opt to continue sending CO_2 up smokestacks and into the atmosphere; or we could select technologies that seek to conserve energy and reduce demand; or, finally, we could really tackle the problem and take on the difficult, but ultimately effective, task of limiting the continuing growth of our population and the number of humans who demand electricity. Which of these technologies and policies, or what mix of these technologies, should we prefer? Technology may, in certain cases, open up options, but technology by itself cannot make those choices for us. To think technology can spare us from having to make profound political, social, and ethical choices is a misguided view of both technology and what it means to be human.

This leads to the fourth, and perhaps most important, reason why we should pause before relying too much on technology as our future redeemer. Technology cannot give us the ethical awareness and moral sensitivity to feel a concern for the impact we are having on our environment and the surrounding ecosphere. It cannot give us a love of Nature and the will to protect the planet. In too many instances, technology serves as an emotional barrier between humans and the Earth we inhabit. Science might provide us with the knowledge to understand what we are doing, but technology often gives us an excuse not to care. The result, on display throughout America and the rest of the world, is knowledge without wisdom. If moral purpose and the collective will to act are the essential missing ingredients in our environmental discussion, then technology, in and of itself, cannot provide us with what we need the most. The connection between technological enthusiasm and the numbing of our moral sense seems to be particularly evident among writers within the cornucopian narrative. When reviewing the cornucopian perspective, two curious facts become apparent: the arrogant optimism of this discourse is without a basis in either rational argument or scientific evidence, and yet, it is an outlook that is, no doubt, widely held by a great many Americans. Since its support cannot be explained by either logic or science, it can only be understood by its connection to deeply held values in our culture. When the facts point one way and our values point in a different direction, many Americans apparently choose to follow their values and ignore or deny the facts of reality. Something in American habits and traditions gives us the predilection to see the continuous, exponential domination and control of Nature by humans as both possible and desirable. In later chapters, we will go looking for the source of those values, habits, and traditions.

The Libertarian Free Marketeers

Libertarians and "free market environmentalists" are normally fellow travelers with the cornucopians. They all want less environmental regulation, more economic growth, and less concern over issues such as global warming, energy depletion, pollution, or population increase. Libertarian organizations such as the Cato Institute, the John Locke Society, the Independent Institute, Laissez Faire Books, and the Federalist Society promote authors who extol the benefits of free capitalist markets, decry political regulation, and assail most environmentalism as an unnecessary and dangerous assault on individual liberty. There are some subtle differences between cornucopinans and libertarian free marketeers, however, which may make dealing with them in a semi-separate fashion worthwhile. Whereas cornucopians attack the substance of environmental concerns, free marketeers concentrate more on the process by which decisions are made. Cornucopians focus on the resources of Nature and argue they are virtually limitless. Free marketeers take the existence of resource scarcity as a given and explore what is to be done given this ecological fact of life. They focus not on Nature but on human nature.

Free marketeers belong to the libertarian wing of neo-classical economics and reflect the premises and logic of that movement. It is seldom, for instance, that one reads a free market tract without seeing the name Friedrich Hayek used repeatedly. Other proponents of the Austrian school of economics such as Murray Rothbard and Ludwig von Mises also appear with some regularity. Although elements of this libertarian economics have been part of our culture since the American Founding, they became specifically linked with environmental skepticism in 1991 when Terry L. Anderson and Donald R. Leal published <u>Free Market Environmentalism</u>.[16] Anderson and Leal would be classified as "conservatives" by most contemporary understandings of that term, but they are "liberals" in as much as they adhere rather strictly to the fundamental tenets of seventeenth and eighteenth century Liberal thought.

The linchpin of this perspective is the view of human beings as self-interest, happiness seeking (the economists call it "utility maximizing"), essentially insatiable, individuals. That last word is the critical one. From the free market perspective, the individual human uses instrumental reason in the pursuit of objects and services; the acquisition of which it is believed will provide the most happiness. Since it is assumed individuals are unique in their values and desires (referred to in the literature as "preferences"), there exists no independent, neutral method for determining one set of values to be better, higher or lower, or objectively more desirable than another. Accepting this subjectivity of preferences, free marketeers build strong theoretical walls against any group that might claim possession of impartial standards of judgment. They reject the idea that scientific managers, economic modeling, or democratic decision-making can or should be able to make informed decisions about environmental policy.[17] According to libertarians, the only result of these intrusions is to restrict the liberty of individuals to choose what they want and what will make the them most happy. In scholarly circles, this view of humans is called *homo economicus*—economic man.

Since all forms of collective or expert judgment are ruled out, free marketeers place their faith—their complete faith—in the one institution they believe will maximize individual freedom of choice—economic markets. It is argued that economic markets take all the subjective, individual preferences and convert them into a set of objective social priorities. Markets do this by reducing each individual's values and preferences to a single unit; the ability and willingness to pay a price for the desired good or service. "Because each of us places different values on environmental amenities, there must be some way of quantifying and aggregating those values," assert Anderson and Leal. "In a marketplace, prices provide an objective measure of subject preferences and are therefore an important source of information about subjective values."[18] Markets, through their price mechanism, permit us to discover what people want, how much of it they want, and how intense the desire for these products might be. The result, according to free market environmentalists, is the rational, efficient allocation of scarce environmental goods and services. Anderson and Leal take the next step and inquire into the institutions and practices necessary to guarantee the effective working of a capitalist market system. For them, the key to the entire free market approach is the enforcement and transferability of private property rights. Free marketeers can be viewed as allies with groups such as the Sagebrush Rebellion and the Wise Use Movement who argue against politically based environmental regulations as an infringement on private property rights. Private property—the ability to assert exclusive use of a resource—gives the owner a vested interest in maximizing the worth of his or her holdings. The belief is that there is no antagonism between environmental "stewardship" and profit maximization, that self-interested profit seekers will rationally and logically be lead to engage in those "tradeoffs" that provide protection and access to Nature while also providing a sound return on investment, and that what is good for the bottom line is also good for ecological processes. Anderson and Leal make this logic explicit. "At the heart of free market environmentalism is a system of well-specified property rights to natural resources .... [A] discipline is imposed on resources users because the wealth of the owner of the property rights is at stake if bad decisions are made .... With clearly specified titles—obtained from land recording systems, strict liability rules, and adjudication of disputed property rights in courts—market processes can encourage good resource stewardship."[19]

Given this view, it is not surprising free marketeers believe "government is not the solution to our problems, government *is* the problem" (to quote one adherent—Ronald Reagan). Their preferred policy is to sell off publicly owned land and leave the management of those resources to profit-seeking individuals. "There are several policy reforms that could further encourage the private sector in its recreation and conservation endeavors. The least politically palatable, but perhaps the most effective, would be to privatize or lease some federal lands and let the private sector manage them for recreational and environmental amenities."[20] This program of privatization would seem to run into an intractable barrier when the issue became those portions of the ecosystem not easily converted into private property such as "common pool resources" like wildlife, air, streams, rivers, and oceans. Even here, however, Anderson and Leal are willing to consider privatizing the environment and letting the profit motive determine use "A stream owner who can devise ways of charging fisherman can internalize the benefit and costs and gain an incentive to maintain or improve the quality of his resource."[21] The authors do not discuss what would happen if toxic polluters were willing to outbid the fisherman. Perhaps, the *reductio ad absurdum* of the free market approach is the suggestion that it might be possible to privatize whales in order to charge people for viewing them.[22]

Issues of poverty and the effects of an unequal income distribution present problems for a discourse that places faith in consumers' purchasing environmental goods and services in an open market. Initially, Anderson and Leal brush aside this critical problem by asserting that it is impossible to know what the poor actually want since "this is not verifiable through voluntary trading," but then, in a remarkable rejection of their own argument, they put forth the possibility that "because poor people do not have access to many amenities, there may be an argument for redistributing income in their favor ..."[23] The founders of free market environmentalism suddenly have to consider the possibility that the intervention of the welfare state may be necessary to save their system from logical bankruptcy! At other moments, Anderson and Leal acknowledge that not all medium can be converted into private property, and where this cannot occur, there are strong, instrumentally rational reasons for polluters to contaminate these areas. "Everyone accepts that managers in the private sector would dump production wastes into a nearby stream if they did not have to pay for the cost of their action."[24]

During his 2008 campaign for the Republican Party's nomination for President, Texas Congressman Ron Paul attempted to reconcile his libertarian principles with the concern for the environment felt by many Americans. In <u>The Revolution: A Manifesto</u>, Congressman Paul states, "Some people falsely believe that advocates of the free market must be opponents of the environment .... But a true supporter of private property and personal responsibility cannot be indifferent to environmental damage, and should view it as a form of unjustified aggression that must be punished or enjoined ...."[25] In his subsequent discussion, however, the congressman devotes only three paragraphs articulating his environmental perspective and deals with the question of this "unjustified aggression" by favorably quoting economists Walter E. Block and Robert W. McGee and their advocacy of nuisance and trespass suits against polluters. "If a firm creates pollution without first entering into an agreement, or if the parties cannot come to an agreement fixing the cost and degree of pollution, then the court system could be used to assess damages."[26] As it turns out, Ron Paul's free market America is going to be a highly litigious society.

Trespass and nuisance cases, however, have significant drawbacks in dealing with environmental issues. They assume the damage has already been done (they cannot be used to prevent what would be obvious destruction before it occurs); they place the burden of proof on the defendants; and they instill strong incentives on polluters to mask their responsibility for injuries they might be creating. In the case of air pollution, for instance, polluters might build (and have built) taller smoke stacks to carry their noxious waste further downwind, mix it with pollutants from other geographic areas, and thereby make successful nuisance suits all but impossible to successfully prosecute. Paul's response to this difficulty is to recommend the creation of "an environmental forensics industry" that would allow us to identify those responsible for pollution. The argument for *laissez-faire* becomes blurred by a state that is monitoring and enforcing property rights, redistributing income in order for the poor to have access to environmental "amenities," providing an extensive court system, and using technology to place tracers on pollutants to hunt down their source.

Contrary to Representative Paul's pronouncements, however, libertarian ideology has an established record of anti-environmentalism. Ron Paul's son, Kentucky Senator Rand Paul, has attacked the Environmental Protection Agency (EPA)'s wetlands safeguards and the Lacey Act's protection of wild animals as "government bullying" that need to be repealed[27] Libertarian organizations such as the Cato Institute and the Federalist Society argue against most sincere attempts at ameliorating environmental destruction.[28] The libertarian free market narrative on environmental politics is a confusing blend of the denial of science, strangely joined with active courts, persistent civil actions, and technology, while coupled with the dismantling of the environmental protections this nation has enacted over the years. Government in a free market society is going to be much more active than first thought, but one is left wondering if this state is protecting individual rights or the privileges of a wealthy elite.

Reform Environmentalism

The anti-environmentalism of the cornucopians and free marketeers are only two forms of environmental politics. Ecological catastrophes that occurred during the late 1960s and into the 1970s—such as Times Beach and Love Canal—stimulated reenergized interest in environmental issues. Americans demanded action by their government. Most Federal legislation during the "environmental decade" of the 1970s reflects the view that pollution is simply another public policy issue that can be solved by reforming the nation's laws and regulations. This "reform environmentalism" became expressed in legislation ranging from the requirement for environmental impact statements contained in the National

Environmental Policy Act of 1969 (NEPA); to broad-ranging laws such as the Clean Water and Clean Air Acts; to the regulation of various pollutants in statutes such as the Toxic Substance Control Act (TSCA), Resource Conservation and Recovery Act (RCRA), and the "superfund" provisions of the Comprehensive Environmental Response, Compensation, and Liability Act (CERCLA). Reform environmentalism now represents the middle ground of mainstream environmentalism.

A fundamental reproach reform environmentalists make against economic-based decision-making is that this perspective is flawed in taking as its basic assumption the self-interested individual or the profit-seeking corporation. Environmental decision-making often requires that judgments be made on the basis of what is best for the community and not simply what the self-interested individual might prefer, it may at times require that sacrifices be made because of some higher ethical standard, and it may require that the common good, or the public interest, be allowed to override more narrow and materialistic concerns. Environmental choices often (reform environmentalists would say "usually" or "always") involve decisions among competing values, and contrary to the position of free marketeers, these choices cannot (reform environmentalists might add "should not") be reduced merely to the ability to pay. They are decisions best made by public, political processes rather than by the methods associated with economics.

Mark Sagoff, in <u>The Economy of the Earth,</u> makes a specific effort to highlight the distinctions between a legal/ethical/political approach and the assumptions of libertarian free market environmentalism.[29] While never denying the existence of economic man, Sagoff calls on us to acknowledge another view of humans. This is *zoon politikon* (the political animal), or, alternatively, *homo civitas* (civic man). This civic man is the citizen, and Sagoff draws sharp distinctions between this aspect of humans and their behavior as consumers. Whereas consumers see themselves primarily as individuals, citizens recognize and celebrate their membership in communities. Although citizens obviously have personal interests also, they acknowledge there are moments when that self-interest must be set aside, or sacrificed, to a higher standard—usually called "the public interest" or "the common good." According to this perspective, laws articulate and codify what we, as citizens, believe to be right and what we, as citizens, will not tolerate without the imposition of sanctions. Legislation does something that markets cannot; it expresses the collective ethical will of the community.[30] In a not too subtle criticism of free market approaches,

Sagoff states, "If you envision political relationships in terms of cooperation and competition among self-interested individuals, you will never fathom American regulatory or environmental law." The struggle, furthermore, is not properly viewed as being conducted between different segments of society nor is it even between different roles humans may choose to perform. The truth is that the conflict between the consumer and the citizen is one that goes on perpetually within each person. "Thus, this conflict pits the consumer against himself as a citizen or as a member of a moral community. The conflict, in other words, arises not only *among* us but also *within* us."[31]

The question is what sort of political process is most conducive to articulating and implementing a vision of a public interest. John Dryzek, in The Politics of the Earth draws attention to different methods democratic nations use to define the public interest, set national standards, enact laws, and enforce environmental regulations. One policy approach used to respond to concerns over resource scarcity and pollution control Dryzek calls "administrative rationalism." This tactic of "leaving it to the experts" entails "a public policy tradition which accorded substantial status to scientific expertise as harness by the administrative state. This sort of nexus of science, professional administration, and bureaucratic structure has been used in many policy settings ...."[32] Augmented by the creation of pollution control organizations such as the EPA in 1970, administrative rationalism can be traced back to the Progressive Era and Gifford Pinchot's attempts to centralize resource management within the US Forest Service during the presidential administration of Theodore Roosevelt. The logic of this approach is that these are highly complex, interconnected, and often exceptionally technical challenges must be dealt with by agencies specially created to manage such concerns. Robert Paehlke, in Environmentalism and the Future of Progressive Politics, advocates this blending of expertise with a political movement that articulates and guides a national commitment to a common good. "We must learn to guide scientific research and application," he asserts, "by political decision-making that is sensitive both to the sciences of limits and to a broad conceptualization of the best future for all."[33]

The other possibility put forth by Dryzek is for regulatory policy formulated and implemented through established democratic channels. Calling this "democratic pragmatism," Dryzek portrays the approach as "interactive problem solving within the basic institutional structure of liberal capitalist democracy."[34] Fundamental to this approach are methods such as public consultation and environmental impact assessment as mandated by the National Environmental Policy Act, transparent public dialog, and right-to-know legislation (built into the Superfund Amendment and Reauthorization Act of 1986). Methods of experimentation, trial-and-error, incrementalism, and compromise—it is believed—will produce policy options that can command both wide public legitimacy and support. Paehlke also defends this democratic pragmatism. "I would stress what I think is obvious: environmental politics, especially in North America, must be a centrist and democratic politics."[35]

The actual implementation used by both subsets of reform environmentalism employs the technique of direct regulation, often called "command-and-control." This is particularly true regarding laws aimed at pollution control (for instance, the Resource Conservation and Recovery Act of 1976). Under this system, Congress sets broad regulatory goals and (usually) charges an executive agency with implementation authority. The legislature may designate which pollutants are to be regulated or they can delegate this responsibility to the executive branch. The next step involves the relevant Federal agency (e.g., the EPA) setting the standards for the permissible limits of pollution, specifying which polluting sources are to be controlled, and the manner in which the controls are to be applied. Often, the Federal agency will work with state governments at this stage since the final step in the command-and-control process entails establishing of a permit system to enforce the pollution standards.[36] This "end-of-the-pipe" system of direct regulation assumes that a certain level of pollution is inevitable in an industrialized society and merely seeks to control discharges to a level that is determined to be tolerable. "Permits to pollute," as they are sometimes called with a degree of cynicism, have been criticized from a number of directions. In establishing maximum standards for pollution discharges, permitting systems send the ecologically undesirable message that it is okay to pollute up to that point. When industry is told that they cannot pollute more than a specified amount, they sometimes proceed under the premise that it is permissible to pollute up to that standard. Other problems exist. As advances in toxicology are made and as scientific measurements become increasingly capable of determining miniscule quantities of pollutants, it becomes more evident that cumulative effects of very small amounts of toxics can result in long-term adverse risks. This phenomenon of the "disappearing threshold" makes it difficult to set hard-and-fast standards, and it further points out the negative consequences of the short-term bias.[37] It became evident to environmentalists that serious weaknesses in command-and-control existed, at the same time anti-environmentalists increased their attack on practically all forms of environmental protection.

Toward the end of the 1970s, there arose a considerable reaction among significant numbers of Americans toward the manner in which environmental policy was being written and implemented. It appeared that a backlash against environmental regulation was underway. Following the 1980 election, the Reagan administration began attacking many of the environmental policies that had been passed in the 1970s. Through the appointment of hostile administrators such as Anne Gorsuch and James Watt, budget cuts to programs such as Superfund and promises to open Federal lands to increased development, Republicans tried to deliver on their campaign promise to provide "regulatory relief" from years of command-and-control rule making. Their significant campaign successes seemed to signal that Americans felt they were being over-regulated and would react sympathetically to the repeal of much previous policy. Yet at the same time, public opinion polls reported a surge in popular support for environmental protection.[38] What Americans apparently wanted was pollution control, but not through the technique of command-and-control regulation. They regularly displayed both a strong commitment to ecological health and an almost equally intense animosity toward governmental regulation. The moment was ripe for an alternative strategy for environmental policy.

Environmental Economics

As the relationship between reform environmentalists on one side and cornucopian and free marketeers on the other became more hostile, a group of economists began to put forward a perspective that sought to combine the basic principles of neo-classical economics with a recognition of serious environmental challenges. Environmental economists, as they became known, attempted to adapt the concepts, language, and perspective of traditional economics to environmental issues such as pollution control, resource scarcity, and policy evaluation. Today, their courses can usually be found on college campuses as an elective offering within most economics departments.[39] Environmental economics remains in the mainstream of academic economics by holding to the fundamental beliefs in the instrumental value of Nature as "resources," viewing humans as *homo economicus*, accepting as given the efficiency, freedom, and participatory decision-making process of the market and proclaiming unquestioning faith in the benefits of a growing economy.

Environmental economists distinguish themselves by acknowledging the existence of serious, socially destabilizing ecological problems. One major source of stress, according to this view, is the improper handling of the industrial and commercial by-products of economic activity. Unlike free marketeers, environmental economists do not believe that unhindered and unregulated markets will automatically create an efficient allocation of industrial resources. Business managers will certainly move to more efficient production techniques if this alternative presents itself and if it is the most cost effect decision available to them, but environmental economists know that managers have other options in addition to becoming more efficient. If the costs of treating industrial discharges or effluent from smoke stacks or if the costs of new, more efficient technologies can be avoided and, instead, these costs of production can be passed onto other individuals not involved in the purchase of the product, then corporate executives can lower their total costs, reduce the price of their product, be more competitive on the economic market, and make higher returns on their investor's capital. Since this is what they are hired to do, managers find it built into their incentive structures to "externalize" as many costs as they possibly can. Note very well, this does not mean these executives are evil, unethical people; nor does it mean they are necessarily ignorant of the consequences of their actions. They are simply behaving as rational economic actors in accordance with the goals of their institution and the incentive structure built into the economic market. From this perspective, polluting the environment makes complete economic and business sense. In the now popular parlance, pollution is described as either a localized or pervasive "externality" of the consumer-based economic market.[40] An externality is any negative consequence of an economic exchange (impaired health, reduced air or water quality, diminished potential for future productivity) that is not accounted for in the market transaction. Tom Tietenberg, an early exponent of environmental economics explains, "As long as the costs are external, no incentives to search for ways to yield less pollution per unit of output are introduced by the markets …. The externalities concept is a broad one covering a multitude of sources of market failure."[41] Here, the distinction between the free market approach and environmental economics is razor sharp and can be summarized in two words; "market failure."

Since the free market approach of privatizing all common pool resources is unlikely to occur and would prove ineffective in any case, a better approach, argue environmental economists, is to place a fee, or tax, on the pollution. If pollution taxes are set high enough, corporations will do simple accounting and conclude being efficient, treating their waste before it is released, or altering their production processes makes more business sense than fouling the environment, and will, therefore, on their own accord (granted, after some nudging by the government) shift to less destructive practices. The externality will have been internalized. Rather than the almost exclusively negative sanctions involved in direct regulation, environmental economists argue that market-based incentives provide superior efficiency and efficacy through more positive incentives such as pollution taxes, tradable discharge permits (now called cap-and-trade), deposit-refund systems, or other techniques while allowing producers and consumers individual freedom to choose how they will react to the market pressures. Examples here include the permit, or allowance, system built into the 1990 and 1995 Clean Air Acts as well as recent proposals to place a fee, or tax, on CO_2 emissions. Unlike the more laissez-faire assumptions of the cornucopians and free marketeers, environmental economists grant a significant role for governmental interference in the economy, but that role is generally restricted to guiding or structuring decisions that are still made within the overall framework of the market (it might be said that the government "rigs" private decisions and then permits the market to operate at will). The premises of the economic approach is that if pollution externalities were internalized, if resources were properly priced at fair value, if private decision-making was made more flexible by techniques such as bubbles and offsets, and if certain resources were privatized, then the inherent rationality and efficiency of the market would provide the greatest good for the greatest number.

Another source of environmental problems is the allocation of scarce resources. This is expressed in a multitude of ways that extend from concerns over the availability of timber, mineral, energy, or water supplies, to apprehension over the apportionment of space for residential, commercial, or recreational use. Here, the second characteristic that defines and distinguishes environmental economics from other policy options presents itself. In addition to the utilization of market incentives, environmental economists heavily depend on cost-benefit analysis to define and evaluate policy. Supporters are broadly united in advocating this approach either in the policy formulation stage (for instance, to compare various policy options or to set environmental quality standards) or in the policy evaluation stage to determine if the selected policies are the most efficient. One example of this approach is President Reagan's Executive Order 12291, which mandated cost-benefit analysis of all environmental regulations. A. Myrick Freeman, an important voice in environmental economics (and purported author of EO 12291), summarizes what he sees as the major applications of this method: "Society should undertake environmental protection and pollution control only if the results are worth more in terms of individual's values than what is given up by diverting resources from other uses. This is the underlying principle of the economic approach to environmental policy. Benefit-cost analysis is a set of analytical tools designed to measure the net contribution of any public policy to the economic well-being of the members of the society."[42] The strength of this argument was reaffirmed in 2015 when the Supreme Court ruled in *Michigan v E.P.A.* that the EPA must calculate the costs and benefits of proposed regulations when setting mercury and air toxic standards (MATS) under the Clean Air Act.

There are clearly many instances where cost-benefit analysis points toward the most efficient policy and aids us in evaluating the policies that have implemented. For cost-benefit analysis to fulfill the sweeping role that many environmental economists have given it, however, the following eight conditions would have to pertain all or most of the time: (1) those who receive the benefits of a particular set of environmental policies would also have to be those who bore the costs of those policies (costs and benefits would have to be shared by the same community),[43] (2) fundamental values or issues of intrinsic worth would not be involved (all involved values would have to be instrumental values), (3) the objective of the analysis would be the efficient achievement of commonly held goals (cost-benefit does not appear to be a good method for determining communal or collective purposes),[44] (4) the issues involved would be relatively local and short-term (since that would afford the best opportunity for consensus on goals),[45] (5) monetary costs could be reasonably assigned (attempts to monetize costs and benefits through such measures as "contingent valuation" have met with difficulties),[46] (6) resources would have to be seen as essentially finite (there would be a need for trade-offs in an essentially zero-sum game), (7) complete, or near complete information would have to exist, and (8) those who conducted the analysis would have to be objective, fair, and unbiased (they could not slant the analysis toward some pre-established ideological position). Tietenberg warns against over exuberance in the use of cost-benefit. "It is fairly easy for most people to accept the general premise that benefits and costs of actions should be weighed prior to deciding on a policy choice," he observes. "The technique becomes more controversial, however, when specific numbers are attached to the anticipated benefits and costs and specific decision rules for translating these numbers into a decision are followed."[47] Observation number 6 is what distinguishes environmental economics and free market economics from cornucopian thought, while observation 7 differentiates environmental economics from the free marketeers. Observation 5 has been called the "fatal flaw" of economic studies. Economists Lisa Heinzerling and Frank Ackerman are convinced, "Cost-benefit analysis cannot overcome its fatal flaw: it is completely reliant on the impossible attempt to price the priceless values of life, health, nature and the future .... Cost-benefit analysis has not enriched the public dialogue; it has impoverished it, covering the topic with poorly understood numbers rather than clarifying the underlying clashes of values."[48]

Still, since there exists a strong chance that at least some of recommendations of the environmental economists will be implemented as official policy, it is important to inquire whether environmental economics is a "practical solution" in the second sense of that term. Can a workable, effective solution to our ecological difficulties rest upon a program that takes so many of our current assumptions for granted? How much real environmental protection can we expect in a society where no one emotionally cares about anything except their own immediate self-interest? The issue is not whether economic approaches to environmental problems are useful. The question is whether or not such a narrow economic outlook is complete, and, therefore, capable of addressing deeper concerns such as the role of common values in environmental disputes, the proper sphere and function of social organizations and government, and objectives that humans may find worthwhile that extend beyond the efficient allocation of instrumental resources. Structured market incentives, pollution taxes, cap-and-trade schemes, and deposit/refund systems have their place, but since environmental economists have opened the door to a large measure of political action—from taxing pollution, to monitoring industrial emissions, to coordinating a system of waste stream recycling, to administering allowance trading schemes—it seems appropriate to inquire into the exact character of that political process, the assumptions that underpin its approach to environmental concerns, and the proposals that emerge from a focus, not only on markets but also the collective good of the community.

Mainstream Sustainable Development

As the Reagan era began to wind down, America, and the rest of the world, looked for some means of combining the goals of reform environmentalism with the means of environmental economics. Reports such as the Brundtland Commission's <u>Our Common Future</u> (1987), the Rio Conference's <u>Agenda 21</u> (1992), and <u>Choosing a Sustainable Future</u> by the (US) National Commission on the Environment (1993), articulated a new discourse in environmental politics, which following the lead of the Brundtland Commission, was dubbed "sustainable development." In 1993, President Bill Clinton impaneled a blue-ribbon Presidential Council on Sustainable Development that produced a study with the long, but revealing title, <u>Sustainable America: A New Consensus For Prosperity, Opportunity, and a Healthy Environment For the Future</u>. Private studies such as Lester R. Brown's <u>Building a Sustainable Society</u>, Al Gore's popular <u>Earth in the Balance</u>, and Thomas Friedman's <u>Hot, Flat, and Crowded</u> have also done much to popularize the idea of sustainable development.[49] From the perspective of environmental politics, sustainable development blended prior concerns regarding resource use and pollution control to a new sensitivity regarding issues of poverty and social justice. The old faith in the value of economic growth remained, while it was assumed that a "demographic transition" would address the crisis of a rapidly expanding human population. The preference for legal, statutory remedies was modified to accommodate a system based primarily on economic motivation, but one where these incentives had been "structured" by the government. In addition to the substantive calls for a reduction in poverty and an increase in technological efficiency, sustainable development carried with it important implications for the policy process.

The new, alternative approach to policy making became institutionalized in laws such as the Pollution Prevention Act of 1990 and the National Energy Policy Act of 1992. Instead of terms such as "command and control" or "direct regulation," the phrase most often heard is "collaboration" or "collaborative planning." According to the President Clinton's Council on Sustainable Development in a section titled "Moving Forward: From Conflict to Collaboration," "More collaborative approaches to making decisions can be arduous and time-consuming ... and all of the players must change their customary roles. For government, this means using its power to convene and facilitate, shifting gradually from prescribing behavior to supporting responsibility by setting goals, creating incentives, monitoring performance, and providing information."⁵⁰ Collaborative planning substitutes the idea of communities working toward a common goal for private groups competing to achieve their own ends. In the area of pollution, for instance, collaborative planning moves away from end-of-the-pipe regulation and toward the idea of reducing of the amount of pollution actually manufactured. Rather than a regulatory system of standards, permits, and sanctions, collaborative planning works toward alternative compliance agreements where industry, government, and (perhaps most importantly) citizens-at-large affirm goals, prescribe methods, and articulate evaluation tools. The aim is to overcome the incremental tinkering of liberal policy making by substituting an approach that more closely resembles the corporatist decision-making found in Japan and many European nations.⁵¹

The Pollution Prevention Act promised to take environmental policy beyond the confines of traditional rule making. In direct opposition to laws such as RCRA and CERCLA which had come to symbolize Federal attempts at "waste management," the Pollution Prevention Act sought to decrease the amount of toxic and hazardous material being manufactured in the first place. The emphasis was shifted from the end-of-the-pipe to the front of the pipe, as "source reduction" became the latest policy motto. An Office of Pollution Prevention was created within the EPA and charged with setting priorities. The result was the creation of a "risk reduction hierarchy" that placed source reduction at the top of the list, followed by recycling, with environmentally safe treatment and storage now made the least desirable method of dealing with pollution.⁵² In addition to new priorities, an alternative set of incentives was put into place. The governmental carrot would replace the bureaucratic stick. Now, in addition to a system of operating permits that could possibly be withdrawn if noncompliance with standards was demonstrated, the EPA offered data collection, information sharing, and, in some instances, appropriations for state grants to aid in programs designed to integrate projects, adapt innovative technologies, and seek long-range solutions.⁵³

The new system of collaborative planning and source reduction required an alternative method of policy evaluation. Traditional techniques, such as outcomes evaluation or cost-benefit analysis, were to be augmented by comparative risk assessment. Risk assessment, it was thought, would balance the economic analysis of the (presumably) more conservative economists in the Office of Management and Budget with the scrutiny of scientists located in the EPA.⁵⁴ When economic evaluation of policy was necessary, proponents of sustainable development/collaborative planning advocated the incorporation of extended product responsibility, sometimes called the life-cycle approach, where everyone involved with a manufactured good—designers, producers, suppliers, users, and disposers—would be answerable for the environmental impact of that product. The bookkeeping side of extended product responsibility is total cost accounting or full-cost pricing. Here, the attempt is made to bring in the external environmental costs that might have been withheld from the cost of the product or costs that are so far in the future that they escape inclusion in the product price. To be sure, even proponents of this different policy approach see there may be problems in what they are suggesting. There exists the threat, for example, that alternative compliance agreements might become sweetheart deals between industry and the government with political regulators becoming more spokespersons for industry than representatives of the public's interest. When this occurs, or is rumored to have occurred, the phrase "captured agency" is used to describe the relevant government agency, and public trust, a key component to any collaborative process, is almost certain to degenerate.[55] In order to avoid the possibility of agency capture, proponents of collaborative planning advocate "transparency" or the broad use of public participation and openness in virtually all phases of the collaborative process.[56]

This idea of a consensus on fundamental goals is critical to the alternative public policy process. Sustainable development requires the belief in, and existence of, a common set of aims in order to move beyond a course of action built around conflict. Hence, the repetitive use of expressions such as "our common future" (Brundtland Commission), "a new consensus" (President's Council on Sustainable Development), "a new synthesis" (National Council on the Environment), and "a new common purpose" (Al Gore). The movement away from conflict-based policy making and toward collaborative planning on a national level is premised on the assumption that there is a countrywide agreement on the goals of environmental policy. Those goals are built upon the belief that there exists no fundamental conflict between the protection of the environment and a strong, growing, productive capitalist economy, and that lack of conflict is based on the conviction that most of our environmental problems can be solved by increased efficiency, better technology, and an equitable distribution of the products of an ever-expanding market. To the extent that conservative capitalists on the right and radical environmentalists on the left, see scale, and not efficiency, as the defining issue, then the right will view environmental protection as a hindrance to further economic growth, and radical environmentalists will view the continued pursuit of exponential economic growth as the major cause of ecological devastation.

At the international level, those who come from a sustainable development perspective describe the current malaise using language that depicts not environmental *problems* but a widespread *crisis*. Yet even while acknowledging the near apocalyptic scope of the threats to human and ecological well-being, proponents of mainstream sustainable development hold fast to their central assumption that there is no fundamental incompatibility between environmental protection and the economic growth.[57] From this transnational perspective, the critical factor contributing to environmental degradation is the extreme inequality in wealth, technology, and institutional structures between the so-called "more developed countries" and the "less developed countries." This inequality, exacerbated by high population growth rates in the third world, produces a series of interlocking and interdependent crises, where lack of development produces negative pressures on the environment (e.g., destruction of forest and soil reserves, pollution of water and air, devastation of native species, crippling governmental debt, and overcrowded cities with inadequate energy, food, and employment). In turn, the lack of development hinders the movement through the "demographic transition," and overpopulation becomes a serious component retarding economic development. While not denying that industrialized nations can have environmental problems unique to their circumstances, the sustainable development approach tends to focus on the issues that are created from extremes of *poverty* rather than those created from extremes of *wealth*. Specific policy proposals, such as technology transfers from the first to the third world, or integrated decision-making among national governments, nongovernmental organizations, businesses, and the United Nations focus on bringing the level of third world development *up*, rather than on redistributing global wealth, deindustrializing the first world, and bringing their level of consumption down.[58] Thus, economic growth remains a central component of sustainable development, and the connection between this perspective and the more conventional economics is maintained.

Contrasts appear when sustainable development distinguishes between different *forms* of economic growth. As Brundtland notes, "harmful growth" tends to "erode the environmental resources" upon which it is based while "[s]ustainable development seeks to meet the needs and aspirations of the present without compromising the ability to meet those of the future."⁵⁹ In other words, sustainable development defines "bad development" or "harmful growth" as growth that is unmanaged or unplanned. It is development that is produced exclusively by the workings of economic markets and by the pursuit of individual self-interest. Good growth, or "smart growth" as it is often called, requires not the abandonment of the market but its management and control in order for the "common good" of "future generations" to be taken into account.⁶⁰ On the global level, these calls for managed growth have become reflected in environmental "side agreements" to free-trade agreements such as the General Agreement on Tariff and Trade (GATT), the World Trade Organization (WTO), and regional trade agreements such as the North American Free Trade Agreement (NAFTA). In actual practice, the lack of international will, or consensus, on the fundamental question of achieving equity through wealth redistribution, has resulted in most international and national efforts at sustainable development becoming little more than pleas for more economic growth and faith in newer, cleaner technologies.

To more than one critic, this resurrected faith in economic growth and technology marks mainstream sustainable development as simply a new form of cornucopian thought. John Dryzek, for instance, notes that mainstream sustainable development has been more effective as a rallying discourse that seems to appeal to practically everyone than as an agenda that shapes actual policy decisions. In 2005, Dryzek commented, "[T]he twenty years that have seen sustainable development establish itself as the leading transnational discourse of environmental concern have seen much less in the way of the wholesale movements in policies, practices, and institutions at global, regional, national, and local levels which its advocates regard as imperative."⁶¹ David Carruthers is even harsher in his criticism. Writing in the Journal of Third World Studies, Carruthers releases acerbic scorn on the "Brundtland-Rio wordsmithing" and notes that this "defanged version of sustainable development" is little more than "a bundle of policies, myths, and faith" that "is premised on precisely the same economic injustices and biophysical impossibilities as the dominant discourse it once rose to oppose." He concludes by asking rhetorically, "Can we really construct a global political economy in the shape of a pyramid, and then hold out as a goal for everyone to occupy its apex?"⁶²

Criticisms of mainstream sustainable development can be gathered around the following claims: (1) It has become a meaningless cliché: "sustainable" and "sustainability" are used to describe processes that cannot possibly be sustained, or they become misapplied synonyms for "efficiency" and "resource conservation." (2) In suggesting that environmental protection can be a "win-win" or we can "do well while doing good," it blinds us to the fact that there are very difficult political (social/communal) decisions which need addressing. We need to have a collective debate on what a "just society" or "harmonious world" will look like. Asserting that debate is not necessary moves us in exactly the wrong direction. (3) Because of the failures listed earlier, the need for social equity gets shunted aside and never really given serious consideration. Environmental devastation created by over consumption by the rich and resource exhaustion by the poor is the predictable result. (4) Because mainstream sustainable development will not, or cannot, address the political and ethical issues stated earlier, it falls back on the old economic fantasy that more exponential economic growth will allow wealth to "trickle down" to those below. But since any environmental discourse that seeks to be practical or realistic must begin by recognizing the essential finiteness of resource inputs and output sinks, "sustainable growth," is an unachievable oxymoron. To suggest that technology and modernization can "delink" or "decouple" economic and ecological processes is simply collective social myopia.

In the end, mainstream sustainable development becomes merely another argument for managed economic growth instead of market-driven growth, another apology for "free trade" and globalization, another plea for more efficiency while acknowledging that efficiency doesn't tell the whole story, another dream that some technological miracle will come to save us, and another desperate delusion that the future will be better while doing pathetically little to bring that better future into existence. Small wonder David Caruthers cites Julian Simon and the Brundtland Commission in the same paragraph. They belong together. In truth, mainstream sustainable development has degenerated into nothing more than cornucopian thought for the politically correct and environmentally chic. It is environmental politics for the liberal brie cheese and chardonnay set.

Conclusion: Beyond the Mainstream

For the past half-century, America and the rest of the world have lurched and stumbled from one set of environmental agendas to another. At one level, there appears to be a concern for humanity's relationship to the Earth that is genuine and shared by almost all except the most recalcitrant cornucopians and free marketeers. At the same time, however, uncertainty and indecision are met when the proper balance between regulation and markets, expert analysis and citizen opinion, or compassion for the Earth and fraternity with our fellow humans is sought. If the direction we want to move is fairly definite, the same cannot be said for the means we feel most confident will get us there, or, indeed, exactly what the "there" is that should stand as our collective objective. Somehow, old aspirations of "more," "progress," "civilization" or even "sustainability" appear less easy to define and defend. Exactly what do we want more of, where is the pursuit of progress leading us, and what are we trying to sustain? Each of the five environmental discourses previously discussed falls short of articulating an agenda that is inclusive enough and deep enough to genuinely comprehend the economic, political, and ethical dilemmas we face. A major reason for that failure may be the fact that none of these mainstream discourses analyzes and challenges the premises and logic of our basic economic system, the political institutions and practices we use to address problems, or the value premises and ethical assumptions that precondition our thinking on all social concerns. Moving ahead socially may entail moving down analytically, down to a level of investigation that isolates and challenges those received assumptions that we too often blindly accept because we fear their exploration takes us into the world of "radicalism." As practical people, we have been so worried about agendas and policies that are feasible that we have ignored those that may actually be workable. Increasing numbers of scientist, economists, political theorists, and ethicists have challenged us to think creatively about our condition, to call into question the intellectual rules under which we operate, and to set our imaginations free in the exploration of alternatives. This group of radical environmentalists is still numerically small, but their numbers grow as our problems worsen and the logic of their arguments becomes seen as being unassailable. As biologist Paul Ehrlich has astutely noted, "It must be one of the greatest ironies of the human species that the only salvation for the practical men now lies in what they think of as the dreams of the idealists. The question now is: can the realist be persuaded to face reality it time?"[63] To those realistic but radical environmentalists, attention now needs to be directed.

Notes

1. See, for example, Ralph Huitt, "Political Feasibility," in Ira Sharkansky, ed., <u>Policy Analysis in Political Science</u> (Chicago: Markham Publishing Company, 1970).

2. Garry Brewer and Peter deLeon, The Foundations of Policy Analysis (Monerey, CA: Brooks/Cole, 1983). For specific application of policy cycle or "stages" to environmental policy, see, Norman J. Vig and Michael E. Kraft, "Environmental Policy from the 1970s to the Twenty-first Century," in Vig and Kraft, Environmental Policy: New Directions for the Twenty-first Century (Washington, D.C.: Congressional Quarterly Press).
3. E.E. Schattschneider, The Semisovereign People: A Realist's View of Democracy in America (Hillsdale, IL: The Dryden Press, 1960), 66.
4. James Lester and John S. Dryzek, "Alternative Views of the Environmental Problematic," in J. Lester, ed., Environmental Politics and Policy: Theories and Evidence, 2nd edition, (Durham, NC: Duke University Press, 1995).
5. John S. Dryzek, The Politics of the Earth: Environmental Discourses (Oxford and New York: Oxford University Press, 1997), Chapter 3.
6. A review of anti-environmentalism can be found in Paul and Anne Ehrlich, The Betrayal of Science and Reason: How Anti-Environmental Rhetoric Threatens Our Future (Washington, D.C. and Covelo, CA: Island Press, 1996), and Michael E. Mann, The Hockey Stick and the Climate Wars: Dispatches From the Front Lines (New York: Columbia University Press, 2012).
7. Norman Myers and Julian Simon, Scarcity or Abundance? A Debate on the Environment (New York: W.W. Norton and Company, 1994), 39–40.
8. North Carolina House Bill 819, reported by Alon Harish, "North Carolina Bans Latest Scientific Predictions of Sea-Level Rise," ABC News Internet Ventures, August 2, 2012.
9. Council on Environmental Quality and Department of State, The Global 200 Report to the President, 2 vols. (Washington, D.C.: US Government Printing Office, 1980).
10. Julian L. Simon and Herman Kahn, eds., The Resourceful Earth: A Response to Global 2000 (Oxford: Basil Blackwell, 1984).
11. Ibid., 7.
12. Julian Simon, "There Is No Crisis of Unsustainability," in G. Tyler Miller, Jr., Living In The Environment, 8th edition (Belmont, CA: Wadsworth Publishing Co., 1994), 24.
13. Julian L. Simon and Herman Kahn, eds., The Resourceful Earth: A Response to Global 2000, op. cit., 65 and 64.
14. John S. Dryzek, The Politics of the Earth: Environmental Discourses, 2nd edition (Oxford and New York: Oxford University Press, 2005), 57.
15. John M. Polimeni, Kozo Mayumi, Mario Giampietro, and Blake Alcott, The Myth of Resource Efficiency: The Jevons Paradox (London and Sterling, VA: Earthscan, 2009).
16. Terry L. Anderson and Donald R. Leal, Free Market Environmentalism, revised edition (London: Palgrave, 2001).

17. Ibid.
18. Ibid., 15.
19. Terry L. Anderson and Donald R. Leal, Free Market Environmentalism, 1st edition. 2.
20. Anderson and Leal, Free Market Environmentalism, revised edition. 74.
21. Terry L. Anderson and Donald R. Leal, Free Market Environmentalism, 1st edition. 21.
22. Anderson and Leal, Free Market Environmentalism, revised edition. 35.
23. Ibid., 25.
24. Ibid., 10–11.
25. Ron Paul, The Revolution: A Manifesto (New York and Boston: Grand Central Publishing, (2008). 105.
26. Ibid.
27. Senator Rand Paul, Government Bullies (New York: Hachette Book Group, 2012).
28. Indur M. Goklany in Clearing the Air: The Real Story of the War on Air Pollution, (Washington, D.C.: The Cato Institute, 1999) argues that federal air quality regulations are unnecessary and ineffective since air quality has been steadily improving through unrestricted market forces. Patrick J. Michael in Climate of Extremes: Global Warming Science They Don't Want You to Know (Washington, D.C.: The Cato Institute, 2009) and Meltdown: The Predictable Distortion of Global Warming by Scientist, Politician, and the Media (Washington, D.C.: The Cato Institute, 2005) shifts from denying the facts of human-induced climate change to admitting the reality of the global phenomenon and then claiming the consequences will be socially insignificant. Michaels, it should be noted, is a senior fellow at the Cato Institute, who while chastising those scientists who receive governmental funding for allegedly bending to the agenda of their sponsors, makes no such assertion regarding his own studies which are financially backed by the oil and gas industry. Libertarian organizations such as the Federalist Society have adopted judicial positions in opposition to most sincere attempts at ameliorating environmental destruction.
29. Mark Sagoff, The Economy of the Earth (Cambridge: Cambridge University Press, 1988).
30. Mark Sagoff, The Economy of the Earth. See especially Chapter 2, "At the Shrine of Our Lady of Fatima; or, Why political questions are not all economic." 24–49.
31. Ibid., 127 and 65, emphasis in original.
32. John S. Dryzek, The Politics of the Earth, 63.
33. Robert C. Paehlke, Environmentalism and the Future of Progressive Politics (New Haven, CT: Yale University Press, 1989) 117.
34. John S. Dryzek, The Politics of the Earth, 85.

35. Robert Paehlke and Douglas Torgerson (eds.), Managing Leviathan, as cited in John S. Dryzek and David Schlosberg, Debating the Earth, 2nd edition, (Oxford: Oxford University Press,2005), 173 and 175.
36. For a review and critique of command-and-control, see A. Myrick Freeman, "Economics, Incentives, and Environmental Regulation," in Vig and Kraft, Environmental Policy in the 1990s, 3rd ed., pp. 194–195. Frederick R. Anderson, Daniel R. Mandelker, and A. Dan Tarlock, Environmental Protection: Law and Policy, pp. 65–71.
37. For a further discussion, see Walter A. Rosenbaum, Environmental Politics and Policy, pp. 161–181; and Richard N.L. Andrews, "Risk-Based Decision making," in Vig and Kraft, Environmental Politics in the 1990s.
38. Riley E. Dunlap, "Public Opinion and Environmental Policy," in James P. Lester, Environmental Politics and Policy, 3rd ed., pp. 63–114.
39. A. Myrick Freeman, "Economics, Incentives, and Environmental Regulation," in Norman J. Vig and Michael E. Kraft (eds.), Environmental Policy in the 1990s, 2nd edition (Washington, D.C.: Congressional Quarterly Press, 1994). Tom Tietenberg, Environmental and Natural Resource Economics, 3rd edition. Eban S. Goodstein, Economics and the Environment, 4th edition (Hoboken, NJ: John Wiley & Sons, Inc., 2005).
40. See Tietenberg, Environmental and Natural Resource Economics, Chapter 3, and Goodstein, Economics and the Environment, Chapter 3.
41. Tietenberg, Environmental and Natural Resource Economics, 52 and 56.
42. A. Myrick Freeman, "Economics, Incentives, and Environmental Regulation," in Norman J. Vig and Michael E. Kraft (eds.), Environmental Policy in the 1990s, 191.
43. Eban S. Goodstein, Economics and the Environment, 205.
44. This criticism is raised by Mark Sagoff, The Economy of the Earth (New York: Cambridge University Press, 1988).
45. M. Jacobs, The Green Economy, (London: Pluto Press, 1991).
46. On the difficulties encountered with contingent valuation, see Walter A. Rosenbaum, Environmental Politics and Policy, 3rd edition (Washington: D.C.: Congressional Quarterly Press, 1995).
47. Tietenberg, Environmental and Natural Resource Economics, 73.
48. Lisa Heinzerling and Frank Ackerman, Pricing the Priceless: Cost-Benefit Analysis of Environmental Protection (Washington, D.C.: Georgetown Environmental Law and Policy, 2002), cited Eban S. Goodstein, Economics and the Environment, 205.
49. World Commission on Environment and Development (Brundtland Commission), Our Common Future (Oxford and New York: Oxford University Press, 1987); Daniel Sitarz, ed., Agenda 21: The Earth Summit Strategy to Save Our Planet (Boulder, CO: Earthpress, 1994). The President's Council on Sustainable Development, Sustainable America: A

New Consensus for Prosperity, Opportunity, and a Healthy Environment For the Future (Washington, D.C.; U.S. Government Printing Office, 1996). The Report of the National Commission on the Environment, Choosing a Sustainable Future (Covelo, CA: Island Press, 1993); Al Gore, Earth In the Balance: Ecology and the Human Spirit (Boston: Houghton Mifflin Co., 1992). Thomas L. Friedman, Hot, Flat, and Crowded: Why We Need a Green Revolution—And How It Can Renew America (New York: Farrar, Strauss, and Giroux, 2008).
50. President's Council on Sustainable Development, Sustainable America, p.7.
51. See Dryzek, The Politics of the Earth, Chapter 8.
52. Norman J. Vig and Michael E. Kraft, "Conclusion: The New Environmental Agenda," p. 139. See also Walter A. Rosenbaum, "The Clenched Fist and the Open Hand: Into the 1990s at EPA," in Vig and Kraft, Environmental Politics in the 1990s, 2nd ed.
53. Ibid., trial projects include the Green Lights Program to reduce energy consumption; the 33/50 Program that sought a 33 % reduction in the releases of 17 of the most toxic chemicals by 1992 and another 50 % reduction by 1995; and the Water Alliances for Voluntary Efficiency (WAVE) that sought water reduction by the lodging industry.
54. On risk assessment and policy evaluation, see Vig and Kraft, Ibid.; Richard N.L. Andrews, "Risk-Based Decision making," and Robert V. Bartlett, "Evaluating Policy Success and Failure," both in Norman J. Vig and Michael E. Kraft, Environmental Policy in the 1990s, 2nd ed.
55. For a case study of agency capture, see Sara Singleton, "Cooperation or Capture? The Paradox of Comanagement and Community Participation in Natural Resource Management and Environmental Policymaking," Environmental Politics, June 2000.
56. See Julia M Wondolleck and Steven L. Yaffee, Making Collaboration Work (Washington, D.C. and Covelo, CA: Island Press, 2000); Philip Brick, Donald Snow, and Sarah Van De Wetering (eds.) Across the Great Divide: Explorations in Collaborative Conservation and the American West (Washington, D.C. and Covelo, CA: Island Press, 2001). Daniel Press, Democratic Dilemmas in the Age of Ecology (Durham and London: Duke University Press, 1994). John DeWitt, Civic Environmentalism: Alternatives to Regulation in States and Communities (Washington, D.C.: CQ Press, 1994). Benjamin Barber, Strong Democracy: Participatory Politics for a New Age (Berkeley and Los Angeles: University of California Press, 1984).
57. In this assumption, sustainable development is markedly different from the "limits to growth" position. See D. Meadows, J. Randers, and W.W. Behrens, The Limits to Growth (New York: Universal Books, 1972). M. Mesarovic and E. Pestel, Mankind at the Turning Point (New York: Dutton, 1974). Council on Environmental Quality and the Department of

State, Global 2000 Report to the President, 2 volumes (Washington, D.C.: Government Printing Office, 1980). J. Stockdale, "Pro-growth, Limits to Growth, and the Sustainable Development Synthesis," Society and Natural Resources, vol, 2 number 3, 1989, 163–176.
58. The Brundtland Commission, for example, believes that first world nations can continue to have economic growth rates of between 3–4 %, while third world countries need to have economic growth rates in the range of 5–6 %. World Commission on Environment and Development (Brundlatnd Commission), Our Common Future, 49–52.
59. Ibid., 39 and 40.
60. Robert Repetto, The Global Possible: Resources, Development, and the New Century (New Haven, CT: Yale University Press, 1985).
61. John S. Dryzek, 2nd edition, The Politics of the Earth, 160.
62. David Carruthers, "From Opposition to Orthodoxy: The Remaking of Sustainable Development," Journal of Third World Studies, 18: 2 (2001), 93–112. Reprinted in John S. Dryzek and David Schlosberg, Debating the Earth, 2nd edition (Oxford and New York: Oxford University Press, 2005), quotations at 294, 290, 291, and 294.
63. Paul Ehrlich, Population, Resources, Environment (San Francisco: W.W. Freeman, 1972), 444.

CHAPTER 3

Radical Environmentalism: Challenging Our Institutions and Beliefs

C. Wright Mills, professor of Sociology at Columbia University and inveterate iconoclast, revolutionized the study of power in America. In his two most important books, <u>White Collar</u> (1951) and <u>The Power Elite</u> (1956), Mills encouraged his readers to look beneath the surface of their society at the business, governmental, and military elites who control decision-making in this country. Mills saw a degree of pluralism in the elite, but there was a uniting force that defined them as being fundamentally different from the masses they managed. That characteristic was an elite consensus on the perspective through which they viewed the world and a basic agreement on the core values that governed their behavior. Because the elite came from similar social backgrounds and educational experiences, and because they tended to cooperate much more than they competed with each other, this shared perception became reinforced. It became an elite version of groupthink. Two years after the publication of <u>The Power Elite</u>, C. Wright Mills wrote another book with the intriguing title, <u>The Causes of World War Three</u>. Although it never achieved the fame of Mills' other work, <u>The Causes of World War Three</u> contains a takeoff on the theme of the elite consensus that might have consequences for the field of environmental politics.

 Mills never implied that the existence of a shared outlook and set of values meant that every member of the elite thought exactly the same. They agreed that some issues and perspectives were beyond the pale, felt many issues were open to variety of reasonable stances, and yet held strongly to certain core values. Outside the elite consensus are those perspectives and concerns that the elite removes from its thought process. These are issues and values considered too absurd, weird, impractical, or unethical to be given serious attention. They are taboo. For this reason, they are never deliberated or debated by the elite, and anyone who extends earnest consideration to these points of view labels himself as an intellectual outcast radical. Inside the parameters of the elite consensus are those topics that the elite is willing to debate, where solutions have not been agreed upon, and where, therefore, the elite agrees to disagree. Mainstream politics, as it is normally defined, takes place in this area. It consists of cooperative competition within the elite where they vie for solutions to problems they have agreed are the only ones deserving of significant attention.

At the very center of the elite consensus are the essential beliefs and values that define the elite's vision of the world. These are the sacred cows, the viewpoints and values the elite have been socialized to accept by their parents, clergy, friends, teachers, media, employers, and peers. While taboo subjects are not disputed because of an overwhelming rejection of their significance, the perspectives and principles of the sacred cows are also not debated, but in this case, for reasons opposite those of taboo subjects. The elite extends to this perspective unquestioning approval, not because they have scientifically or rationally explored it but because they were conditioned to defer to the people who taught it to them. The effectiveness and fairness of our political and economic institutions, the sanctity of private property, the requirement for economic growth, a boundless faith in technological optimism, and the value of individual humans all come to immediate mind. Recognition of this situation, however, gives rise to interesting questions: what is there to guarantee that the elite consensus—the elite's perception of reality—bears any relationship whatsoever to the actual physical, biological world as it truly exits? Remembering that it is the elite consensus that guides their management of society, is it not possible that the elite consensus could be nothing more than a shared, reinforced pattern of group delusion?

It is here that the relevance of <u>The Causes of World War Three</u> can be seen. Contained within Mills' argument is a theoretical concept that can be put to uses beyond the 1950s Cold War context. For Mills, the elite consensus on the inevitability of a thermonuclear exchange between the USA and the Soviet Union was the true "cause of World War Three." On one hand, the shared perception of the higher circles defined the world in which we all lived and was, in that sense, a critical component of reality.

Yet, at the same time, planning for and constructing the weaponry to execute the complete annihilation of the civilized world was irrational madness in its clearest, most objective form. The consensus of the elite, therefore, was both absurd and an accurate description of the world we occupied. Mills coined the colorful expression "crackpot realism" to describe an elite consensus that was widely shared yet completely at odds with normal notions of social health or the world as it actually exists.[1]

Crackpot realism is the self-delusion of the elites that they impose on the world through their power over the central institutions of politics, economics, the military, academia, and the information media. Mills describes crackpot realism as containing "the official definitions of the world reality clung to by the elites" and "the mask behind which elite irresponsibility and incompetence are hidden ...."[2] Although he rejected the idea that crackpot realism was evidence of a conspiratorial plot, Mills denounced both the unthinking, dangerous, incompetent power of the elites and the consensus upon which they operate, as well as the cowardly, unprofessional, complacent passivity of academics and intellectuals who fail to speak out and, instead, often became its associates and apologists in the "pretentious triviality of much that passes for social science." Never one to mince language, Mills decried the institutionalized power that put the elites in a position to control the affairs of the nation and argued "[t]hey are allowed to occupy such positions, and to use them in accordance with crackpot realism, because of the powerlessness, the apathy, the insensibility of publics and masses; they are able to do so, in part, because of the inactionary posture of intellectuals, scientists, and other cultural workman."[3] <u>The Causes of World War Three</u> is as compelling as anything C. Wright Mills wrote, and although its apocalyptic predictions have fortunately not come to pass, the concept of crackpot realism may remain analytically applicable today.

The idea of an elite consensus in defiance of the truth, a crackpot realism that the elites believe and the masses are taught, a collective pattern of social self-delusion that defies all attempts at refutation in spite of its repeated failures to adequately or morally guide human actions, remains a useful intellectual tool for describing and decrypting our social situation, for in their essence, all forms of radical environmentalism are united in accusing mainstream environmentalism of being merely so many variations on a common theme of crackpot realism. For radical environmentalists, the notion that our political, economic, and ethical systems are not major causes of environmental destruction is absurd nonsense.

The idea that a sustainable economic system can be built upon the pursuit of unending exponential growth on a finite planet is crackpot thinking in its most glaringly obvious form. For all forms of radical environmentalism, there exists something profoundly wrong with our institutions, our patterns of behavior, and our values, and the devastated natural systems we observe are, in one sense, simply outward signs of a dysfunctioning set of human creations. Both mainstream and radical environmentalists are looking at the same instances of environmental degradation, but they are defining, interpreting, and seeing them very differently. It is not the facts that are in question, but rather the interpretation and analysis of the data that is causing the disparity.

The point is not that radical environmentalists think being efficient, or reducing pollution, or recycling cans, bottles, and paper are wrong; in fact, they generally support such efforts. Radical environmentalists believe, however, that by themselves, such efforts are shallow and superficial. The word "radical" should, therefore, be understood in its most basic, almost medical, sense of "going to the roots" (as in the idea of "radical surgery").[4] The different forms of radical environmentalism then go on to propose solutions and an alternative social consensus that seeks to be genuinely realistic rather than crackpot. Radical environmentalism needs to be surveyed with an eye to the attack, the deeper roots that are explored, the redefinition of the causes of the crisis, and the solutions put forth as the only workable options to a world adrift toward ecological devastation. Finally, since environmental theories seldom spring into the world *ex nihilo*, comprehensiveness requires that the historical and intellectual origins of each form also be explored. There are at least five versions of radical environmentalism.

ECOLOGICAL ECONOMICS

The early formulation of ecological economics is generally considered the work of four men: Nicholas Georgescu-Roegen, Kenneth Boulding, Robert Costanza, and Herman Daly. Contemporary ecological economists include John Cumberland, Robert Goodland, Richard Norgaard, and Joshua Farley.[5] Of these, Daly has proven to be the most prolific author and the ecological economist with the widest impact. In Beyond Growth, he recounts an incident that happened while he served as an economist at the World Bank. In 1992, the World Bank produced a report entitled Development and the Environment, which like other mainstream reports assumed the economy was independent of the ecosystem and consisted simply of "inputs" and "outputs." Daly believed, however, that since economic activity took place on planet Earth, the economy was a subset of the environment. He wanted to raise the question regarding how long exponential economic growth could continue on a planet with limited resources and limited space. He remembers discussing one of the report's diagrams at a meeting with the Bank's chief economist:

> During the question-and-answer time I asked the chief economist if, looking at that diagram, he felt that the question of the size of the economic subsystem relative to the total ecosystem was an important one, and whether he thought economists should be asking the question, What is the optimal scale of the macro economy relative to the environment. His reply was immediate and definite: "That's not the right way to look at it."[6]

In addition to illustrating the essential difference between ecological economics and all forms of mainstream economics, this story also says something about the relative political posture of the two perspectives. Shortly after the incident related earlier, Herman Daly resigned from the World Bank in frustration over their lack of solidly based environmental policies. The World Bank chief economist who thought recognizing that economic activity takes place inside an ecosystem is "not the right way to look at it" was Lawrence H. Summers. After leaving the World Bank, Summers went on to become Bill Clinton's Secretary of the Treasury, President of Harvard University, managing partner of a Wall Street hedge fund, and then joined the Obama administration in 2009 as director of the National Economic Council where he presumably went on denying that the economy is a subset of the ecosystem.[7] Those who challenge the prevailing elite consensus meet with frustration, while those who support the current version of crackpot realism become Cabinet Secretaries, Ivy League presidents, and administrative insiders.

Ecological economics believes recognizing the economy as a subset of the ecosystem *is* the right way to look at it. If, as traditional economics (used here to include both neo-classical economics and environmental economics) believes, the economy is freestanding and not a subsystem of the ecosystem, then it can grow forever as long as "inputs" keep being supplied and a place exists to consume, discard, or reuse all the "outputs." In traditional economics, there is no consciousness of limits because nothing is recognized that could do the limiting. By contrast, acknowledging the economy as a subset of the environment changes everything. It becomes possible to transfer the notion of "carrying capacity" from ecology and incorporate it into the study of economics. Carrying capacity, the limits imposed on population by finite space and finite resources, is an essential intellectual tool for ecologists, yet until the advent of ecological economics, it has seldom been used in any rigorous, interdisciplinary manner. It is almost never mentioned in articles or textbooks, and it is this oversight that allows traditional economists to focus almost exclusively on issues of allocation and distribution while ignoring the critical subject of scale. Of course, scale, or the size of economic activity, is not ignored completely by economists, but the discussion is typically reserved for courses in microeconomics, where one hears talk of relative scarcity, zero-sum tradeoffs, and Pareto optimums.[8] When attention shifts to macroeconomics, however, the focus becomes the "win-win" benefits of free trade, globalization, input substitutability, the ability of technology and capital to solve all problems, and the infinite, exponential growth of the economy. Viewed in relationship to these issues, ecological economics can be seen as taking the insights of microeconomics and applying them at the macrolevel by introducing the concept of "scale" to the standard topics of allocation and distribution. With the introduction of the idea of carrying capacity into economics, it can be seen that not only can economies be efficient or inefficient; just or unjust; they can also be *sustainable* or *unsustainable*.

Although ecological economics is a relatively new, interdisciplinary field of study, the idea of moving to a no-growth economy can be traced back at least to the middle of the nineteenth century. Herman Daly has long acknowledged his debt to the thought of John Stuart Mill, who, in the 1848 edition of his Principles of Political Economy, wrote "[I]n contemplating any progressive movement, not in its nature unlimited, the mind is not satisfied with merely tracing the laws of the movement; it cannot but ask the further question, to what goal? ... It must always have been seen, more or less distinctly, by political economists, that the increase of wealth is not boundless: that at the end of what they term the progressive state lies the stationary state ...."[9] Advocating the shift to a no-growth, steady-state, stationary economy places ecological economics at loggerheads with traditional, mainstream economists. John Stuart Mill's assertion that political economists understand the impossibility of infinite growth has not proven to be the case for most economists. In spite of the lessons from thermodynamics and ecology, neo-classical and environmental economics continue to preach the possibility and necessity of infinite, exponential growth. Economic growth remains one of the most sacred of crackpot realism's sacred cows.[10]

Although they seldom reveal their radicalism with straightforwardness, ecological economists do not always shy from intellectual fights with the mainstream. Prominent has been their criticism regarding recent trends in the globalization of the international economy, free trade, and institutions such as the North American Free Trade Agreement (NAFTA), the World Bank, the International Monetary Fund (IMF), the General Agreement on Tariffs and Trade (GATT), and the World Trade Organization (WTO). Ecological economists are not opposed to all trade—they often voice their approval for balanced or fair trade—but as currently structured by these institutions, globalization and free trade stand as intractable obstacles to an environmentally sustainable economic system.

Once their underlying logic is exposed, the differences between the globalized free trade and ecological economics could not be more stark. From a social, economic, and ecological perspective, free trade advocates want national specialization, international economic interdependence, globalization, and continuous, unending economic growth.[11] Ecological economists and opponents of free trade want national (and even local) self-sufficiency, the opposite of specialization (which may be called economic diversification or community omni-competence), local control of economic decision-making, within a stationary, or steady-state economy. Ecological economists argue that our current system of free trade carries with it a number of very negative political and ecological consequences, and that the system is based upon an economic logic that does not fit with the realities of world production and finance. Free trade, in other words, is both environmentally disastrous and logically flawed.

In order for environmental protection to be compatible with free trade, a global system of strict environmental regulation would have to be in place, and increasing efficiency would have to be the only way private corporations could enhance per unit profit margins. Unfortunately, for supporters of economic globalization, neither of these conditions exists. The environmental problem free trade presents is that globalization gives corporations multiple opportunities to lower costs by reducing benefits or externalizing costs. Wide disparities exist among nations with regard both to policies designed to protect the environment and the enforcement of those policies officially on the books but seldom implemented. In fact, several countries openly tout their anti-environmentalism as a way to encourage corporations from the developed world to relocate.[12] Absent incentives to do otherwise, there is no reason to believe businesses will not take advantage of those opportunities. Global free trade and national specialization provide poor countries with a strong incentive to attract corporations by lowering their social and environmental standards with a resulting "race to the bottom." As Daly observes, "All these social and environmental measures raise costs and cannot withstand the standards-lowering competition induced by free trade with countries that have lower standards."[13] The empirical signals, therefore, are mixed. Occasionally, globalization results in increased efficiency, a reduction of waste, and a cleaner environment. At other times, free trade leads to "competing down" protection standards, threats to move jobs if existing regulations are strictly enforced, and a dirtier, less healthy environment. Opponents of free trade contend that instances of the latter type far outnumber those of the former, but where there seems to be little debate is over the fact that globalization produces increased specialization and economic growth.

Ecological economics, then, is the radical alternative to environmental economics. Ecological economics explores deeper into the sources of ecological destruction and pictures exponential growth, increased economic scale, and market globalization as the root causes of a fiscal system that has breached the Earth's ecological carrying capacity. In response to this market-driven, growth-oriented capitalist economy, ecological economists have drawn a fairly clear picture of the type of economy and society they would like to see implemented. Again, using Herman Daly as the spokesman for this approach, his recommendations can be grouped into three categories or sets of institutions: a socially established method for maintaining a constant population; an organized set of policies for maintaining a constant stock of physical wealth; and an institutional scheme for limiting the degree of inequality in the distribution of the steady-state wealth to the steady-state population.[14] Daly is skeptical that individual choice alone will lead us to a situation of zero population growth. The optimal size of a nation's population should be a collective decision and can, in all probability, only be achieved by using some system of socially regulated incentives and sanctions.[15] The second component of Daly's alternative is the maintenance of a constant stock of physical wealth. This stationary-state economy would encourage a shift to renewable resources (particularly energy resources), promote recycling, and, where possible, provide incentives to improve the efficiency of production. Explicit policy recommendations here include a system of protective tariffs, depletion quotas, the removal of governmental subsidies for certain industries (such as nuclear power), and a strict enforcement of anti-trust laws. Behind these policies (the working and implications of which Daly draws out in some detail) lie a few basic principles or guidelines to direct and shape the process: a commitment to invest in natural capital, as opposed to human-made capital; a redirection of economic activity away from global or national scale toward local or regional self-sufficiency; and an alternative accounting systems that would replace current Gross Domestic Product (with its focus on measuring only quantitative throughput) with an Index of Sustainable Economic Welfare that seeks to unite a realistic appraisal of the economy's performance with a standard for judging qualitative impacts on distributive justice and ecological sustainability.[16]

One of the perverse consequences of economic growth is that it permits society to temporarily avoid the discussion of distributive justice. Once the long-term impossibility of exponential growth becomes apparent, a serious communal consensus on economic justice and wealth distribution would become essential. Early in his work, Daly recognized, "Distribution is the rock upon which most ships of state, including the stationary state, are very likely to run aground."[17] He recommends a tax system that affords a negative income tax for anyone earning below the minimum and a confiscatory progressive tax on all incomes above a pre-determined level. Inside this range, personal incomes would be allowed to fluctuate depending upon the initiative, ambition, or luck of the individual, but everyone in the society would know that a collective decision had been reached on the broad outlines of economic equality and that a cap had been set on the scale of human financial aggrandizement.[18]

With deeper reflection, ecological economics can be recognized as a particular manifestation of a long-established trend in Western thought. The writings of ecological economists are filled with language that depicts the need for individual interests to be made subservient to the common good, for the rights of the individual to be balanced against the necessities of the community, and for the corrupting influence of personal acquisitiveness to be subsumed beneath the duties of the civic-minded citizen. This is the discourse of civic humanism or the classical republican tradition and demonstrates the logical incompatibility with the self-interested individualism of economic liberalism. It is an alternative to neo-classical capitalism and environmental economics that still remains within the established traditions of American political thought. As he is always eager to point out, Herman Daly is no friend of socialism. His point is not to abolish the private ownership of property (in fact, he sees himself as defending private property) but simply to make its institutionalized practice ecologically sustainable. In the history of economic thought, Daly et al. are best seen as advocates of corporatist economics—neo-physiocrats or neo-mercantilists—with a distinct preference for small-scale operations.

Although ecological economists implicitly charge mainstream economics with being crackpot realism, there are certain aspects of the elite consensus to which ecological economics continues to adhere. While they are highly skeptical of free market, individualistic capitalism and although their commitment to democracy displays a preference for decentralized populism rather than national representative liberalism, in the area of ethics, there is no appreciable disagreement between ecological economics and the sacred ethical cows of mainstream elitism. The belief that *Homo sapiens* are ethically distinct from and morally superior to the rest of Nature is just as much a part of Herman Daly's thought as it is part of the environmental politics of Julian Simon, Myrick Freeman, or Al Gore. Daly and co-author John Cobb (who teaches theology) reaffirm their commitment to Christian morality and recognize that such a commitment is incompatible with any form of ecocentric ethics. At the close of For the Common Good, Daly and Cobb strain to make ecological economics compatible with Christian ethics even while acknowledging, "There is no doubt that the biblical call for dominion has been responsible for much unjustified cruelty and destruction."[19] Still, the two men find themselves intellectually incapable of carrying their radicalism into the area of ethics. Ecological economics, then, is radical in its economic stance, reformist in its politics, and mainstream in its approach to environmental ethics. Exploring radical environmentalism at a deep level entails moving beyond economic theory to more foundational levels of politics and morality, and that, in turn, means moving beyond or beneath ecological economics.

Social Ecology

Although all other environmental discourses are the products of multiple minds, social ecology is so much the creation of one man that the movement is virtually synonymous with his name. The man is Murray Bookchin, and for over 40 years, he articulated and vigorously defended the perspective he created and named.[20] Bookchin's social ecology is important because it provides penetrating insights into the possible sources of environmental destruction, offers a theoretical and historical critique of how we arrived at our current predicament, and points the general direction toward a more healthy, sustainable, and liberated existence for both humans and Nature. Whereas the thought of ecological economics places emphasis on economic theory, Bookchin stresses the importance of political theory. Readers familiar with classical Western philosophy dating back to the Greeks and Romans or the theoretical work of the early Enlightenment will easily place Bookchin in this tradition. Where he takes Western humanism, however, earns Bookchin a unique place in environmental and political thought. Bookchin's writings go a long way toward fulfilling the need to "stir the imagination into creating radically new alternatives to every aspect of daily life."[21]

When Bookchin writes of environmental concerns, there is remarkably little discussion of actual physical or biological issues. For social ecologists, environmental destruction, the upsetting of the balance found in natural ecosystems, and the undermining of the resource base needed to sustain life processes are profound but still superficial signs of deeper, more rudimentary dysfunctions. According to Bookchin, environmental destruction is the outward, tangible manifestation of social systems that are incapable of functioning in a free, participatory, egalitarian fashion. Hence, there is an aptness in Bookchin's choice of the term "social ecology." As the founder of the movement expresses it, "ecology alone, firmly rooted in *social* criticism and a vision of *social* reconstruction, can provide us with the means for remaking society in a way that will benefit nature *and* humanity." The questions that need to be probed, therefore, are "What factors have produced ecologically harmful human societies? And what factors could yield ecologically beneficial human societies?" Bookchin's perspective can be summarized in the assertion that "*all ecological problems are social problems.*"[22] His intricate and elaborate theory is a blend of Hegelian dialectics, evolutionary theory, social history, nineteenth century anarchist thought, and ethological explanations of human behavior.

One writer toward whom Bookchin acknowledges indebtedness is the nineteenth century German philosopher, Georg Wilhelm Friederich Hegel. Incorporating Hegelian dialectics within environmental theory permits Bookchin's social ecology to blend rationalism into a dynamic theory of evolution and history. At first glance, the whole process might come across as being incredibly and unnecessarily cerebral, but after Bookchin has had his say, the complete vision of social ecology fits neatly into place and becomes, in a way, elegant. Dialectics provides a method of understanding and a method of judging where events can be comprehended and the course of their development can be evaluated. Rational explanation and normative judgment—facts and values—become blended or transcended. "Whether a society is 'good' or 'bad,' moral or immoral," claims Bookchin, "can be objectively determined by whether it has fulfilled its potentialities for a rational and moral society."[23] If the foundation of Hegel's philosophy can be characterized as dialectical idealism, and Karl Marx's theory can be seen as resting on dialectical materialism, Murray Bookchin accurately labels his thought as "dialectical naturalism."[24]

One of Bookchin's major contributions to environmental politics is his recognition that environmental destruction goes beyond individual lifestyle choices. In modern societies, human behavior is largely channeled through institutions and the impact those institutions have on the character of the humans operating within them cannot be ignored. For all of its celebration of the individual, the Enlightenment depended, and modernity continues to depend, on political, economic, educational, and social institutions in order to structure thought, values, and behavior.

In order to fully comprehend what it means to be a human being, Bookchin argues, it is necessary to have a dialectical understanding of humans as biological entities, humans as social creatures, and humans as evolving life forms exploring and actualizing the full potential of these characteristics. Human society grows out of the biological facts of our instinctive makeup. Many animals have genetically derived behavior patterns blended with learning from their surroundings, but Bookchin emphasizes that humans are unique in our ability—indeed in our requirement—to produce cultural traditions built around speech, elaborate conceptual powers, and the institutionalized ability to reshape purposively our surroundings.[25] This "prolonged degree of human plasticity, dependency, and social creativity," he maintains, produces two very important results. First, all humans have a high level of interdependence that draws them to other humans in order to satisfy a wide range of physical and psychological needs. Second, this human interdependence can be manifested in a wide variety of highly structured forms. Humans produce much more than innately gregarious communities; we produce very complex and ordered societies, and it is the very complexity of those societies that allows them to be built around behaviors that are oppressive or those that are liberating.[26]

If human behavior represents an unfolding dialectical process of instincts and learning; of first nature and second nature; of individual adaptation and institutionally structured conduct, then to what extent can our knowledge of evolutionary processes serve as a guide to how we should shape and build social institutions? Can evolution be used as a model for society?

The applicability of ethological principles to society is a controversial topic, but one that can be traced back over a century and a half to the reactions generated by the publication of Charles Darwin's Origin of Species. The movement to apply Darwinian theory to society became known as Social Darwinism. In the intervening years, so many Social Darwinists used evolutionary theory to support politically conservative, if not outright reactionary, agendas that Social Darwinism has become almost synonymous with this right-wing variant.[27]

There is another side to the argument, however. The Russian anarchist prince, Peter Kropotkin constructed a powerful rebuttal to the dominant conservative interpretation. Although he also accepted Darwin's theory, Kropotkin pointed out that mutual aid was every bit as important in species' survival as competitive struggle and that applying evolutionary principles to human politics could just as easily result in a defense of cooperative, anarchist communities as in the justification of capitalist warfare. It is certainly intellectually fair to say that Kropotkin was as much a "Social Darwinist" as Herbert Spencer or T.H. Huxley, but Kropotkin was a Social Darwinist of the left.[28] Kropotkin pointed out that species who help their young, who warn others of approaching danger, who collectively defend the weak, and who resist forces that would tend to tear apart the group are truly the "fittest to survive." In human societies, there is very much a "struggle for existence," but it is not between one individual and another, it is between cooperative, communal, free communities and those who would destroy those free associations by imposing hierarchy, domination, oppression, and authority. These two paths, the path of cooperation and freedom, and the road of competition and oppression can be traced throughout the historical evolution of the human species. In two master works, Kropotkin outlined these mutually antagonistic but historically coexisting trends.[29]

Bookchin's social ecology produces an analysis of history that closely parallels Kropotkin's. As human institutions developed, they could have been built upon, and reinforced, our species' innate needs for cooperation and solidarity. In too many instances, however, institutions were constructed around the assumption that human instincts needed to be dominated, that chaos would result if hierarchical authority was not imposed, or that the process of oppressing nonhuman Nature was the necessary prerequisite for the liberation of humanity. Bookchin emphasizes, however, that history is not determined. It could have gone one way or the other, but the evaluation of these alternatives is not a relativistic individual preference.

Because of our instinctive first natures and the liberating potential of our second natures, some patterns of social interaction can be classified as objectively alienating, and some are liberating in a manner subject to ethological and historical verification. In his essay, "Post-Scarcity Anarchism," Bookchin declares, "the *preconditions* for freedom must not be mistaken for the *conditions* of freedom. The *possibility* of liberation does not constitute its *reality*."[30] In the conceptually broadest terms, social ecology seeks institutions and habits that mimic ecological systems. For Bookchin, this means processes that are complex, diverse, and resilient. Nature becomes not an enemy to conquer and dominate, but a partner in a joint project. Over the years, this picture of an alternative society has been given the name "libertarian municipalism" to indicate both the non-hierarchical diversity of its human relationships and the communal, collectivist social arrangements that blend an ecological sensitivity and a pastoral setting.[31] Bookchin, who died in 2006, left a legacy of voluminous writings spanning social theory, ecology, the Spanish Civil War, and anti-Marxist radicalism; recognition as a founder of American "ecoanarchism;" and a committed group of followers in Vermont and elsewhere who carry on his cause for decentralized "appropriate technology" and free communities.

Bookchin's belief that the future will be one of either escalating hierarchy and oppression, or a revolutionary movement toward the liberation of humans and Nature, begs comparison with another visionary inquiry into the human prospect. Francis Fukuyama's 1992 book, The End of History and the Last Man, became an instant classic in neo-conservative thought which has fascinating similarities with the work of Murray Bookchin.[32] Both men are committed Hegelians, both use dialectics to rationally explain and predict social history. Both are Social Darwinists (although neither uses the expression). For Fukuyama, however, the basic instincts or drives that precondition human social institutions are the need for material economic security and the spiritual need for respect, dignity, or what Plato called *"thymos."* Fukuyama argues that these needs are universal to our species and the quest to satisfy them has driven history up to this point. The consequence, he believes, is the creation of science and technology committed to the domination of Nature contained within liberal societies who acknowledge and protect human rights. Technological capitalism and liberal democracy, therefore, completely fulfill human's most basic needs. For Fukuyama, "all of the really big questions have been settled" and, as a result, "we have trouble imagining a world that is radically better than our own, or a future that is not essentially democratic and capitalist."

Social regression is, of course, an unfortunate possibility. "But we cannot picture to ourselves a world that is *essentially* different from the present one, and at the same time better."[33] Is this true?

Contrasting social ecology and neo-conservatism compels us to ask some critically fundamental questions: is liberal, technological, capitalism (the dream of cornucopians, free marketeers, and mainstream sustainable development) really the best society we can imagine? Have humans truly reached the "end of history?" Are there ecological and ethical negations within liberal, technological capitalism that necessitate the social evolution to a different system? Investigating some of those possible contradictions drives us even deeper into radical environmentalism.

The Neo-Malthusians

Progressive liberals and radical environmentalists are most likely to part company over the issue of human population control. To call for halting or reversing the growth of the human population, to suggest that giving birth is not a decision best left to individuals or families, seems to many to be oppressive and misanthropic. To suggest policies for halting the expanding American population—expansion most demographers acknowledge is mostly attributable to legal and illegal immigration—is often labeled nativist, xenophobic, or racist. The intensity and intractability of these conflicts are all signs that analysis and debate have moved beyond economic and political institutions, and have entered a more basic realm of ethical perspective. With the neo-Malthusians, radical environmental theory crosses a threshold and moves beyond, or beneath, the critique of institutions. For neo-Malthusians, the impacts of human overpopulation are pervasive and pernicious. "Too many cars, too many factories, too much detergent, too much pesticide, multiplying contrails, inadequate sewage treatment plants, too much carbon dioxide—all can be traced easily to *too many people*."[34] In less developed countries, the effects are normally felt as the degradation of the environment, while in developed countries (it would be more accurate to call them "overdeveloped countries"), the impacts manifest themselves by the depletion of renewable and nonrenewable resources caused by the excessive conversion of resources into consumable products. Perhaps the clearest summary of the neo-Malthusian position was stated by Paul Ehrlich in his book, The Population Explosion, "what ever your cause, it's a lost cause without population control."[35]

The harm caused by exponential population growth, the impossibility of sustaining exponential population growth, and a discussion of the means available to halt exponential population growth, were all first enunciated by the Reverend Thomas Robert Malthus in <u>An Essay on the Principle of Population</u> (1798).[36] Although expressed in terminology sometimes unfamiliar to the modern reader, Malthus outlined the argument that would become fundamental to many forms of contemporary environmentalism. What Malthus brought to the world's attention was the remarkable fecundity of the human species. Given that males and females are capable of reproductive activity from their early teens until their mid-forties, and given that females ovulate year round, the biological potential for increased numbers of humans is extraordinary. We are a species genetically poised to overpopulate.

Even while recognizing fertility (the actual number of children produced by the average women during her reproductive years) is much lower than the potential fecundity, any fertility rate greater than zero will generate an exponentially increasing population. In <u>How Many People Can the Earth Support?</u>, Joel Cohen notes, "if human populations continued to grow, in each major region of the world, at the rate presently observed in each, then the population would increase more than 130-fold in 160 years, from about 5.3 billion in 1990 to about 694 billion in 2150."[37] Clearly, these events will never occur. Something—a limiting factor—will intervene and keep the human population well below its theoretically possible maximum. As expressed by Leibig's Law, the limiting factor is the resource a given population needs that happens to be available in the lowest per capita abundance. The limiting factor helps define "carrying capacity." Writing at the close of the eighteenth century, Malthus was being both logically reasonable and factually correct to consider food the primary limiting factor for the size of the human population.

Although Malthus believed starvation the ultimate, brutal curb on population growth, he considered a number of alternatives that would also serve as checks on population expansion. He grouped potential limiting factors into two categories, which he named "misery and vice," or, in less inflammatory language, "positive checks and preventive checks."[38] A preventive check was anything that stopped potential children from being born. Here, Malthus made two important observations: first, the motivating factor behind preventive checks was the ability of couples to look into the future and practice self-restraint in their sexual activities, and second, in what would later become an important consideration in population issues, Malthus opined that while preventive checks operate throughout the various social classes, they seem to be most employed by members of the middle and upper classes.[39] Positive checks, on the other hand, consist of all those factors that lead to the early death of humans after they are born. These checks most frequently operate on the lower classes and among the poorer (less developed) nations. In the terminology of contemporary demography, preventive checks are associated with birth rates, and positive checks are classified with death rates. Regardless of the altered language, nothing has changed over the years to refute the basic correctness of Malthus' position—increasing population growth will be restrained either by decreases in the birth rate, or by increases in the death rate. Over the long term, there are no other alternatives.

Strategies for responding to population increase consist, therefore, of relying on higher death rates, encouraging lower birth rates; or short-term, intermediate tactics for expanding the human carrying capacity. Since higher death rates and lower birth rates both seem to challenge the dominant ethic of individual humanism, it is not surprising that, over the years, most attention has been directed toward methods for increasing the number of humans the Earth can support. Malthus was aware of, and discussed, these intermediate tactics, which he referred to as "palliatives." The total number of humans inhabiting the Earth can be enlarged (temporarily) either by adding additional portions of the planet predominantly dedicated to producing the means of subsistence for our species (now referred to as "takeover"), or by using technological innovation to discover new means for supporting humans (now called "drawdown" or "environmental modernization").[40] It is important to recognize, however, that there is a sharp difference between a temporary tactic that will postpone the inevitable time when population growth must stop and a fundamental refutation of Malthus' population theory. In other words, a "palliative" must not be confused with a "solution." There is a clear distinction between delaying the day of reckoning and "proving Malthus wrong."[41]

Any resolution of the population crisis; that is, any approach that extends beyond the "palliatives" of takeover, technological efficiency, and drawdown, must be built around either decreasing birth rates or increasing death rates. Recalling Malthus' observation that preventive checks were most often adopted by the middle and upper classes, and assuming that preventive checks are morally preferable to positive checks, then it is a small step to propose that if everyone could be made middle or upper class, these people would choose to reduce the size of their families through the voluntary use of birth control. The appeal of this approach is that it meshes the biological need to control population with political liberalism's preference for individualism and freedom of choice. Today, this is mainstream environmentalism's preferred method for addressing population issues. It is supported by a line of reasoning known as "the demographic transition."

The expression "the demographic transition" is used to signify the generalized population trend as societies shift from population explosions found in "the demographic trap" to stabilized population growth through a combination of low death rates and low birth rates.[42] It is argued that attaining essentially zero population growth is accomplished through a combination of factors including reduce infant mortality, higher educational levels for the general public (but especially for women), improved social services, and an economy characterized by higher skilled industrial labor. The cumulative effect of these factors is to create a more socially and economically developed nation where children become less of an economic necessity, where multiple births are not required to assure the survival of familial lineages, where access to and knowledge of birth control techniques are commonly available, and where women have an active, effective voice in making reproductive decisions. The result, according to the demographic transition thesis, will be a reduced wish for larger families and the joint determination by husbands and wives to limit the size and spacing of their families. From the Bruntland Commission, to Agenda 21, to President Clinton's Commission on Sustainable America, to the Worldwatch Institute, to Al Gore, to most mainstream environmental studies textbooks, hope for reaching sustainable human population levels is placed in the demographic transition and "family planning" (as the policy is normally called). It is here where the rupture between mainstream advocates and radical neo-Malthusians occurs. The issue, put simply, is that the mainstream accepts and trusts the demographic transition while neo-Malthusians explicitly reject such convictions. Recognizing the critical need to control human population growth is what separates environmentalists from anti-environmental cornucopians. Acknowledging the failure of the demographic transition and the need to move to a strong, more explicit population policy is what separates mainstream from radical environmentalists.

Beyond the historical fact that the demographic transition may be culturally bound (Saudi Arabia, for example, experienced a dramatic population explosion at the same time burgeoning oil revenues were creating economic affluence, social security, and increased, albeit still very limited, opportunities for women), reliance on the demographic transition overlooks the critical fact that reduced fertility rates are purchased by increases in individual affluence. That affluence means more consumption of resources, more conversion of land to agriculture and development, more per capita accumulation of material goods, and more creation of the waste and pollution associated with industrial production. The negative environmental impact associated with increases in human population is merely transferred to negative environmental impacts of affluence. Overall, the total negative impacts may not be reduced. In truth, moving through the demographic transition may actually increase the aggregate environmental damage being done. Paul Ehrlich has expressed these relationships as an equation, $I = P \times A \times T$, where I represents total environmental impact, P stands for population, A symbolizes affluence, and T signifies the relevant level of technology being applied (i.e., how efficiently the society is allocating resources; what level of toxic or hazardous waste is being produced; what degree of recycling and reuse is built into the applicable technology).[43] Once the correlation between population and affluence is recognized, then the devastating impact of the American life style can be seen and the essential ecological requirement for limiting the number of Americans must be acknowledged. Given the multiplier effect of population times affluence in determining environmental impact, the most overpopulated nation on the planet is not China, India, or Indonesia, but the United States.[44]

Neo-Malthusians advocate a shift in policy from a family planning agenda based upon individual choice to a national population policy based upon a communally designated system of incentives and sanctions. Like any effective policy, a national population policy must begin with setting a goal, or target, and a prospective time-line for reaching benchmarks on the way to that goal. In a democratic nation, that target should be the product of an open, transparent debate. What is the optimum population of the United States? Should we try to hold America's population at its current 310 million, or should we adopt policies with the intention of reducing American population to a more biologically sustainable and ethically defensible number of 150 million (America's population in 1940)? Reestablishing biological diversity, wilderness restoration, and respect for indigenous cultures would probably necessitate an American population living at a more simple level of affluence, using renewable resources for their energy demands, and numbering between 50 and 100 million people.[45] Simultaneous with this national debate on population size would be an ongoing discussion of the most appropriate and acceptable means of moving toward the proposed population target. Some preventive checks are still compatible with an overall voluntary strategy of family planning, and include wide-ranging reproductive education programs; inexpensive, easily accessible contraception; and the empowerment of women. These every society should adopt. Beyond these simple programs, more controversial, but more effective, policies would include free access to sterilization; safe, available, publicly-funded abortions; and options for the elderly to pursue death with dignity through such instruments as living wills.[46]

There is one population control measure that would make an important, immediate contribution in preserving American wildlife and open space. This is the control of our borders, the restriction of illegal immigration, and the deportation of those people who have entered the United States illegally. An effective population policy for America—one that recognizes our disproportionate negative environmental impact and also takes into consideration the ecological needs of other species and ecosystems—would include a strong system of incentives and sanctions designed to reduce the birth rate, coupled with an immediately implemented immigration policy based upon tough border security, stringent sanctions for employers who hire illegal immigrants, repeal of trade agreements that lower living standards in other countries, and a system of deportation that encourages voluntary repatriation.

This is the point where the sharply ethical component of the population debate becomes evident. Many might respond that given the choice between more humans in America and fewer wild species, why should we not opt for more humans? Anything broader than family planning appears to confront the fundamental individual humanism built into our culture. The apparent unlikelihood that any effective population policy or immigration control measures are forthcoming seems to signal that Americans are, as yet, unwilling to break out of their dominant moral perspective and include natural landscapes and other species in their ethical calculations. Garrett Hardin, for example, in a much maligned essay titled "Lifeboat Ethics" has suggested that Americans should act as if we are stranded passengers on a lifeboat—we should protect our own resources first, repel would-be borders (immigrants), and severely restrict the freedom of those on our small craft.[47] Hardin's lifeboat ethic suggests we constrict our moral horizon, but perhaps the opposite is true: we may need to expand our ethics, beyond white, black, brown, male or female *humans* to include other animals, plants, streams, mountains, plains, and wetlands. American environmentalism may need to search through its philosophic roots to find values and viewpoints built upon a culture of limits, simplicity, and lower consumption coupled with a sense of place and a love of the Earth that could act as the foundation for a sustainable biosphere and an expanding moral sense.

Bioregionalism

The cleverly subversive element in bioregionalism is that, on the surface, it appears within the tradition of American Populism, pragmatically moderate and, therefore, likely to find wide acceptance; while, at the same time, at a deeper intellectual level it holds the promise for a radical shift in our perceptions and our ethics. Contemporary American writers such as Kirkpatrick Sale, Wendell Berry, Daniel Kemmis, Wes Jackson, David Orr, Lynn Miller, Scott Russell Sanders, and a host of others have articulated an image of human relations and human/Nature relations premised on decentralized agrarian communities with a reduced scale of production and consumption, and a strong civic base in small, participatory democratic communities.[48] Among certain bioregionalist, there is a simultaneous articulation of an alternative environmental ethic built around the thought of Aldo Leopold that extends the concept of "community" to include other species, plants, watersheds, and mountains to eventually encompass all of "the land."

In <u>Community and the Politics of Place</u>, Daniel Kemmis reasons that we Americans have difficulty solving our political, social, and environmental problems because we lack the language necessary to find solutions to *any* public issue. Too often, legislative or open meetings are characterized by "shrillness and indignation ..., which is a symptom that something is profoundly wrong with the way we make 'public' decisions." Kemmis calls the pattern of individualism, self-interest, competition and conflict, private rights, and the isolated, autonomous utility-maximizing consumer America's "first language" of politics.[49] He draws attention, however, to the existence of a "second language" in American political discourse. Here, people speak of themselves not as isolated individuals but as members of communities. They talk about their neighbors, the history and traditions of their place, the need to cooperate toward some common objective, and their obligations as citizens. Throughout our history and all across our country, Americans have persistently spoken both languages. Kemmis calls for the rediscovery and resuscitation of that civic tradition, and links the development of active, virtuous citizens with attachment to a particular geographic location. "The kind of values which might form the basis for a genuinely public life, then, arise out of a context which is concrete in at least two ways. It is concrete in the actual things or events—the barns, the barn dances—which the practices of cooperation produce. But it is also concrete in the actual, specific places within which those practices and that cooperation take place."[50] This bond between citizenship and place—this "inhabitation"—is a concept that Kemmis traces to the work of Wendell Berry.

Along with Gary Snyder and Wes Jackson, Wendell Berry is the preeminent spokesman for agrarianism in America.[51] Berry's contribution to the bioregional perspective is to outline and delineate the values and practices that contribute to a culture being healthy or unhealthy, nurturing or exploitive, while recognizing that the two elements are, in present America, interwoven. "A healthy culture is a communal order of memory, insight, value, work, conviviality, reverence, aspiration. It reveals the human necessities and the human limits. It clarifies our inescapable bonds to the earth and to each other"[52] Our most basic relationship with Nature is the fact that we eat it—hence the importance of agriculture. Our most fundamental link with tools and techniques is the appreciation that nothing can do everything—hence the importance of appropriate technological diversity. And our most essential association with other humans is the simple fact that without them our lives lack direction, purpose, or meaning—hence the importance of communities.

For bioregionalists, any comprehension of environmental concerns must begin with a recognition of the connections and interdependencies, what Holmes Rolston calls the "entwined destinies," between humans and humans, living creatures and their food supplies, and humans and Nature.[53] Lack of connection leads to lack of restraint, and lack of restraint leads to exploitation. Berry argues that the modern person "assumes that there is nothing that he *can* do that he should not do, nothing that he *can* use that he should not use. His 'success'—which at present is indisputable—is that he has escaped any order that might imply restraints or impose limits."[54] Wendell Berry has no doubt regarding the basis for this value of restraint: only a powerful ethic, a community morality built into the culture, can provide the recognition of limits that will keep the human and ecological connections intact. He cites an example. "[T]he basic cause of the energy crisis is not scarcity; it is moral ignorance and weakness of character. We don't know *how* to use energy, or what to use it *for* .... If we had an unlimited supply of solar or wind power, we would use that destructively too, for the same reasons." Civic morality can only exist where there is the recognition of a commitment and accountability to other citizens. Beyond that, nothing exits except the relativistic, individual preferences of the market and the shallow quest for limitless consumption. For bioregionalists, the environmental crisis can finally be distilled down to a clash of cultures between the yuppie world where "whoever dies with the most toys, wins," and a community of ethical citizens. "People whose governing habit is the relinquishing of power, competence, and responsibility, and whose characteristic suffering is the anxiety of futility, make excellent spenders. They are the ideal consumers." On the other side, "The moral argument points toward restraint; it is a conclusion that may well be in some sense tragic, but there is no escaping it .... [W]e will have either to live within our limits, within the human definition, or not live at all."[55]

All of this points to the final unifying element in Berry's thought. For connections to be recognized, for limits to be appreciated and practiced, and for communal morality to hold corrupt individual values in check; economics, politics, and social life must be conducted on a scale that permits and fosters a healthy culture. "[W]e must address ourselves seriously, and not a little fearfully, to the problem of scale. What is it? How do we stay within it? What sort of technology enhances our humanity? What sort reduces it? The reason is simply that we cannot live except within limits, and these limits are of many kinds: spatial, material, moral, spiritual."[56]

Among bioregionalist writers, perhaps no one has dedicated more time to the issues of scale and transition than Kirkpatrick Sale.[57]

Sale contends that as a general rule, the bioregion should be "not so small as to be powerless and impoverished, not so large to be ponderous and impervious, a scale at which human potential can match ecological reality."[58] Regarding the critical issue of transition, Sale claims bioregionalism meets all three "conditions of an effective political project ...: it is irrevocably grounded in historical realities ... it accords will with the apparent patterns of the present ... and its visions of the future seem practical and real ...."[59] Bioregionalism "is capable of uniting many different kinds of people: the National Rifle Association hunter in Pennsylvania with the environmentalist in Colorado ..., the woman in the Virginia commune and the housewife on the Iowa farm ..., the activist in Vermont who has been fighting a nuclear plant and the farmer in Minnesota who has been resisting an electric power line on his land ...." The bioregional project has "the potential to join what are traditionally thought of as Right and Left in America" in a movement of centrist Populism built upon a commitment to "local control, self-reliance, town-meeting democracy, community power, and decentralism, all basic elements in what are thought of as the traditional American—at least Jeffersonian—values."[60]

Since "community" is such a key term in bioregional theory, and since ethical or moral restraint is put forth as one of the essential elements in maintaining proper scale, the relationship between community and ethics pleads for further exploration. On this topic, no one is more influential than Aldo Leopold, and his <u>A Sand County Almanac</u> remains one of the foremost studies (many would say *the* foremost study) of environmental ethics. Leopold's ethical theory links community, citizenship, and place, and promises to add a radically subversive element to the bioregional discourse. His concept of ethics moves the discussion beyond the usual bounds of anthropocentric morality, individualistic humanism, and universal cosmopolitanism; thereby challenging the dominant assumptions of mainstream moral theory. The objective or fixed portion of the ethic consists of the claim that a community is comprised of a collection of interdependent parts. For Leopold, the fact of interdependence, once recognized by humans, will lead us to expand our homocentric notion of community to include other members of the biotic system, but human interdependence with the biotic system exists whether or not humans recognize and respond to it. We humans, through ignorance or insensitivity, can refuse to be good *citizens* of our ecological communities, but we cannot alter the fact of our *membership*. To be good citizens of our ecological communities requires three things: there must be an ethic founded in our instinctive moral sentiments; an esthetic that connects and sensitizes us to our surroundings; and experience to teach us to expand our environmental awareness while remaining rooted to the primitive origins of our social conscience. Final inclusion in the biotic community comes when the person becomes emotionally rooted to a place that they can call their own.

Whether he was conscious of it or not, Leopold's ethical theory is part of an historical tradition of some repute known as the "moral sense" school or the "theory of moral sentiments."[61] Leopold asserts that an ecological conscience rests on "something we can see, feel, understand, love, or otherwise have faith in."[62] The place of esthetics in this scenario becomes clear if we accept a definition of an "esthetic experience" as being simply the opposite of an "anesthetic experience." Viewed in this light, to be ethically anesthetized is to be made insensitive to both pain and pleasure; it means becoming dulled, desensitized, and emotionally numb. On the other hand, being "esthetized" (to coin a word) would indicate the process of increasing our level of emotional reaction. The esthetic experience is one that elevates our perceptual acuity. While objects that are usually viewed as "pretty" may serve this "esthetizing" function, Leopold states that this is only the beginning of a process that "expands through successive stages of the beautiful to values as yet uncaptured by language."[63] The ethical experience and the esthetic experience are both linkages to basic emotions. The boundaries of these passions can be expanded when training, education, and reason lead us to feel esthetic moral sentiments toward entities previously excluded from our consideration. This, for Leopold, is the relevance of an expanding ethical community. If the ethical community is the range of entities toward which the individual feels an esthetic, moral relationship, then ethics, esthetics, and community are simply three strands in an ever expanding web of moral and emotional relationships where humans become not only more ethical, but also more alert and alive. The potentially positive role of education and moral improvement consists in this expansion of the community. With civilization comes an associated increase in our understanding of ecological processes and the potential enlargement of our ethical/esthetic perception. Comparing the potential of our twentieth century sensitivity to the nineteenth century perspective of Daniel Boone, Leopold says, "The incredible intricacies of the plant and animal community—the intrinsic beauty of the organism called America ...—were as invisible and incomprehensible to Daniel Boone as they are today to Babbitt." Leopold then asserts the ethical value of this scientific training, "The only true development in American recreational resources is the development of the perceptive faculty in Americans."[64]

But Aldo Leopold was no unrestrained advocate of progress and civilization. There is both a progressive and a primitive element within his work as indicated by the many contemptuous references to the individual who is "supercivilized," a "shallow-minded modern," a Babbitt, or a "mechanized man, oblivious of floras, [and] proud of his progress."[65] For Leopold, there are both positive and negative effects of civilization. On the one hand, civilization holds forth the possibility of expanding our sense of community by allowing us to see, and more importantly to *feel*, the web of connections that unite the elements of ecosystems. We are capable of loving and caring about other species, rivers, mountains, or bioregions. On the other hand, the artifacts of civilization—the gadgets, specialized occupations, and massive institutions of industrialized society—threaten to disconnect us from the primal and atavistic origins of our ethical sentience. We become numb to the ugliness and environmental destruction we see (but do not feel) all around us. For Leopold, when we become disconnected, we become desensitized. Leopold fears that our species is becoming *too* civilized as the world is made increasingly "artificial." "Your true modern is separated from the land by many middlemen, and by innumerable physical gadgets. He has no vital relation to it... Synthetic substitutes for wood, leather, wool, and other natural land products suit him better than the originals. In short, land is something he has 'outgrown'."[66] The function of the "Conservation Esthetic," therefore, is to preserve and expand the link between humans' environmental conduct and the moral sense rooted deep in the human conscience. The key to expanding our environmental ethics is the enlargement of our moral sense beyond humans, and beyond even other sentient creatures, to include the entire biotic community, or, as Leopold puts it, "the land."

In the 1947 Foreword to his book, Leopold describes the event that began his journey of ethical transformation. "My first doubt about man in the role of conqueror arose while I was still in college. I came home one Christmas to find that land promoters, with the help of the Corps of Engineers, had dyked and drained my boyhood hunting grounds on the Mississippi River bottoms."[67] There is anger in Leopold's voice as he relates how his territory was defiled. The passions that the land ethic summons forth can be not only those of attachment and love, but also those of moral outrage. This call for activism can be found toward the end of the book in the combative exhortation directed toward "Defenders of Wilderness." "[A] militant minority of wilderness-minded citizens must be on watch throughout the nation and vigilantly available for action." Those who are willing to defend wilderness are juxtaposed to "the shallow-minded modern who has lost his rootage in the land..."[68] Leopold recognizes that resisting the spread of environmental destruction will almost certainly require individuals to defend land that is not "theirs" in the legal sense of property. Indeed, there is every reason to believe that protecting the environment will often entail defending the land *against* the actions of the people who *do* own it. It requires moving beyond interactions with the land that are economic and individualistic to a relationship with a place that is built upon the ethical obligations that one member of a community has toward another.

If "radical environmentalism" is defined as those theories that reject our current political, economic, and ethical systems, then bioregionalism certainly deserves to be counted among these radical discourses. Bioregionalism rejects representative democracy in favor of direct, participatory democracy. It calls for a shift from our current globalized, specialized, highly consumptive economy to one that is decentralized, self-sufficient, and sustainable. Perhaps most fundamentally, bioregionalism moves us away from an anthropocentric, humanistic, individual ethic toward a moral system based upon ecocentric interdependent communities. Yet this radicalism is put forward in a way that allows the transition to be accomplished gradually and without dramatic social convulsion. Transitioning to the bioregional alternative can begin as simply as attending the planning committee of your local government, buying food from your local farmer or co-op, joining a volunteer or civic organization, consuming locally and consuming less, or defending your community from the onslaught of chain stores, fast food, and sprawl. Most importantly, moving toward bioregionalism begins with learning the history, traditions, and ecology of the place where you live. It means getting to know your neighbors and taking the good of your community—*all* members of your community—into consideration. It means, in the succinct phrase of Wendell Berry, that you "find a place and dig in your heels." In contemporary America, all of these actions help undermine capitalist consumer society, the power of the central government, and the expanding global economy. These alternative actions are all subversive—and they are all good.

As powerful as bioregionalism's vision appears, and as attractive as its transition seems to many people, it is not without its drawbacks. These need to be acknowledged. For example, even someone as sympathetic as David Orr has pointed out that rural people are not always the paragons of environmental virtue pictured by writers such as Berry and Sale. Life on farms can be tenuous, built on the displacement of native people, and unjustly hard on women and children. "In any case," Orr contends, "agrarian life was never as good as we sometimes imagine." Without a strong environmental ethic, "we start mostly from ruins" with the need being, not the return to a prior lifestyle but the creation of a "new agrarianism."[69] Part of that new agrarianism must be an unflinching recognition of the need for population control. Without a firm population policy, the bioregional vision is undermined. Yet there is nothing in bioregionalism that prevents communities from having a degree of diversity or from helping members of other communities address their environmental problems;

and exploring the boundaries between communal morality and individual autonomy is something every society must bear.

Something like bioregionalism, then, will certainly have to form part of the vision of any comprehensive approach to the environmental crisis. Concerns for citizenship, decentralized production and consumption, the scale of our activities, and the boundaries of our ethics are unavoidable. To the extent bioregionalism's weaknesses exist, they may result from a failure to recognize how the agrarian lifestyle itself is an anti-environmental assault on wilderness, and to appreciate that resolving our predicament must start, not only just with our communities but also deep within the Western psyche. The real problem with bioregionalism might not be that it is extreme to the point of xenophobia or misanthropy, the essential predicament may be that it is not radical enough.

Deep Ecology

Deep ecology shifts our focus from farms to wilderness. It moves from a critique of our cultural habits to an exploration of our fundamental assumptions regarding ourselves as one species in the biosphere. Reading the wide array of authors within the discourse, one gets the impression of a single movement, but one with two faces. One face gives the appearance of scholarly, intellectual analysis, while the other is the visage of passionate, emotional resistance. Both sides of this environmental Janis are components of a single perspective—the discourse that has been called "green romanticism."[70] The two faces of deep ecology were evident from the beginning when Norwegian philosopher and activist, Arne Naess, named the movement and set out eight principles that defined the discourse. The critical propositions are those dealing with ecocentric ethics, population reduction, and direct action.[71]

Unlike all mainstream approaches, as well as ecological economics and social ecology, deep ecology is willing to forthrightly challenge the central premise of Western ethics. As Naess and George Sessions phrase the first principle of deep ecology: "The well-being and flourishing of human and non-human life on Earth have value in themselves (synonyms: intrinsic value, inherent worth). These values are independent of the usefulness of the non-human world for human purposes."[72] This is the principle of ecocentric intrinsic worth as opposed to anthropocentric, or humanist, utilitarianism. Australian Robyn Eckersley defines "anthropocentrism" as "the human-centered orientation" and contrasts this with an ecocentric, or biocentric, orientation, which holds "there is no valid basis to the belief that humans are the pinnacle of evolution and the sole locus of value and meaning in the world."[73] For her, the "litmus tests" for ecocentrism are control of human population growth and the preservation and protection of wilderness.[74] Writers such as Thomas Berry and Lynn White have traced the source of anthropocentrism to human "alienation from the natural world" and the emerging ethic of the late Middle Ages in Europe when science and technology were joined to a variation of Christian theology which pictures the nonhuman environment as part of God's plan where "no item in the physical creation had any purpose save to serve man's purpose." For White, the fact that Christianity is "the most anthropocentric religion the world has seen" promotes the view that "it is God's will that man exploit nature for his proper ends."[75]

Theodore Roszak, in <u>Where The Wasteland Ends,</u> argues an important corner was turned in humanity's relationship to Nature with the beginning of the scholarly tradition that we now call the Enlightenment. From an ecological point of view, there was a terrible price paid for the objective analysis, empirical methodology, and humanist philosophy that characterize the thought of men such as Francis Bacon, Isaac Newton, and John Locke. That cost was the loss of spiritual wonder, intrinsic value, and communal purpose attributed to Nature by earlier animistic and Gnostic patterns of awareness.[76] Philosophers from Bacon and Newton to Descartes and Kant released enormous quantities of human intellectual power by alienating human moral sensitivity from any guilt or obligation in our dealings with the nonhuman. "They found the great truth," argues Roszak in describing these early proponents of the Enlightenment; "break faith with the environment, establish between yourself and it the alienative dichotomy called objectivity, and you will surely gain power. Then nothing—no sense of fellowship or personal intimacy or strong belonging—will bar your access to the delicate mysteries of man and nature. Nothing will inhibit your ability to exploit."[77]

Parallel to the development of Enlightenment values ran a counter set of ideas now generally grouped around the name of Romanticism. The logic, thrust, and style of Blake, Wordsworth, or Goethe proceeded from fundamentally alternative assumptions about knowledge, value, and reality: that truth cannot be reduced to discreet, autonomous facts; that facts cannot be divorced from values; that scientific research and technological innovation possess profound moral implications; that knowledge is inseparable from wisdom; and that to purchase material "progress" by ignoring these essential connections is not merely to build a civilization upon an alienated mode of existence, but to open the way for environmental destruction which must ultimately become suicidal.[78] Romanticism believes the way out of the wasteland is through reintegration of human inquiry with meaning found in a vibrant, living Earth, the respiritualization of Nature, and the freeing of the human imagination from its technocratic cage in order to realize creative, nondestructive, and joyous other options. Romantic deep ecology comes quite close to, if not actually achieving, the animistic worship of Nature.

Roszak believes the reconciliation of the Enlightenment and Romanticism might be at least partially realized in the science of ecology. Ecology has been called "the subversive science" for good reason.[79] In ecology, animal, biological, and chemical interactions are not portrayed as mechanized, blind "forces," but rather as interactive, symbiotic interchanges where ever-changing patterns shape and define each component. Ecology also permits the bridging of the fact/value dichotomy and the realization of an empirical standard for ethical judgments. Roszak declares, "ecology is through and through judgmental in character. It cannot be value-neuter." He explains that ecologists "prescribe a standard of health. What violates the natural harmony must be condemned; what enhances it, be endorsed."[80] Ecology promises to serve this higher purpose; to see facts as embedded in ethics; and to return the concept of justice to the classical meaning of communal harmony. "Ecology is the closest our science has yet come to an integrative wisdom. It, not physics, deserves to become the *basic* science of the future."[81]

The brilliant, prolific, interdisciplinary studies of Paul Shepard explore how the study of ecology might be joined to anthropology, genetics, ethology, art history, and psychology in the new fields of human ecology and sociobiology while attempting to answer the most basic environmental question, "Why do men persist in destroying their habitat?"[82]

In all of his voluminous work, Shepard leads the reader through an inquiry into the human consciousness. Epigenisis, the simultaneous unfolding of instinct and consciousness—our perceptions, categories, and maturing awareness—came out of and can only be explained by our genetic, evolutionary, historical journey of millennia as *Australopithecus* slowly transformed into *Homo sapiens*. The process where we slowly acquired this ontogeny—the genetic code determining the life cycle from infancy, to childhood, to adolescence, to maturity—evolved over hundreds of thousands of years of primate evolution where behavior was tested in a particular ecological setting and genetic strains best adapted to survival in those settings emerged. In <u>Thinking Animals</u>, <u>Man in the Landscape</u>, <u>The Tender Carnivore and the Sacred Game</u>, and other writings, Shepard emphasizes and explores the fact that our genetically determined ontogeny became fixed during those eons as hunters and gatherers.[83] He salutes this hunting culture as a time when humans were more complete, when the process of maturation was encouraged to develop fully, and when humanity's relationship with the nonhuman was built around difference, respect, and limits.[84] For Shepard's wing of deep ecology, "man's environmental crisis signifies a crippled state of consciousness as much as it does damaged habitat …."[85] A world where humans have difficulty establishing relationships with their environment that are not built upon domination and destruction reflects an inner world where humans struggle to discover their own identities and their own symbiosis with their surroundings. In this world, we face what Shepard calls "an irrational (though not unlogical) and unconscious … failure in some fundamental dimension of human existence, an irrationality beyond mistakenness, a kind of madness."[86]

In <u>Nature and Madness</u>, Shepard points out and explores the obvious, if startling, similarities between the collective environmental actions of modern societies and adolescents experiencing arrested development in the process of maturing to adulthood. Shepard's analysis is not restricted to any particular country, but it is difficult to read his book without drawing parallels to current America. No other nation on Earth so glorifies youth, panders to pubescent cravings, or establishes the juvenile as the cultural role model. This culture of the brat has several defining characteristics: there is the boisterous, frenetic assertion that each and every wish must be instantly gratified and the equally rapid swing to some new fad in the never exhausted rummaging for fresh toys. Impatience, scattered attention, and a low threshold of boredom seem inextricable from a narcissistic individualism that refuses to acknowledge the existence of anything bigger or more important than themselves. Adolescents of whatever chronological age affirm their ability (and their right!) to determine for themselves the world in which they choose to live. It follows that Nature can have little significance for people who are convinced that they can "make up their own minds." All of this is built upon the firm, absolute, and completely groundless belief that these adolescents live in a world without exterior or interior limits: that they can believe, want, and do anything and everything in the world and that the only momentary restrictions are their individual wills and the temporary deficiency of their personal fantasies.

In itself, there is nothing inherently neurotic about being an adolescent, as long as the stage is limited to a particular period in a person's aging process, and as long as these adolescents are not put in charge of making important social decisions, but contemporary society seems to have violated both of these checks, so we now confront a situation where "the only society more frightful than one run by children, as in Golding's Lord of the Flies, might be one run by childish adults."[87] Shepard is emphatic in maintaining that human beings are "a perfectly good species" that has all the instinctive skills necessary to survive.[88] The fact that we existed for at least 250,000 years at home in wilderness is proof that we do not have to destroy our environment in order to succeed biologically. The present ecological crisis is the result of our cultural choices and not our genetic endowment. Our salvation lies in coming to accept and appreciate who we are and not in the impossible and unworthy task of completely remaking our species. Simply put, we, as members of American and Western culture, need to grow up; to recognize that others exist who are just as important as we are; to understand there are limits to what we can want, have, and be; and to appreciate that being a contented, but active, member of a community is much more fulfilling than the endless acquisition of material possessions.

In a series of provocative and imaginative essays, John Rodman, Shepard's colleague at Pitzer College and Claremont Graduate School, articulates the motivational impulses that might bridge the gap between theory and action; and transition ecocentrism from a philosophic critique to a source of social opposition. One of his strongest efforts was an essay published in Arne Naess' Inquiry, titled "The Liberation of Nature?" Ostensibly, Rodman's piece is an extended review essay of Peter Singer's Animal Liberation and Christopher Stone's Should Trees Have Standing?, but Rodman goes far past the normal duties of a book reviewer. Singer and Stone attempt to construct a nonanthropocentric ethic around animal rights (Singer) or the inclusion of nonhuman and nonsentient entities in our legal system (Stone). Rodman challenges this liberal extension of moral rights and legal jurisprudence, arguing that it is an inappropriate, insufficient, and undignified response to our ecological crisis. For many deep ecologists, liberalism is more a source of our ecological predicament than the solution to our problems.[89]

In language balanced on a thin edge between philosophical exegesis and emotional derision, Rodman holds up the logic underpinning the liberal line of reasoning and exposes the patronizing, demeaning, pretentiousness of its putative ethic. "[T]he process of 'extending' rights to nonhumans conveys a double message," Rodman summarizes. "On the one hand, nonhumans are elevated to the human level by virtue of their sentience and/or consciousness; they now have (some) rights. On the other hand, nonhumans are by the same process degraded to the status of inferior human beings, species-anomalies: imbeciles, the senile, 'human vegetables'—moral half-breeds having rights without obligations (Singer), 'legal incompetents' needing humans to interpret and represent their interests in a perpetual guardian/ward relationship (Stone)."[90] It is here that the disagreement between liberal environmentalism and deep ecology is exposed. "Is this, then, the new enlightenment—to see nonhuman animals as imbeciles, wilderness as a human vegetable?" Rodman rhetorically and sarcastically queries. "As a general characterization of nonhuman nature it seems patronizing and perverse." The perversity of liberalism is then enunciated.

> It is not so much that natural entities are degraded by being represented in human legal actions, or by not having us attribute to them moral obligations. They are degraded rather by our failure to respect them for having their own existence, their own character and potentialities, their own forms of excellence, their own integrity, their own grandeur—and by our tendency to relate to them either by reducing them to the status of instruments for our own ends or by 'giving' them rights by assimilating them to the status of inferior human beings. It is perhaps analogous to regarding women as defective men who lack penises ....[91]

Extending civilized concepts such as rights and law is now seen as an act of species imperialism instead of a projection of benevolent paternalism. Human ecological imperialism has taken this shape: rather than animals, rivers, or landscapes developing and acting as they will, humans impose our own agenda, our own rules, and our own techniques onto those processes. This, in essence, is what imperialism means; it is the substitution of the aggressor's priorities for the goals, objects, or lifestyle of the subjected entity. The negation of this oppression is to be self-determined, autonomous, or self-willed. This is the origin of the world "wilderness." The emancipation of wilderness from human subjugation becomes the objective of the movement, and ecological resistance becomes the means to achieving this end. It is the equivalent of colonial liberation struggles to overthrow oppressive imperialist regimes.[92]

Rodman takes the parallel of human and ecological liberation to even deeper and more profound implications. The act of liberation has two components. Part of the process involves resisting and fighting against the further spread of the intruding power, but another, no doubt much more difficult, task is the self-liberation and self-emancipation from the "introjected self-image of inferiority and impotence." Seen this way, human technological and economic domination of Nature is essentially an outward manifestation of the pathological lack of security on our own part. We humans have convinced ourselves that we are too weak, too stupid, too fragile, and too lacking in instinctive skills to survive unless we turn to technology, economics, law, and anthropocentric ethics to rescue us from our pitiful condition. The drive to dominate Nature originates in, and reflects, an unarticulated need within ourselves to control some terrible inner demon: the need for reason to conquer passion, for Christian righteousness to subdue satanic evil, and for civilized propriety to suppress savage bestiality. From this view, conquering wilderness reflects the perceived need to have the ego and superego suppress the id, and hating Nature becomes a neurotic reflection of self-loathing. The liberation of Nature, then, goes hand in hand with self-liberation. If limits are perceived as internal, if a person or a society can become overly civilized and desensitized to the undomesticated animal within themselves, then deep ecology is at least as much about liberating this internal wildness as it is about managing external "resources."[93] The interplay between our behavior and our self-image and the connection between our perspective on Nature and our philosophies regarding human social and political actions are so entangled that it is probably impossible to change one without changing all or to declare one to be more foundational than the other. This is not so much an issue of substructure and superstructure, as it is a matter of metaphorical mirroring among multi-facetted images.[94] Acts of resistance against the encroachment of civilized "development" upon wilderness become, then, liberating attempts to preserve some personal inner wildness against the threat of becoming overly civilized, of losing all spontaneity, passion, or creativity, of becoming merely a machine. In this struggle, saving Nature becomes part and parcel of saving our own souls. This is ecocentrism in its deepest and most radical sense.

When we come to see wild animals and wild places on an equal ethical footing with humans, defending those species and places from destruction will seem no more strange than shielding a fellow human from an immoral thug. Indeed, an ethical obligation will be attached to such a resistance, and any action or force used to stop the attack will be viewed as the appropriate behavior of a mature, moral, concerned member of the community. Only the most unreflective pacifist would deny that a person whose house has been broken into by hooligans intent on molesting their child and ransacking their habitat has not the right, and the duty, to forcibly stop that invasion. For deep ecologists who understand that the Earth is our home, this vandalism is now under way on an enormous scale, and they feel under the same ethical authorization, and the same ethical duty, to use available means to end the devastation. Groups such as Greenpeace, Sea Shepard, and Earth First!; and activists such as Edward Abbey, Christopher Manes, Dave Foreman, and Derrick Jensen have moved resistance out of the arena of intellectual debate and into the actual ecological setting where attack and defense is an everyday occurrence.

Inspired by the wilderness preservation work of men such as Bob Marshall and Howard Zahniser and heeding the call for direct action by early advocates of ecological resistance such as Aldo Leopold and Edward Abbey, Dave Foreman and a small group of deep ecologist formed Earth First! in 1980. Earth First! laid out a program of radical resistance to any and all future destruction of American wilderness.[95] Forman sagely points out that resistance to oppressive and immoral laws is as much a part of American culture as the Boston Tea Party, the Underground Railroad, or Martin Luther King's program of civil disobedience. The ecological resistance he advocates is as American as John Muir and Henry David Thoreau. It is aimed against the real "environmental extremists" who seek to convert every last stand of old growth forest to timber, dam the last free-flowing river, or build roads into the last tract of undisturbed wilderness. The tactics selected to halt this destruction can be as diverse as the people willing to fight for salvation of the planet. "Our most fundament duty is that of self-defense," contends Foreman. "Our self-defense is damage control until the machine plows into that brick wall and industrial civilization self-destructs as it must. Then the important work of bioregionalism begins." Foreman supports a diverse assortment of means that environmentalists of all different types can use to prevent ecological destruction and promote a healthy, diverse, sustainable environment for all species and all ecosystems. "Every available tool needs to be employed," he argues, "every style, from business suits and laptops to camouflage and tree spikes, needs to be encouraged." Foreman counts off a partial list of tactics, including "purchasing land for conservation purposes, lobbying Congress and agencies on ecological issues, filing environmental appeals and lawsuits, conducting scientific and economic research into the value of wild nature, and developing alternative soft-path lifestyles are all valid and necessary methods, just as are the hard-ass, court-of-last-resort avenues of the Sea Shepard Conservation Society and Earth First!"[96] All of these tactics are legitimate parts of American political culture. The legal forms of environmental politics derive their legitimacy from the structures and practices of our Constitutional system. Those methods of ecological resistance outside the bounds of our current laws derive their moral authority from the values, ethics, and practices enshrined in our political traditions and by the biological, evolutionary requirements for a healthy, sustainable, diverse planet. We need to investigate and understand the role of the US Constitution in this process of shaping responses. Then, our attention can turn to the perspectives and values found in American political culture.

Notes

1. C. Wright Mills, The Causes of World War Three (New York: Simon and Schuster, 1958), Chapter 13.
2. Ibid., 82.
3. Ibid., 85 and 89.
4. Since the events on September 11, 2001, the term "radical" has acquired such a negative connotation that one is tempted to shift to a more neutral expression, such as "foundational environmentalism." Such a shift, however, would by neither more enlightening nor ingenuous.
5. Nicholas Georgescu-Roegan, The Entropy Law and the Economic Process (Cambridge, MA: Harvard University Press, 1971); Kenneth Boulding, The Meaning of the Twentieth Century (New York: Harper and Row, 1945); Robert Costanza, John Cumberland, Herman Daly, Robert Goodland, and Richard Norgaard, An Introduction to Ecological Economics (Boca Raton, FL: St. Lucie Press, 1997): Herman Daly, Beyond Growth (Boston: Beacon Press, 1996). Herman Daly and Joshua Farley, Ecological Economics: Principles and Applications, 2nd ed. (Covelo, CA: Island Press, 2011).
6. Ibid.
7. Summers has said, "There are no… limits to the carrying capacity of the earth that are likely to bind any time in the foreseeable future. There isn't a risk of an apocalypse due to global warming or anything else. The idea that we should put limits of growth because of some natural limit is a profound error." Quoted Bill McKibben, Deep Economy: The Wealth of Communities and the Durable Future (New York: Times Books/Henry Holt and Company, 2007). 24.

8. For an excellent look at traditional economics from an environmental perspective, see Herman E. Daly and John B. Cobb, Jr., For the Common Good: Redirecting the Economy Toward Community, the Environment, and a Sustainable Future (Boston: Beacon Press, 1989), especially Part One.
9. John Stuart Mill, Principles of Political Economy, edited with an introduction by Donald Winch (Middlesex, U.K.: Penquin Books, 1970), 111. Early references by Daly to Mill are found in Herman E. Daly, ed., Essays Toward a Steady-State Economy (San Francisco: W. H. Freeman, 1972), and Herman E. Daly, "Institutions Necessary for a Steady-State Economy: Three Suggestions," IDOC—North America, Number 47, October 1972, 22-27.
10. Bringing thermodynamics into economics is credited to Nicholas Georgescu-Roegen, The Entropy Law and the Economic Process (Cambridge, MA: Harvard University Press, 1971).
11. David Vogel, Trading Up: Consumer and Environmental Regulation in a Global Economy (Cambridge, MA: Harvard University Press, 1995).
12. See the essays in Jerry Mander and Edward Goldsmith (eds.), The Case Against the Global Economy (San Francisco: Sierra Club Books, 1996), as well as John J. Audley, Green Politics and Global Trade: NAFTA and the Future of Environmental Politics (Washington, D.C.: Georgetown University Press, 1997), and Joshua Karliner, The Corporate Planet: Ecology and Politics in the Age of Globalization (San Francisco: Sierra Club Books, 1997).
13. Daly, Beyond Growth, 147. Daly's most complete argument against free trade is found in For The Common Good, Chapter 11.
14. Daly, "Institutions Necessary for a Steady-State Economy: Three Suggestions," 22.
15. Daly and Cobb, For the Common Good, Chp. 12. Daly, Beyond Growth, Chps. 8 and 9.
16. Ibid., passim.
17. Daly, "Institutions Necessary for a Steady-State Economy," 26.
18. Ibid.; Daly and Cobb, For the Common Good, Chp. 17.
19. Daly and Cobb, For the Common Good, 388 and all of Chp. 20.
20. A subset of ecofeminism has formed around the concepts of social ecology, which they have adapted and labeled "social ecofeminism." Major proponents include Ynestra King, Machina Ex Dea: Feminist Perspectives on Technology (New York: Pergamon Press, 1983), and Janet Biehl, Rethinking Ecofeminist Politics (Boston: South End Press, 1991).
21. Published in 1963, Bookchin's, Our Synthetic Environment, predates Rachel Carson's Silent Spring by a few months. The quotation is from Bookchin, The Ecology of Freedom, 334.

22. Murray Bookchin, Remaking Society: Pathways to a Green Future (Boston: South End Press, 1990), 13, 17, and 24. Emphasis in this, and all subsequent Bookchin quotations, are from the original.
23. Murray Bookchin, The Philosophy of Social Ecology: Essays on Dialectical Naturalism (Montreal: Black Rose Books, 1990), 35.
24. Bookchin explains his use of this term in a chapter titled "Where I Stand Now," in Murray Bookchin and Dave Foreman, Defending the Earth: A Dialogue Between Murray Bookchin and Dave Foreman, Steve Chase, ed. (Boston: South End Press, 1991), 131. See also Murray Bookchin, The Philosophy of Social Ecology: Essays on Dialectical Naturalism.
25. Bookchin, Remaking Society, 24–28.
26. Ibid., 28–29.
27. Richard Hofstadter, for example, only discusses the conservative version of Social Darwinism, see his Social Darwinism in American Thought, rev. ed., (Boston: 1955). The remarkable historical irony, as Donald Worster points out, was that Darwin's original theory was itself a projection onto Nature of the social relationships he saw in Victorian England. See Donald Worster, Nature's Economy: A History of Ecological Ideas 2nd ed. (Cambridge: Cambridge University Press, 1994), 169.
28. Key readings in the thought of Peter Kropotkin are Mutual Aid: A Factor of Evolution; "The State: Its Historic Role," in Selected Readings on Anarchism and Revolution, edited by Martin A. Miller (Cambridge: the M.I.T. Press, 1970); Memoirs of a Revolutionist (New York: Grove Press, 1968); "Anarchism: Its Philosophy and Ideal" (London: Freedom Pamphlets, No. 10, 1897); and "Anarchist Communism: Its Basis and Principles" (London: Freedom Pamphlets, No.4, 1891).
29. In addition to the sources cited earlier, the fascinating life and thought of Peter Kropotkin is also revealed in his autobiography, Memoirs of a Revolutionist; his historical analysis, The Conquest of Bread (New York: Vanguard Press, 1926); and his picture of an alternative, anarchist society, Factories, Fields, and Workshops (New York: Harper Torchbooks, 1974).
30. Murray Bookchin, "Post-Scarcity Anarchism," in Post-Scarcity Anarchism (and other essays), (San Francisco: Rampart Press, 1971), 34. Again, emphasis in original.
31. Murray Bookchin, The Philosophy of Social Ecology: Essays on Dialectical Naturalism (Montreal: Black Rose Books, 1990), 35. Further reflections on libertarian municipalism are found in Murray Bookchin, The Limits of the City (New York: Harper Torchbooks, 1974).
32. Francis Fukuyama, The End of History and the Last Man (New York: Free Press, 2006).
33. Ibid., xii and 46.

34. Paul R. Ehrlich, The Population Bomb (Binghamton, NY: Sierra Club, 1968), 57. Emphasis in original.
35. Paul R. Ehrlich and Anne H. Ehrlich, The Population Explosion (New York: Touchstone/Simon and Schuster, 1990), 23. Other neo-Malthusians include Robert L. Heilbroner, An Inquiry Into the Human Prospect (New York: W.W. Norton and Company, Inc., 1975), 31. Garrett Hardin, Living With Limits (New York and Oxford: Oxford University Press, 1982). William R. Catton, Overshoot: The Ecological Basis of Revolutionary Change (Urbana and Chicago: University of Illinois Press, 1982).
36. Thomas Robert Malthus, An Essay on the Principle of Population, edited by Philip Appleman (New York: W.W. Norton and Company, Inc., 1976), 19. At its core, Malthus' essay was a heresy against the dominant philosophic orthodoxy of the time, which held that progress—as defined by increasing material consumption for increasing numbers of humans—was both possible and desirable. Malthus never directly took up the task of challenging the *desirability* of material consumption (that mission would be assumed by the Romantics)
37. Paul R. Ehrlich and Anne H. Ehrlich, The Betrayal of Science and Reason: How Anti-Environmental Rhetoric Threatens Our Future (Covelo, CA: Island Press, 1996), 66. Joel E. Cohen, How Many People Can the Earth Support? (New York: W.W. Norton and Company, Inc., 1995. Another observation: if the world's population is currently seven billion and increasing at an annual rate of 1.7 % (or a doubling time of approximately 41 years), in 18 doublings (738 years), there will be 10 human beings for every square meter of ice-free land on the surface of the Earth.
38. Malthus, An Essay on the Principle of Population, 34–42, *passim*.
39. Ibid., 33 and 34.
40. In a chapter titled "New Colonies," Malthus showed that wherever Europeans found sparsely inhabited portions of the globe, they were able to transport a portion of their excess population to these areas and promote a quickly increasing population in both the mother country and the colonies. He particularly focused on "the English North American colonies, now the powerful People of the United States of America." Essay on Population 45–46. Malthus' second "palliative" builds upon the Enlightenment faith in science and capitalism's devotion to exponentially increasing wealth. But again, a palliative is not a solution. The long-term effect of economic and technological growth is merely to increase the total number of humans who will suffer when the limiting factors on carrying capacity are eventually met. On environmental modernization, see Paul Hawken, The Ecology of Commerce (New York: Harper Collins, 1993). Ray Anderson, Mid-Course Correction: Toward a Sustainable Enterprise: The Interface Model (White River Junction, VT: Chelsea Green Publishing,

1999). Paul Hawken, Amory Lovins, and L. Hunter Lovins, <u>Natural Capitalism: Creating the Next Industrial Revolution</u> (Boston and New York: Little, Brown and Company, 1999). William McDonough and Michael Braungart, <u>Cradle to Cradle: Remaking the Way We Make Things</u> (New York: North Point Press, 2002). Thomas L. Friedman, <u>Hot, Flat, and Crowded</u> (New York: Farrar, Straus and Giroux, 2008).

41. Examples of writers who claim to "prove Malthus wrong" include Everett E. Hagen, <u>On The Theory of Social Change: How Economic Growth Begins</u> (Homewood, IL: Dorsey Press, 1962) and Mark Blaug, "Malthus, Thomas Robert." In David L. Sills, ed., <u>International Encyclopedia of the Social Sciences</u>, IX (New York: Macmillan and Free Press, 1968). 549–552. Of course, virtually all cornucopian economists would also fit into this category. For reviews of this literature, see Garrett Hardin, <u>Living Within Limits</u>, Chapter 4; Paul and Anne Ehrlich, <u>The Betrayal of Science and Reason</u>, Chapter 5, and William Catton, <u>Overshoot</u>, Chapter 8.

42. It is argued that phase one of the demographic transition is characterized by high birth and high death rates and is typically found in primitive and undeveloped countries. It is in the second demographic stage where populations explode. Typically found in underdeveloped countries and former colonies, population trends in phase two are exemplified by high birth rates and low (or at least declining) death rates. The upshot is that phase two is generally pictured as large, overcrowded populations which are too poor to develop economically and demographically growing at rates that make overcoming the poverty impossible. Overpopulation is both a cause and a consequence of poverty. This is referred to as "the demographic trap."

43. Paul and Anne Ehrlich, <u>The Population Explosion</u>, 58.

44. One set of projections estimates that if current trends in birth rates and legal, but mostly illegal, immigration continue, the population of America would reach 422 million by the year 2050. Paul and Anne Ehrlich, <u>One With Nineveh: Politics, Consumption and the Human Future</u> (Covelo, CA: Island Press, 2004), 107.

45. Cornell University Professor David Pimentel estimates the long-term sustainable population of the world is one to two billion people, with America's optimal population at one hundred to two hundred million. See David Pimentel, "How Many Americans Can The Earth Support?" Population Press, 1999.

46. Some countries have adopted national population policies. China and India implemented far-reaching schemes combining incentives and sanctions to move toward population targets. The all-too-often biased and heavy-handed manner in which many of these plans were implemented has included strong governmental pressure for sterilizations and gender-based abortions in societies where male offspring are culturally preferred. See

Paul and Anne Ehlich, The Population Explosion, op.cit., 205–210. Ecologist Raymond B. Cowles has proposed direct subsidies to all young women for *not* having children. Such an "infertility-reward program" would, no doubt, be expensive, but probably no more costly than the social and ecological costs of overpopulation. Perhaps, the boldest suggestion has come from economist Kenneth Boulding who proposed that every women be given a socially determined number of "baby permits" (probably one or two) that she could sell, give to someone she thought would be a deserving parent, or cash in herself when she reached an age to make that decision. Enforcement could be achieved by a sterilizing agent added to the water or food supply and the antidote administered at the time the woman redeemed her permit. Boulding's system has many virtues, including the fact that it keeps a maximum of individual choice in birth planning, it would lead to more equality of incomes, and it would give societies resolve with flexibility in meeting their population goals. The obvious fact is that neither Cowles nor Boulding's plans are likely to be adopted, or even given a fair public hearing in the USA in the near future. For a further discussion of the Cowles and Boulding plans, see Garrett Hardin, Living Within Limits, 272–274. See also Herman Daly and John Cobb, For the Common Good, 236–251.
47. Garrett Hardin, "Lifeboat Ethics," Bioscience 24 (1974).
48. Edited volumes of bioregional thought include William Vitek and Wes Jackson (eds.), Rooted in the Land: Essays on Community and Place (New Haven: Yale University Press, 1996), and Eric T. Freyfogle (ed.), The New Agrarianism: Land, Culture, and the Community of Life (Covelo, CA: Island Press, 2001).
49. State legislator, mayor of Missoula, and associate at the University of Montana's Department of Environmental Studies Daniel Kemmis, Community and the Politics of Place (Norman, OK: University of Oklahoma Press, 1990), 62. Kemmis traces his use of the expression "first and second languages" of Robert Bellah, et al., Habits of the Heart: Individualism and Commitment in American Life (New York: Harper and Row, 1985).
50. Ibid., 79.
51. Wendell Berry, The Unsettling of America: Culture and Agriculture (San Francisco: Sierra Club Books, 1977).
52. Ibid., 222, 66, and 43.
53. See Holmes Rolston, Conserving Natural Value (New York: Columbia University Press, 1994), 6–9, for Rolston's discussion of "entwined destinies: nature supporting culture."
54. Ibid., 53.
55. Ibid., 94 and 24.

56. Ibid., 222.
57. Kirkpatrick Sale, Human Scale (New York: Coward, McCann, and Geoghegan, 1980) and Dwellers in the Land: The Bioregional Vision (Philadelphia: New Society Publishers, 1991).
58. Ibid., 55 and 64. For human settlements, Sale recognizes the necessity for some variation, but suggests that history demonstrates the best size for the "basic village or intimate settlement" is between 500 and 1000 inhabitants, while the "larger tribal association or extended community" should be between 5000 and 10,000 citizens.
59. Ibid., 136.
60. Ibid., 162, 161, 169, and pp. 167–168.
61. J. Baird Callicott, A Companion to A Sand County Almanac (Madison: The University of Wisconsin Press, 1987), 190. For an early version of this argument, see Franklin A. Kalinowski, "Aldo Leopold as Hunter and Communitarian," in Vitek and Jackson, Rooted in the Land, 140–149.
62. Aldo Leopold, A Sand County Almanac (New York: Balantine Books, 1970), 251.
63. Ibid., 102.
64. Ibid., 291.
65. Ibid., 227, 279, 291, 119, and 50.
66. Ibid., 261–262.
67. Quoted in Callicott, A Companion to A Sand County Almanac, 282.
68. Leopold, A Sand County Almanac, 279.
69. David Orr, "The Urban-Agrarian Mind," in Eric T. Freyfogle (ed.), The New Agrarianism, 96 and 97.
70. John S. Dryzek, The Politics of the Earth: Environmental Discourses, second edition (Oxford and New York: Oxford University Press, 2005), Part V.
71. The other principles are diversity, recognition of limits, a rejection of consumer society, the acknowledgment that ecological conditions are growing worse, and a call for radical change. Arne Naess, "The Deep Ecological Movement: Some Philosophical Aspects," originally published in Philosophical Inquiry, 8, nos. 1–2 (1986). Page references are to the reprint found in George Sessions, ed., Deep Ecology for the Twenty-first Century, (Boston and London: Shambala Publications, Inc., 1995). 67. See Arne Naess, "The Shallow and the Deep, Long-range Ecology Movements," originally published in Inquiry, 16 (1973). See also, Bill Devall and George Sessions, Deep Ecology: Living as if Nature Mattered (Salt Lake City, UT: Gibbs Smith for Peregrine Smith Books, 1985).
72. Ibid., 68.
73. Robyn Eckersley, Environmentalism and Political Theory: Toward and Ecocentric Approach (Albany, NY: State University of New York Press, 1992). 26 and 28. Similar efforts are made by David Ehrenfeld, The

Arrogance of Humanism (Oxford: Oxford University Press, 1978); Robert E. Goodin, Green Political Theory (Cambridge, UK: Polity Press, 1992): and Holmes Rolston, III, Environmental Ethics: Duties to and Values in the Natural World (Philadelphia: Temple University Press, 1988).
74. Eckersley, Environmentalism and Political Theory: Toward and Ecocentric Approach, 51, 50, and 29. All emphases in original. Other deep ecologists have been radically specific about population. In 1987, Christopher Manes, writing in the Earth First! Journal under the pseudonym "Miss Anne Thropy," created an intellectual firestorm by arguing the AIDs epidemic may be salvation of the planet.
75. Thomas Berry, "The Viable Human," in George Sessions, ed., Deep Ecology for the Twenty-first Century, 9, 11, 12, and 16. Lynn White, Jr., "The Historical Roots of our Ecological Crisis," appeared originally in Science, Vol. 155, March 10, 1967, 1203–1207. Page references are to reprinted version in Ian G, Barbour (ed.), Western Man and Environmental Ethics, (Reading, MA: Addison-Wesley Publishing Company, 1973), 20, 24, 28.
76. Theodore Roszak, Where the Wasteland Ends: Politics and Transcendence in Postindustrial Society, (Garden City, NY: Anchor Books, 1973), especially Part Two "Single Vision and Newton's Sleep: The Strange Interplay of Objectivity and Alienation."
77. Ibid., 155 and 157.
78. Ibid., especially Part Three, "A Politics of Eternity."
79. Paul Shepard and Daniel McKinley, The Subversive Science: Essays Toward An Ecology of Man (Boston: Houghton Mifflin Company, 1969).
80. This shift to an integrative, ethically driven science of intrinsic value is not foreordained, however. "Ecology stands at a critical threshold," Roszak recognized over three decades ago. "Is it to become another anthropocentric technique of efficient manipulation, a matter of enlightened self-interest and expert, long-range resource budgeting? Or will it meet the mystics on their own terms and so recognize that we are to embrace nature as if indeed it were a beloved person in whom, as in ourselves, something sacred dwells?" Deep ecology calls for this transformation of knowledge and this unification of truth and spirit. "The question remains open," observes Roszak, "which will ecology be, the last of the old sciences or the first of the new?" Where the Wasteland Ends, 370 and 371
81. Theodore Roszak, Where the Wasteland Ends, 368. Emphasis in original.
82. Paul Shepard, Nature and Madness (San Francisco: Sierra Club Books, 1982), 1. Given that males are responsible for a great deal of our environmental destruction, Shepard may, perhaps, be forgiven his use of gender-specific language.

83. Paul Shepard, Thinking Animals: Animals and the Development of Human Intelligence (New York: The Viking Press, 1978). Paul Shepard, Man in the Landscape: A Historic View of the Esthetics of Nature (New York: Ballantine Books, 1967). Paul Shepard, The Tender Carnivore and the Sacred Game (New York: Charles Scribner's Sons, 1973). See also Paul Shepard, Coming Home to the Pleistocene, edited by Florence R. Shepard (Washington, D.C.: Island Press/Shearwater Books, 1998).
84. In Paul Shepard, The Tender Carnivore and the Sacred Game, he asks "Can we face the possibility that hunters were more fully human than their descendants? Can we embrace the hunter as part of ourselves as a step toward repairing the injury to our planet and improving the quality of life?" 36.
85. Shepard, The Tender Carnivore and the Sacred Game, xvi.
86. Shepard, Nature and Madness, 4.
87. Ibid., 17.
88. Paul Shepard, The Tender Carnivore and the Sacred Game, xv and 110.
89. John Rodman, "The Liberation of Nature?" Inquiry, Vol. 20 (Spring 1977), 83–145. Besides Robyn Eckersley, another philosopher who acknowledges his debt to Rodman is J. Baird Callicott, "Animal Liberation: A Triangular Affair," in J. Baird Callicott (ed.) In Defense of the Land Ethic: Essays in Environmental Philosophy (Albany, NY: State University of New York Press, 1989). See also Mark Sagoff, "Animal Rights and Environmental Ethics: Bad Marriage, Quick Divorce," in James Sterba (ed.) Earth Ethics, Englewood Cliffs, NJ: Prentice-Hall, 1995.
90. Ibid., 93–94.
91. Ibid., 94.
92. Ibid., 95–98.
93. John Rodman, "Paradigm Change in Political Science: An Ecological Perspective," American Behavioral Scientist, Vol. 24, No. 1, September/October 1980, 73.
94. Ibid., 101–105.
95. The quotation is from Dave Foreman, Confessions of an Eco-Warrior (New York: Harmony Books, 1991), 186. Access to the wonderful works of Edward Abbey can be made through an anthology such as Edward Abbey, The Best of Edward Abbey (San Francisco: Sierra Club Books, 1984). Indispensable reading includes Edward Abbey, Desert Solitaire: A Season in the Wilderness (New York: Ballantine Books, 1968); Abbey's Road (New York: E.P. Dutton, 1979); and, of course, The Monkey Wrench Gang (New York: Avon Books, 1975). Another book that needs to be considered is Christopher Manes, Green Rage: Radical Environmentalism and the Unmaking of Civilization (Boston: Little, Brown and Company, 1990).
96. Ibid., 172.

CHAPTER 4

The Constitution and the Environment: Ecological Principles and Liberal Policy Making

Tens of millions of Americans, many of whom have never read the Constitution and know almost nothing of its background or principles, hold that document in hallowed status because they have been taught Constitution worship from early childhood and have never found it necessary, or worthwhile, to challenge those lessons. Other, more thoughtful students of early American history and political thought have made a careful analysis of our Founding Document, and this group, too, leaves their investigations with deep reverence for what the Constitution has accomplished.[1] America is an economically wealthy, militarily powerful, and politically stable nation that has provided safety, opportunity, and more than a few civil liberties to millions and millions of native and immigrant people. Most Americans have a sincere love of their country, and although they may, at times, have reservations about their government and their elected leaders, they still believe the Constitution is the embodiment of that patriotic nationalism and is, therefore, immune to criticism. A third, numerically small group of Progressive intellectuals and political reformers have studied the Constitution and rendered a less-glorifying verdict. For these analysts, the Constitution is not nearly democratic enough, our history could have been, and should have been, much more egalitarian and participatory, and they call for changes (usually in the direction of making our system more along the lines of parliamentary nations, or more radically decentralist) to what they perceive as a closed, plutocratic, elitist arrangement dominated by special economic interests.[2] The debate on the Constitution's impact on minorities, the poor, and the unorganized segments of the human population will, no doubt, continue, but what are the ramifications of the Constitution for our ecosystem? How does the political structure and decision-making process enshrined in the American Constitution affect choices this society makes regarding the relationship between humans and their geological and biological surroundings? If our rivers, forests, soils, and wild animals could study and render an opinion, what would be their judgment of the Constitution?[3] Answering these questions requires an investigation of the biological requirements for a healthy and sustainable ecosystem and the decision-making process built into the political structure of the American Constitution.

This chapter explores the interconnections and conflicts that arise when attempting to manage complex ecological processes within the constraints imposed by liberal policy making as institutionalized in the American Constitution. The thesis is that policy outcomes are limited to a very narrow set of options, and that this limited set of policy choices often conflicts with the requirements of effective conservation biology. Underpinning this argument is the premise that only certain types of public policies can be formulated and implemented at the Federal level of the American government. By shaping and constraining the policy-making process, the Constitution limits the policy options that can be put into practice. It does this by institutionalizing the individualism of interest-group liberalism, incremental decision-making, policy fragmentation, free market economic growth, and a bias toward short-term remedial action. In the area of environmental policy, these limitations are particularly evident. Only those policies that are acceptable and adaptable to mainstream environmental politics can work their way through the formulation process institutionalized within the Constitution.

Once the full range of possible environmental responses is understood, and once the distinction between mainstream and radical environmentalism is appreciated, questions can be asked regarding the most likely policies to be implemented and the policy options most likely to succeed in resolving our ecological crisis. In order to understand the logic and consequences of this argument, a number of core ecological principles need to be spelled out in simple clarity and then contrasted to the ideological and procedural checks built into the American Constitution. Our analysis will conclude with some reflections on the responses that have been produced by this tension between ecological reality and policy practicality.

Ecological Principles: Things You Can Ignore But Not Change

During the late 1960s and early 1970s, several seminal books were published that helped define that period as the reawakening of environmental thought in America. One of those works was The Closing Circle by noted ecologist Barry Commoner.[4] Commoner begins by defining "ecology" as "the science that studies the relationships and the processes linking each living thing to the physical and chemical environment ..." and goes on to describe four "laws of ecology" that help define the discipline. The idea of taking a few basic principles of ecology and stating them in straightforward, commonplace language was further developed by ecologist Garrett Hardin in Living Within Limits, published 20 years after Commoner's book.[5] Using Commoner and Hardin as guides, we can stipulate seven fundamental ecological principles to which all environmental policy must conform if it seeks to be effective.

Principle Number One: Everything Is Connected to Everything Else

Commoner refers to this as his First Law of Ecology. It corresponds to the third of what Hardin calls his "biological default positions." Hardin's version of the principle states, "we can never do merely one thing," but expressed either way, the basic tenet is that biological or ecological processes never occur in isolation. The ecosystem is so interconnected that whatever happens to one element of it impacts on other elements and causes changes in their processes also. Although the exact nature of these impacts cannot always be predicted beforehand, their occurrence is certain.[6] For example, in a model of a simple carbon cycle, CO_2 is released into the air from living organisms through respiration or through the burning of fossil fuels. This atmospheric CO_2 may be taken up in plants by photosynthesis and converted into sugar molecules, passed on to animals that eat the plants and convert the sugar molecules into energy or more complex molecules, and then released back into the atmosphere by respiration, decomposition, or combustion. Whatever occurs in one sector of the carbon cycle effects what happens in subsequent sectors.[7] If more CO_2 is released into the air than plants can sequester through photosynthesis, then there will be a buildup of CO_2 in the atmosphere, and since CO_2 is a greenhouse gas, this will lead to an overall increase in global temperature.

Nitrogen cycles, water cycles, and predator–prey relationships are all examples of what Commoner calls systems of "multiple interconnected parts, which act on one another."[8] Through a process known as branching, cycles, or portions of cycles, can be further connected to each other and made even more complex. As a general rule, the more complex an ecological process, the more flexible it is and the more resilient to change. Using Commoner's example of the interrelationship between rabbits and lynx, if the lynx kill all the rabbits in their territory, and if no alternative food supply exists, then the lynx will quickly starve and their population will collapse. The presence of a substitute food source will allow the lynx to survive declines in the rabbit population, give the rabbits an opportunity to replenish their numbers, and, as a result, keep the system in an oscillating balance. These ecological webs may be interpreted by humans as being positive (as when complexity leads to resilience) or negative (as when the linkages lead to the collapse of one component precipitating a general disintegration), and Hardin may extend this concept to include economic processes summarized under the "law of unintended consequences" or the sociological postulate of "unanticipated consequences of purposive social action," but the irrefutable point is the fact of interconnection.[9] The belief that reality can be reduced to fragments that can be managed in isolation from each other is utter delusion.

Principle Number Two: Everything Must Go Somewhere

Put simply, the second of Commoner's axioms asserts that there is no such place as "away." To realize the truth that everything must go somewhere is to recognize the self-deception involved in believing that anything ever really disappears or that a mere alteration in the location of a problem is tantamount to its solution. Principle number two is, in some respects, a derivative of principle number one, but it extends the first concept by pointing out that connections can also exist across ecological mediums or historical time. By way of illustration, attempting to remedy difficulties caused by solid waste landfills by incinerating municipal garbage merely shifts the problem from the environmental medium of land and groundwater pollution to an issue of air pollution. It may not even accomplish this minor palliative if we consider that the ash from the incinerator will have to be buried at some location that will lead to further land and groundwater contamination. Any of these methods of temporary "disposal" (now properly placed in quotation marks) is, in fact, a transferal of the problem from the people who produced the garbage (and presumably derived benefits from the initial product) to other individuals who will have to live with the consequences of land, groundwater, or air pollution (and who probably gained no benefit from the consumption of the product). This shift in point of impact without a consequent reduction in the severity of the problem is "tinkering," and, according to the second principle, tinkering is both ecologically ineffective and socially unjust. As Hardin observes, the denial of the second principle continues in clichés such as "the solution to pollution is dilution" or "out of sight, out of mind," but these human self-deceptions cannot alter ecological reality.[10]

Principle Number Three: Nature Knows Best

If ecological processes are interconnected and if it is impossible to avoid the consequences of those interconnections, then there is a strong conservative bias regarding how change in ecological systems takes place. The patterns of interconnection and complexity described in principles one and two evolved over millions of years of trial and error. The resiliency, flexibility, and stability seen in ecosystems are the result, therefore, of countless experiments in Nature. It can be supposed that, over the course of time, a great many possible combinations and relationships were attempted that failed. From all the possible combinations of relationships among populations and their environment, evolution has selected those that are most sustainable. The probability strongly exists, therefore, that any intervention by humans into these evolved systems is much more likely to be harmful than beneficial.[11] This does not mean that all human action is proscribed. Neither does it imply that humans should view themselves as alienated from and hostile to the natural world. Clearly, there are instances when humans need to intercede in order to restore ecosystems that have been previously damaged by human activity. What this third principle emphasizes is that this human intervention must be done cautiously and skeptically. Nature, in the sense of wild, undomesticated individuals, populations, and ecosystems should be the standard by which we judge both the propriety and extent of our interactions.

This line of thought can be extended and, instead of applying it to human manipulation of natural systems, it can be used to focus on the process of change itself. In contemporary German environmental policy, the concept of *vorsorgeprinzip* or the precautionary principle has been adopted which assumes that scientific uncertainty regarding environmental risks will not be taken as a reason for delaying action to address those risks.[12] If there exists a reasonable suspicion that a serious environmental problem is occurring, then corrective policies will be pursued. Garrett Hardin adopts similar thinking in arguing that "guilty" should be the default position of choice when evaluating technological innovations such as the introduction of synthesized chemical compounds into the environment. While the assumption of "innocent until proven guilty" may be reasonable in judicial matters, the cautious belief that humans are more likely to do harm than good when manipulating Nature leads Hardin to advocate a shifting of the burden of proof away from those who think they are being harmed and onto those who believe they are doing no damage. Until new chemicals, practices, or technologies have been proven not to do harm, their production and distribution will be restrained. Stated as "Nature knows best," the precautionary principle or "guilty until proven innocent," this principle acts as a corrective to human arrogance toward Nature. Regarding ecological systems, the rules should be, if it ain't broke, don't fix it; if you must tinker, save all the pieces; and, if you don't know where you're going, slow down.[13]

Principle Number Four: There Is No Such Thing As a Free Lunch

While on very rare occasions there may be actions that benefit everyone involved, this principle asserts that those instances are so extraordinary that they should not be used as a guide. Far more common is the situation where gains to one participant entail costs to another. Sensitivity to ecological science should make us incredulous when we are told that a policy or course of action will result in a "win-win" situation. Far more likely is the possibility that negative costs are being imposed somewhere where their impact cannot be seen, are being ignored, or are being imposed upon some group that is powerless to express their dissent. The environment taken as a whole (which it must be if we recognize the preceding principles) is most accurately pictured as a zero-sum game. Since the no-free-lunch principle is borrowed by ecologists from economists, it is ironic that economists, particularly macroeconomists, are often the violators of the criterion. For example, the winners and losers of free trade and economic growth illustrate this phenomenon. Many macroeconomists are fond of arguing that increases in total economic output, or elimination of protective tariffs, benefit everyone involved in the transactions, but these claims ignore the negative impacts to ecological systems that are created when resources are extracted, energy consumed, or waste generated. The situation is "win-win" if, and only if, the environment is ignored and not considered a "participant" in the transaction. If the natural world were not ignored, it would be clear that economic gains to the humans are purchased by ecological costs to the environment. Remembering the earlier principles that everything is connected and there is no such place as "away," the environmental damage or ecological pollution inflicted in the name of free trade or economic growth, will, in all probability, eventually feed back onto human societies. Principle number four, the last of those outlined by Commoner, warns us to be alert to the negatives that usually accompany perceived positives. In addition to these "laws of ecology" as outlined by Barry Commoner, three additional principles can be added.

Principle Number Five: Individualistic Markets Cannot Manage Common Pool Resources

The ecological principles described by Barry Commoner govern the general operation of all ecosystems. While humans are not exempt from these (see principle number seven), behavior specific to humans requires the stipulation of some further guidelines. Principle number five is a brief restatement of what has become known as the "tragedy of the commons," first articulated by Garrett Hardin.[14] This essay has two expressed theses. First, Hardin seeks to demonstrate that the population problem has "no technical solution," by which he means that the ecological ruin implicit in the unlimited increase in the human population cannot be resolved by science and technology. Attempting to address exponential population growth by an exponential increase in the output of human economic production must eventually flounder upon the fact that the earth, its resources and its ability to absorb the inevitable pollution generated as a by-product of manufacturing, is finite. A solution to the population crisis (to the degree to which a solution exists) can only lie, therefore, in a fundamental extension or reordering of our moral assumptions. The second point of Hardin's argument is his refutation of Adam Smith's "invisible hand." As postulated by Smith in 1776, if each individual in a society pursues their own self-interest, then the workings of a competitive, open-market economy will result in the public interest, or the collective good, being produced. Smith contends that this movement from private interest to public interest, from individual goods to the common good, is achieved without anyone in the system consciously endeavoring to seek it. It is achieved spontaneously, without volition, "as if led by an invisible hand."[15] Hardin's rejection both of Smith's argument and a technical solution to exponential population growth is tantamount to a denial of the assumptions underpinning laissez-faire liberalism.

From the outset, Hardin points to the absence of logic in such liberal goals as Jeremy Bentham's dictum that society should pursue "the greatest good for the greatest number."[16]

According to the principles of mathematics, it is impossible to "maximize" two or more variables at the same time. It may be possible to maximize "the greatest good" or to maximize "the greatest number," but you cannot do both at the same time. Still less is it possible to maximize three variables at once, as was suggested by Gifford Pinchot and the Progressive Conservationists when they sought as a goal "the greatest good for the greatest number over the longest period of time."[17] This social dictum can be made logical by changing it to something like "the *optimum* good for the *optimum* number for the longest time," but this requires that a collective, social (and ultimately moral) decision be made regarding the definition of "optimum." In order for the optimum good, or real sustainability, to be the goal of public policy, there must be a collective realization that the optimum is less, quite possibly considerably less, than the maximum. Beyond these considerations of formal logic, Hardin's tragedy of the commons goes on to attack the idea that resources available to everyone at no cost can ever be managed by a process that relies on individual, voluntary compliance.

Hardin's now famous illustration takes as its starting point the idea of a "commons" or a communally owned pasture such as those prevalent in pre-modern Europe and colonial America. When an individual contemplates exploiting the commons by adding an additional head of livestock, he or she will weigh the potential benefits of the exploitation and discount the potential costs. Assuming the individual to be rational, and defining "rationality" as the instrumental weighing of personal benefits and costs, then it is easy to see that while the benefits accrue totally to the individual, the costs are spread across the entire society. Hardin argues that this privatization of benefits and socialization of costs must inevitably lead the individual to add the additional animal, and since all other members of the society are assumed to behave in the same, individualistically rational manner, the increasing result must be an exponential accumulation of negative impacts that continues until the pasture collapses from overexploitation. The important thing to note is that the individual's behavior cannot be changed by reason or education, for they are acting rationally and with complete information. Given the assumption of *Homo economicus* operating in a liberal society where he or she is free to pursue any action they find egotistically and materially rewarding and which is not prohibited by the law, destruction of open access to "common pool resources" such as air and water, must be the unintended, disastrous, yet inevitable result.[18]

Principle Number Six: What Is No Good For the Hive Is No Good For the Bee

Principle number six resurrects an aphorism crafted by Marcus Aurelius in his Mediations written in second century A.D. It is important now because it articulates the guideline that needs to be used in resolving the tragedy of the commons. If individual self-interest does not lead to collective good, and if a resolution can only be found in a restructuring of decision-making rules, then the formula of Marcus Aurelius indicates the direction the process should take. The lesson being taught here is that the primary focus in deciding collective action must be the interest of the community, population, species, or ecosystem and not the good of the individual. It is easy to imagine actions that are quite beneficial to the individual—perpetual life is an example—that would be disastrous if generalized to everyone. It is much more difficult to picture actions that are favorable to the species that do not ultimately benefit individuals also.

Many people, raised in Western cultures that emphasize the dignity and importance of the individual, may be philosophically offended by a tenet that seems to degrade the significance of the individual as it elevates the value of the collectivity. Perhaps, these misgivings can be partially mitigated by observing that principle six does not require us to ignore completely the good of the individual "bee." It simply reminds us that, in the long run, the health, welfare, or good of individuals ultimately rests in the integrity and sustainability of the environment in which they live. The dictum admonishes us to look first to the health of the environment, to the welfare of the species, and to the long-term sustainability of communities and systems. Instead of moving from the personal to the communal, as suggested in Smith's invisible hand, the approach built into sound ecological science is to move from the health of interdependent wholes to the well-being of the entity. This is true of forests in regard to pines, marshes in regard to geese, and it is true of societies in regard to humans.

Principle Number Seven: Humans Are Not Exempt from the Principles of Ecology

Commoner and Hardin, being trained in the natural sciences, may have overlooked the most important ecological principle of all. Everything that has been said earlier is only of peripheral interest to humans if we are not subject to the laws of ecology. Embedded in the Western intellectual tradition is the faith that the human species is somehow exempt from the principles that guide the rest of the natural world. Carolyn Merchant and William Catton have described this faith as "human exemptionalism."[19] Basically, it argues that since humans are capable of creating such things as technology and culture, our connection to the natural world is tenuous at best. In its crass form, human exemptionalism can be found articulated in the belief that humans study the principles of Nature, not in order to live by them but in order to conquer and control Nature. The idea that we are exceptional and unique flatters our trust in our freedom and ability to choose from among an infinite set of options, and the choices are seen as infinite because we have the power to create them as we wish. Human exemptionalism is so pervasive that it can be found even among well-intentioned individuals who honestly hope to address some of the planet's ecological problems and who consider themselves environmentalists. Vegetarians argue that abandoning a meat-based diet would free land from use as pasture and permit us to support a much larger population. Setting aside the question *why* we should want to have more humans on the planet, the vegetarian argument fundamentally advocates that our species should "live lower on the food chain."[20] Note that implicit in this logic is the assumption that our place on the food chain is up to *humans* to decide. Unlike wolves, deer, mice, or clover whose place on the food chain has been determined by millions of years of evolution, humans, it is assumed, are exempt from this process.

Principle number seven asserts that human exemptionalism is an essentially flawed picture of the affiliation between humans and the rest of Nature. While there is no doubt that technology, culture, dietary choices, and other artifacts of our creation place us in a distinct relationship to Nature, it is even more certain that we need clean air and water, that toxic chemicals and radioactive materials damage our cells, that the numbers of our species is limited by the carrying capacity of the planet, that the laws of thermodynamics apply to our inventions and undertakings, and that the exponential increase in human civilization must necessitate the intrusion into the niches of other creatures. If humans are not exempt from the principles of ecology, the connections may be psychological as well as biological, and human "needs" may extend far beyond those necessary merely to sustain a bare physical existence.

Our connections to the natural world may also extend to such issues as a human need for solitude and wilderness or the psychological pathology that appears to inevitably follow when we lead our lives on the assumption that we have no connections, no communal ties, no rules, and no guidelines to judge our actions beyond our own inventions and constructs.[21]

If these seven ecological principles are widely known and acknowledged, and if legislators are eager (as they often claim) to build public policy on a scientific bases, then why does it often seem so difficult to enact environmental practices that conform to the facts of ecological reality? Since these principles establish an objective, scientific basis for evaluating public policy, it is possible to analyze the decision-making process established by the American Constitution and see how close to effective ecological practice it is capable of moving.

The Liberal Bias of American Policy

In The Semi-Sovereign People, E.E. Schattschneider makes the following observation; "All forms of political organization have a bias in favor of the exploitation of some kinds of conflict and the suppression of others because *organization is the mobilization of bias.* Some issues are organized into politics while others are organized out."[22] To say there is a liberal bias in American politics is to say there exists a pre-disposition in favor of representative government built around the values of individualism, human rights, private property, and ethical relativism.[23] These specific beliefs can be subsumed under the two fundamental principles of liberal society: toleration and compromise. A liberal society can only exist if people acknowledge that others do not have to think like them in order to participate in the political process: citizens are tolerant of diverse opinions. And liberal society also requires that participants are willing to accept something considerably less than complete victory in the policy process: they are willing to negotiate, bargain, and compromise with those holding divergent policy prescriptions. This liberal bias is reflected in the American political system by competing perspectives on how power ought to be distributed and the way power actually is distributed in this nation.

Among political scientists, the most often used explanations of power are pluralism, elitism, and bureaucratic theories. Pluralism and elitism share the premise that certain citizens exert significant influence over policy formulation. They differ, however, on where the focus of power actually exists. Pluralists claim that average citizens can exercise political influence by joining groups. In turn, these interest groups compete, bargain, compromise, and form coalitions with each other. In the end, this competition among countervailing forces permits a degree of citizen participation in a representative democracy and allows the views of the average citizen (or at least the views of citizens who choose to join groups) to be heard.[24] Whatever specific policy citizens want, and can negotiate, is, presumably, capable of being put into practice. It is easy to see the descriptive and prescriptive elements in pluralism. The claim is that public policy *is* the result of this bargaining process, and it is also asserted that this interest-group pluralism *ought* to be the way decisions are made. Interest-group pluralism rests upon and reinforces the liberal principle of compromise. The premise is that the best ideas, the best individuals, and the best policies will win any contest where the rules are fair and assiduously followed. Put differently, liberalism believes that winners are winners because they deserve to be winners. One result is that defenders of the environment become seen as simply another in a long line of interest groups, with no particular factual or ethical superiority over those who contend for exponential economic growth, habitat destruction, resource depletion, or environmental pollution. In a liberal society, it is all a matter of opinion.

Writers who argue from an elitist theory of power, have a different viewpoint. According to them, public policy represents, not the wishes of organized citizens, but rather, the preferences of a small, self-selected elite. Elite theorists make it clear that this is not a process of representation. The elites are not spokesmen (they are almost all males) for popular opinion. Nor is the system competitive. Public decision-making, according to this theory, is characterized by cooperation and collaboration among individuals who hold a basically similar view of the political world. This shared perception, or elite consensus, derives from, and is instilled by, similar class and educational backgrounds, coupled with a complacent commitment to the status quo.[25] One group of elite theorists argues that a critical component of the elite consensus is a commitment to individual rights, social tolerance, and the respect for diversity often lacking among the less-educated and less open-minded masses. According to this view, it is the "irony of democracy" that liberal values are most strongly supported by the upper-class members of the society.[26] The important thing to notice in both pluralist and elitist theories of power is that they both place the power to make decisions in the hands of people and not institutions. Pluralism and elitism both believe that citizens control organizations more than institutions shape policy.

In contrast, all written constitutions, in fact virtually all systems resting upon the rule of law, are built upon the premise that it is possible to construct social institutions that mold the choices made by individuals and groups. Constitutions, then, pay homage to the bureaucratic theory of power. As theorists from David Hume to Max Weber have argued, the genius of politics over the past 300 years has been to rationalize the social system and turn politics into a "science." By creating organizations of power, granting and withholding political authority, and establishing the rules for decision-making, written constitutions influence the possible options of political actors. In other words, the process by which public policies are made helps determine, to a great extent, which policies are made. The process shapes the product. If elites or masses were trusted completely, there would be little need for institutions beyond those necessary to implement the will of the rulers. The mere existence of formal organizations for policy making bears testimony to the masses' lack of faith in elites and elites' lack of faith in the masses.

One interesting attempt at reconciling the competing claims of elitism and pluralism has been made by Peter Bachrach and Morton S. Baratz in an essay titled, "The Two Faces of Power."[27] According to these authors, the key to understanding the two faces of power is to recognize that they exhibit themselves in different stages of the policy cycle as issues move through agenda setting, policy formulation, policy legitimation, policy implementation, policy evaluation, and policy revision.[28] The power of elites is felt in the agenda setting stage. It is here where the political issues to be discussed and decided conform to elite value preferences regarding what problems are important and what solutions will be considered practical or reasonable. Perhaps most importantly, power can be exercised at this stage by keeping issues *off* the public agenda. What is never discussed can never be decided. This is where the full impact of C. Wright Mills' "crackpot realism" is felt. In contrast, interest-group pluralism can be found in the policy formulation stage of the policy cycle. In the actual writing of the laws and setting of policy guidelines, there can be seen the competitive bargaining, the compromise, and the incremental decision-making that characterizes the pluralist theory of power. In other words, the consensus of the elites determines the issues where policy will be formulated, and then negotiating interests make deals that decide which policy will be established. This can be clearly seen in an issue such as economic growth. No matter how ecologically rational, scientifically supportable, and socially necessary a no-growth economy may be, the idea is so inconceivable to the consensus of the elites that it never makes it on to the public agenda for serious consideration. In this sense, America has not two political parties, but only one: the pro-growth, free trade, DemRep coalition.

If we were to ask what determines policy in America, the response must be that elites restrain the viewpoints on the public agenda, interest groups have predominant influence over the formulation of policy, and the Constitution biases the overall system by institutionalizing these views and this process. By setting parameters on how policy proposals become enacted into law, the Constitution limits which bills will make it through the legislative system. The liberal bias of American politics would mean little unless it was actually put into practice.

Liberal Policy Making: Institutionalizing the Bias

When decision-makers formulate public policy, they must follow the procedures established by the Constitution and the law, and when they follow those procedures, the biases built into the system get passed on in the policies they create. Procedural rules become manifested as policy constraints. Reviewing the policy-making process will expose the mechanisms by which the liberal biases are institutionalized and the consequences of those legal mechanisms when their impact on environmental policy is considered. Unfortunately, remedying the situation by eliminating the negative mechanism is not as simple as has been suggested.[29] The techniques by which the liberal bias is built into the policy process are also the methods by which individual freedom, equality before the law, and personal rights are maintained. The real dilemma is how to avoid the "baby and the bathwater" syndrome: from an environmental perspective, it becomes obvious that these mechanisms prevent the formulation of effective environmental policy, but from the social perspective of the nation's citizens, there are a great many positive consequences that follow from the liberal bias. It seems that ridding politics of the negative features could only occur by also removing a great many of the positive outcomes. One approach is to attempt reconciliation between the mechanisms and the liberal consequences by addressing environmental problems in a superficial, tinkering, incremental fashion that never gets at the roots of the problem, but, on the other hand, never seriously offends our liberal principles. It can be argued that this is the path we have followed.[30] Whether or not it is possible to move beyond this shallow response can best be addressed after looking at the particular policy mechanisms and the environmental results that follow from them.

The initial feature of policy making in America that must be recognized is the splintering of authority, problems, responsibility, and information that constantly occurs. The fragmentation of the elements necessary to coordinate any possible response characterizes policy making in America. Problems of pollution are divided according to the medium in which they occur (air, water, land use), administrative responsibility and authority are split among different agencies and different levels of the government, interests and values are splintered among various groups who might be involved, and the information and knowledge necessary to tackle the problem may be divided among multiple interests who do not share their findings or who distort their data to support their particular agenda.[31] Commoner's "first law of ecology," the principle that everything is connected to everything else, is ignored, denied, and contradicted in a process that takes as its organizing assumption the belief that reality can be meaningfully disjointed and responded to in a piecemeal fashion. In many cases, the policy process is the cause of this fragmentation. In other instances, the process encourages the lack of issue coordination that originates elsewhere. While deep at its source, this political fragmentation derives from the liberal faith in specialization, division of labor, and institutional structure, it is put into practice at the Federal level in various ways.

When the Constitution specifies *separation of powers* between the legislative and executive branches, it requires the fragmentation of policy formulation and policy implementation. Those who make the rules are not those responsible for carrying the procedures into effect. One consequence of this is that even in those instances where valuable policies may be written, there can exist little assurance that they will be put into practice in the manner intended by their creators. Implementing agencies may engage in slowdowns, reinterpretations, or noncompliance when faced with Congressional policies they disapprove of or find repugnant to the thrust of the administration's agenda. This is particularly evident when American national politics exhibits "divided government," which means that one political party controls the executive branch and the opposition party controls at least one house of Congress. Divided government is possible only with the American Constitutional system, and although it strengthens our checks and balances, it also permits policy obfuscation and the tendency toward political gridlock. Policy implementers in the executive branch may have a fairly clear idea of the actions necessary, but without a coherently formulated policy and the legislative majorities necessary to get their programs enacted, they may lack the authority to proceed. Over the years, the situation has been partially mitigated by having Congress write broadly stated guidelines in the form of law and then permitting the executive agencies to write more detailed regulations in conformity to the general policy. For example, in writing the Clean Water Act, one of the goals set by Congress was to make the surface waters of America "fishable and swimmable." Beyond the fact that this objective may be considered so vague as to be meaningless, the practical impact was to establish Federal (i.e., Congressional) authority in the area of water quality and then to delegate to the Executive branch—in this case, the Environmental Protection Agency (EPA) working in conjunction with state governments—the task of establishing legally enforceable standards and a National Pollution Discharge Elimination System (NPDES) to implement those standards.[32] Congress writes the authorizing *laws*, and executive agencies write the enabling *regulations*. In the exercise of its oversight function, the legislature can, by using investigations and hearings, keep a rein on the bureaucracy to make sure the regulations are in line with the policy. But if oversight is too tightly applied, it will be claimed that Congress is attempting to micromanage the implementation of policy. If Congress permits the executive agencies wide latitude in the writing of regulations, the call will go out that the bureaucrats are abusing their discretionary authority.[33] In either case, authority and responsibility will be fragmented between legislative and executive functions, making a coherent, rational, long-range course of action less likely.

Within the legislative branch, there are several institutional mechanisms that both further the liberal bias and make effective environmental policy harder to achieve. *Bicameralism* and the reliance on the *committee system* lead to further policy fragmentation. The division of the legislative branch into a Senate and House of Representatives was designed by the Framers to further their goal of checks and balances. The expressed purpose of bicameralism is to assure that policy measures will be given careful, thoughtful consideration before adoption. The (possibly) unintended consequence of this division is that, as policy proposals move through each house separately, the goals of each plan may become significantly different, the definitions of key terms may vary, and the prescribed solutions will lack coherence. Only in the conference committee, after both houses have acted, will there be an attempt to reconcile the House and Senate versions of the bill. In this, the conference committee attempts to unite what the standing committees and subcommittees have disjointed, but the political objective of the conference committee is compromise and not coherence.[34]

Throughout the process of policy formulation, Congressional committees dissect and compartmentalize problems and solutions to the point where it is often difficult to see, let alone address, the overall interconnections. This fragmenting can begin in the very organization of standing committees, as when the Senate creates one committee to handle Energy and Natural Resources, and a different, distinct committee to structure policy in the area of the Environment and Public Works. The Energy and Natural Resources Committee may encourage the building of new electric-generating facilities fueled by natural gas, oil, or coal, in order to address perceived shortages. At the same time, the Environment and Public Works Committee may be considering the implications of American ratification of the Kyoto Convention that would require lowering the emission of the greenhouse gases that the Energy Committee is increasing. Issues of pollution generation are fragmented from issues of pollution reduction. Within a given standing committee, subjects can be further broken up according to medium, geographic region, enforcement agency, level of government, or the sub-content of the proposal. After lengthy hearings, investigations, markups, and votes, precision in these specialized, narrowly defined areas may be attained, but it is a precision purchased at the price of integration, connection, or coherence in the resulting policy.

Most of what has been said thus far applies only to specific policies that empower particular agencies to act in certain legally prescribed ways. Before they actually perform their duties, administrators must be designated, capital expenditures must be made, and enforcement officers must be hired. The best policies in the world are useless until they are implemented, and that implementation invariably requires the spending of money. Under the Constitution, there is required an appropriations bill for all expenditures of public funds, and that, in turn, requires the writing of separate legislation and the passage of distinct appropriations measures. Once again, the policy process is fragmented by the separation of *authorization* and *appropriations,* where "authorization" refers to laws that address the actual substance of the policy and authorize actions to be taken, and "appropriations" refers to laws that designate monies to be spent and appropriates public funds toward those ends. The phrases "unfunded mandates" and "implementation deficits" are sometimes heard in reference to situations where legislatures pass laws that require the adoption of certain policies and then the appropriations committees of those same legislatures fail to allocate the funds necessary in order to accomplish what they require must be accomplished. Superfund legislation is one example where substantive policy has been written to address the issue of toxic and hazardous waste but has been crippled by a lack of sufficient funds to adequately do the job. State and local governments have persistently complained that the Federal Congress is long on statutory requirements and short on appropriations to meet those directives. Another illustration is when municipalities have had to design very expensive systems to deal with both wastewater and storm water in order to comply with provisions of the Federal Clean Water Act but are denied Federal funds to help build those facilities. In appropriations for "Phase II" of the Clean Water Act—the statutory provisions addressing "nonpoint source" pollution—actual Federal allocations have consistently fluctuated between being inadequate and being nonexistent.[35]

These are not the only instances where relations between the Federal government and the states cause conflict in environmental policy. Although state and local governments have justifiably been called the "laboratories of democracy" and have created many innovative environmental programs ranging from recycling systems, deposit/return programs, land-use initiatives, and public disclosure reforms to broader-based schemes such as regulatory integration and growth management plans, it remains the case that serious fragmentation between Federal and state policy often exists. The overall performance of state and local governments with respect to environmental protection remains mixed, with some states innovating and performing much better than the national average, while other states lag behind even the most rudimentary measures of environmental protection.[36] Still, as a Constitutionally required structure of government, *federalism* represents one of those rare instances where an arrangement that promotes policy fragmentation also holds forth the opportunity for policies that are innovative, integrating, and have depth.

Separation of powers, bicameralism, the committee systems, the division of authorization and appropriations, and federalism; all represent institutional and procedural provisions that lead to policy fragmentation. Beyond these, there are other elements of the Constitution that both reflect the liberal bias and impede the likelihood of well-designed environmental policy. In addition to fragmenting policy, which violates the first two of the previously discussed ecological principles, the American political system exhibits a considerable *short-term bias* which works against the design and implementation of policies with more than a shallow, cursory impact. The time horizon of most policies, and the legislators who create them, seldom extends beyond two, three, or five years. Even in those remote instances where policies are formulated with longer perspectives in mind, it is common for them to be revised, modified, redesigned, or terminated before their full negative or positive impacts can be felt. Almost never is policy designed, implemented, and evaluated with a view that extends beyond six years—three terms for a member of the House and one term for a Senator—or eight years at the outside—the Constitutionally limited period that a President can serve. This short-term bias in policy making has several causes and some significant consequences. The most obvious reason for the existence of a short-term bias is that the Framers built it into the system in order to assure that legislators would be held responsible to their constituents. The positive outcome of this frequent electoral tempo is to make the system more democratic, responsive, and individually accountable. But the recognition of these well-intentioned and salutary effects should not blind us to the negative outcomes that accompany such a penchant for the immediate over the extensive.

The rules of Congress have multiple provisions for assuring that policy will be frequently revisited. Periodic reauthorization provisions are built into most laws requiring that they be reviewed and evaluated after a certain period of time—usually five years. Sunset clauses are sometimes included in bills stipulating that the act shall be automatically terminated unless specifically renewed after some, specifically stated, period of time. A two-year budget cycle necessitates the return to the House and Senate Appropriations Committees for reconsideration, refunding, or elimination. Again, the politically beneficial results of these stipulations cannot be ignored. They keep the policy process mindful of constituent demands, sensitive to frequent evaluation, fearful of a loss of justification and legitimacy, and attentive to productive results. They also keep the policy process focused on the short run.

Policy analysts, especially environmental policy analysts, have pointed out other impacts. Michael Kraft and Zachary Smith have noted the untoward side of the short-term bias. Kraft cites a variety of issues from population growth and a national energy policy to global warming and the protection of biological diversity that can only be solved by comprehensive, extended approaches that will, almost without question, require decades and decades to resolve. These are critical issues that do not admit of short-term solutions and will, therefore, probably be ignored or dealt with superficially. The "short-term and narrow view" means that "members of Congress are likely to be more concerned with local and regional economic impacts of environmental and resource policies," with the result that "action on environmental policies is rarely easy. Sometimes it is impossible."[37] Smith echoes Kraft's conclusions and produces an illustrative hypothetical example. Suppose, Smith conjectures, Congress is presented with two policy options. The first bill "is estimated to cost $5 billion a year for each of the first 5 years and then $25 billion for each of the next 20 years." The total cost of this bill is, therefore, $525 billion. Now, suppose the existence of a competing measure that will accomplish the same ends but "has an initial cost of $25 billion for each of the first 5 years with subsequent costs of $5 billion for each of the following 20 years." In other words, the total cost of the second bill is $225 or $300 billion less than the first proposal. Which bill will most likely be adopted? Smith argues that members of Congress will almost undoubtedly support the first measure, not because it is more effective (it is assumed that both bills will accomplish the task), and certainly not because the first bill is more cost-efficient (the total expenditures are over twice as much), but because the initial, short-term economic costs are lower.[38] The examples put forth by Kraft and Smith illustrate the short-term bias of environmental policy, the tendency to weigh immediate costs more heavily than distant benefits, and the fondness of legislators to discount the future.

Hidden in their analysis is another implication that follows from the short-term bias. It is the proclivity to seek out and prefer economic thinking and economic solutions to approaches built upon ethical standards or community consensus regarding what is right and wrong. A bias for the immediate is a bias for the economic over the ethical. This reciprocal attachment between short-term thinking and an economic perspective, on the one hand, and long-term considerations and an ethical viewpoint, on the other, needs to be recognized and appreciated. Perhaps, this can best be seen by viewing the situation in reverse. When we evaluate actions in the long distant past or future, we seldom judge them in personal or economic terms. More likely, we will consider their impact on the community as a whole, the trends they alter or set in motion, and the way they influence the values of those who will come after or reflect the values of those who came before. For instance, when Americans study their Civil War and consider the deeper meaning of what happened, we virtually never conclude that Lincoln's actions represented an efficient use of tax dollars or that Jefferson Davis' decisions produced the best benefits for the marginal costs. We may extend moral approbation to one or ethically question the actions of the other, but at that time distance, our interest in the immediate economics of the situation is tangential or contextual at best. Our overall evaluation of the different policies is framed in terms of the ethical residue and not the monetary outlay.

Robert V. Bartlett, in "Evaluating Environmental Policy Success and Failure," faults our propensity to emphasize short-term economic consequences of environmental policy to the detriment of long-term ethical outcomes. By focusing on the narrowly defined success or failure of specific projects and programs, we lose sight of the broader, longer-ranged impacts that come when institutionalized rules fundamentally alter problem definitions, organized routines, and moral perspectives. "Institutions shape politics not only by reducing chaos and bringing order through rules but also through the construction and elaboration of meaning .... [I]t is only in relation to institutional transformation that any lasting merit or worth of environmental policies can be ascertained, and it is the transformation of institutions to be ecologically rational and sustainable that poses the ultimate test for environmental policy."[39] To limit the evaluation of policy to economic cost-benefit analysis is to tilt the analysis in favor of short-term projects and programs, rather than long-term evaluation of institutions, and it is also to bias the evaluations toward negative conclusions. Economic assessments "using such values as efficiency, cost-effectiveness, and cost minimization ... frequently find environmental programs to be failures .... Because of their strong bias toward findings of no impact, simple outcomes evaluations almost always play into the hands of the do-nothing crowd and the undo-what-we've-done-right wingers."[40] What Bartlett asks us to remember, however, is that "public policy processes 'have an intrinsic moral element which exists apart from their instrumental value.' The Endangered Species Act, for example, can be said to be good because it codifies a collective ethical judgment and declares a responsibility, and because in doing so it creates a cognitive dissonance between moral commitment and political action that forces perpetual reconsideration of both ethics and action."[41] It is not so much that cost-benefit evaluations are useless or that society is misguided in using economic approaches to policy, the point is that these approaches exhibit a bias toward the short-term and programmatic. Since courses of action relating to the environment obviously have both short- and long-range implications, we must recognize and discount that bias in order to more effectively evaluate policies and institutions, but as structured under the Constitution, the formulation of policy favors fragmented decisions with a short-term bias built upon an "economic model of human rationality."[42]

Combining the pre-disposition toward policy fragmentation with the bias toward short-term, economic thinking produces an overall approach to policy that has been labeled incremental decision-making or, more simply, "incrementalism." As the name implies, public policy is formulated in increments, in a piecemeal fashion, bit by bit. Problems that may contain multiple interconnected elements are fragmented and the parts are dealt with separately. Incrementalism means that decisions are made in a reductionistic and reactive manner. In this context, the phrase "if it ain't broke, don't fix it" describes a process that waits until a situation reaches such critical proportions that public pressure demands some response before action is contemplated. This approach to policy assumes, of course, that problems can be meaningfully resolved in a fragmented way, and it assumes that all crises are capable of being reversed. Neither of these premises may be valid in the case of environmental destruction, for as we have seen, fragmentation violates fundamental ecological principles and, in the case of species extinction or global climate change, ecological processes may not be capable of being reversed. In these instances, at least, once it is broken, it cannot be fixed.

Incrementalism is conservative to the extent that it reinforces a basic faith in the status quo. It is built upon the notion that there needs to be no drastic or radical change in our manner of perceiving and reacting to policy issues. Change, when it does occur, is minimal, and usually amounts to no more than the fine tuning of a machine that is assumed to be running tolerably well and headed in the right direction. This rather complacent mode of addressing policy has been called the strategy of "muddling through;" an apt phrase for the process institutionalized within the American political system by the Constitution. The benefits of this muddling incrementalism are clear and should not be overlooked. The process is democratic, it permits multiple points of impact from a variety of citizen interests and lends itself well to collective decision-making, it does not require a complete or accurate understanding of the issue, nor does it necessitate the mobilization of the entire society before action is undertaken. When considered separately, incremental policies are unlikely to do devastating harm, they will not offend large numbers of voters, and they will subtly reinforce a basic commitment to things-as-usual. The process generates policy stability, if not necessarily policy coherence.

In Ecology and the Politics of Scarcity, William Ophuls and Stephen Boyan follow a line of argument quite similar to what has been said here. They acknowledge the advantageous effects of incrementalism and go one step further in suggesting that there may be no real alternative. The substitute for incrementalism—rational decision-making or what Ophuls and Boyan term "synoptic" outcome-oriented decision-making—presupposes clearly articulated communal goals, which may be possible only in small communities or in societies with exceptional political leadership possessed of nearly complete information and flawless reasoning. Rational decision-making may commit society to a course of action that is nearly impossible to achieve, difficult to alter, and disastrous if proven to be a mistake. In the daily administration of a democratic society, therefore, some degree of muddling through seems inevitable and desirable.[43] These positive aspects of incrementalism, however, do not completely outweigh the negative features that also appear.

Incrementalism is inappropriate to policy decisions that require long-range, comprehensive programs that may diverge radically from the status quo. As Ophuls and Boyan observe, "disjointed incrementalism is not well adapted to handling profound value conflicts, revolutions, crises, grand opportunities, and the like—in other words, any situation in which simple continuation of past policies is not an appropriate response." The environment presents numerous examples that fit into these categories. "A perfect illustration of muddling through is our approach to global warming .... We have made compromise and short-term adjustment into ends instead of means, have failed to give even cursory consideration to the future consequences of present acts, and have neglected even to try to relate current policies to some kind of long-term goal."[44] The fundamental problem with incrementalism is that it assumes we are muddling in the right direction, but when incremental decision-making is combined with a myopic inability to see the calamitous results of pursuing the present course of action, muddling through becomes little else than a prescription for disaster on the installment plan.[45] Worse: to the extent that incrementalism is built into the American system through the various mechanisms of the Constitution, we seem placed in a situation where some form of catastrophe appears the only means for altering the general direction of public policy. The substitute approach of a national debate engendering an alternative consensus regarding the public interest seems precluded by the assumptions of liberal politics. A fundamental alternative to incrementalism would require, in other words, that America cease being a liberal society.

Given that incrementalism is "an entrenched reality with which the environmental reformer must cope," it is important to point out some implications that follow from it.[46] When incremental decision-making combines with interest-group pluralism, the tendency is for some groups to assume strategic importance within specific, fragmented issue areas. So-called "iron triangles" will form among Congressional committees, administrative agencies, and clientele interest groups, which will, in turn, put these critical groups in a position where their concurrence is essential for any policy to be effective. The American Medical Association with regard to health policy, the major defense industries with regard to military policy, and the timber and cattle industries with regard to public land-use policy are all examples of groups that hold the power to veto any policy option within their sphere of expertise and influence. When this situation occurs, interest-group pluralism degenerates into what Robert Lineberry calls "hyperpluralism." Seeing their decisive bargaining location, groups so positioned will be inclined to "dig in their heels" and force the policy process into gridlock until and unless they receive major concessions.[47] The consequence of this state of affairs is predictable enough. Incrementalism, hyperpluralism, and the ever-present threat of gridlock coalesce to create a situation where every significant group must be conceded a sizeable portion of its demands. This creates a political situation that is extremely expensive to maintain and that predicament can only be addressed in an economy that is large and growing. Economic growth—the constant enlarging of the available spoils—becomes the only way of partially satisfying all the competing, checking groups and keeping the system out of gridlock. Economic growth thus becomes instilled and encrusted as the *sine quo non* of public policy. There are other factors that compel the American system toward an uncritical commitment to economic growth. The consensus of the elite, divided government that further fragments power and requires additional pork to lubricate the system, and liberal society's inability to define and implement a just distribution of limited resources all contribute to making economic growth, and its concomitant ecological destruction, the biased policy choice.

When one steps back and looks at the overall political system in America, the liberal bias of the Constitution, and the specific policy rules that assure fragmented, short-range, economically slanted, incremental decision-making that is only sustained by exponential growth, the shape that environmental policy takes should not be surprising. Over the years, America has done a fairly good job at addressing specific, narrowly defined problems. In many ways, the water is cleaner, the air is getting worse at a slower rate, and millions of acres of public land are receiving protection that might not be possible if they were privately owned. A solid case can be made that businesses and industries are becoming more efficient at the same time the overall economy is growing in scale. However, ecological issues that necessitate comprehensive, integrated, long-range solutions built upon a stationary economy are not being addressed. In the denial and delusion that surrounds such vital concerns as population growth, species extinction, global climate change, and an energy policy that takes into account peak oil and the laws of thermodynamics, the deep, systemic inadequacy of the Constitution is revealed. The system does well in adapting to shallow, narrow, explicit problems. Its performance is much worse when confronted with larger, deeper, and more fundamental concerns. When there is no fundamental antagonism between the policy process built into the Constitution and the basic principles of sound ecology, the public policy process is able to formulate and implement policies that have at least short-term effectiveness. But in those many areas where the contradictions between sound ecology and liberal policy making are tragically obvious, a fundamental altering of the policy process is clearly necessary. This situation has been apparent to a number of activists, analysts, and scholars over the years. They have proposed changes in the policy process, and many of their suggestions have been adopted. The past record and future prospects of those adjustments shed further light on the state of environmental politics and policy.

Beyond Liberal Policy Making (And Back Again)

As the growth of our economy slows because of increasing energy costs and stagnant consumer spending; as ecological transformations brought on by global climate change require quick, decisive, and dramatic flexibility in our policy responses; and as an ever-increasing population makes more and more difficult demands for social services, the capabilities of our Constitutional system will be severely strained. Given what we now know about ecological principles and liberal decision-making, it is hard to picture how decision makers operating within the parameters established by the American Constitution will be able to meet those challenges. The enshrined process is simply too fragmented, too slow, too incremental, too shortsighted, and too oriented to individual self-interests to provide the type of decisions that will become ever more necessary. America may (or may not) remain politically free, but under the current rules of policy formulation, it will be difficult for America to become ecologically sustainable. Some change in the framework by which decisions are reached will have to be made. The implications of Constitutional arrangements for environmental politics are threefold.

America can continue operating within a free market economic system and a Constitutional arrangement built around checks and balances, separation of powers, fragmentation, incrementalism, and adversarial competition among entrenched groups. We can try to muddle through, making only small adjustments to policy as the immediate circumstances dictate, while ignoring the deeper problems whose resolution is required in order to address the larger causes of environmental disruption. This commitment to the status quo appears to be the preferred choice of many of our economic and political elites. It fits well with the version of crackpot realism to which they have been socialized, it requires little in the way of genuine leadership, and it is compatible with a Constitutional system that makes basic amendment extremely difficult. Its fundamental denial of basic scientific principles assures that it is doomed to ecological failure, but change will probably only come as the result of some rather significant social catastrophe.

The second possible response is to move toward what John Dryzek calls "democratic pragmatism."[48] The hope here is that significant popular support can be raised through extensive public education and, when converted into an influential green political movement, this influence can call for the type of change necessary in American policies and practices. Like free market approaches, democratic pragmatism works from the bottom up. Unlike free marketeers, however, democratic pragmatism assumes that Americans will act as citizens and not exclusively as consumers. Democratic pragmatism seems to work best in constitutional settings that have strong, disciplined political parties; unified legislative and executive functions; and a system of parliamentary supremacy capable of converting popular majorities into comprehensive and coherent national programs. It is probably no accident that the major advocates for this approach are John Dryzek (an Australian) and Robert Paehlke (a Canadian). It is also probably no accident that those nations who have done the best job in protecting their environment (Dryzek lists Germany, Japan, the Netherlands, Norway, and Sweden) are those countries who combine parliamentary constitutions and strongly regulated economic systems. The American Constitutional system is built around the specific rejection of parliamentary government, national majorities, and disciplined political parties. More than a few American political scientists have expressed admiration for parliamentary government, but absent a significant shift in American culture, both the explanation of the necessity for the move and the actual transformation to a parliamentary system are unlikely to occur.

The third alternative is to keep the formal structure of the Constitution but shrewdly subvert it to operate more like a responsible-executive, parliamentary model. Although the time frame still remains fairly short and economic growth is still taken as a given, policy formulation becomes effectively transferred out of the legislative branch and into the executive branch where it becomes the task of experts to adapt law to scientific requirements, and it becomes the mission of appointed cabinet officials to implement an overarching administrative strategy. This is what Dryzek calls "administrative rationalism." Its defining characteristics are a top-down, hierarchical structure; scientific risk assessment; professional resource management bureaucracies, and expert advisory commissions seeking to define and implement rational policy analysis.[49] In America, this has been the approach adopted during the administrations of Progressive presidents. The checks and balances, as well as the separation of powers built into the formal Constitution have been bypassed, as the plebiscitary President becomes the articulator and spokesman for a comprehensive national interest that is then implemented by technical experts operating within specialized bureaucracies. Congress still passes enabling legislation, but checks and balances become reduced to oversight as policy is both formulated and implemented (and usually also evaluated) within the executive branch. Progressive resource conservation is probably as close to parliamentary responsible government as America can get without fundamentally altering the structures and rules of our Constitution.[50]

The elections of 1980 and 2000 seemed to indicate that a majority of Americans wanted to return to a period of classical liberal policy making. In 1980, environmentalists across the nation were initially dismayed, but not surprised, at the actions of the Reagan/Watt administration. After the attempted retreat from environmental protection, environmentalism rebounded with the removal of Watt, Anne Gorsuch Burford, and the promise of George H.W. Bush to become "an environmental president." Those who believed that the anti-environmental Republican right had learned its lesson with the rapid rise and decline of Newt Gingrich's "Contract With America" during the mid-1990s, soon had their misconceptions corrected. After the 2000 presidential election, anti-environmentalism reappeared as George W. Bush made appointments such as Gale Norton to head the Department of the Interior (a James Watt protégé), adopted anti-population control policies to placate the anti-abortion Christian right, relaxed Federal guidelines on environmental quality issues such as arsenic standards, announced an openly hostile rejection of the Kyoto Convention on global warming, and adopted a national energy plan that was strongly supportive of big corporate oil, gas, and nuclear interests while aggressively unfriendly toward conservation and wilderness protection as symbolized by proposals to open the Arctic National Wildlife Refuge to drilling. Environmentalists countered with opposition to all these schemes, Progressives shifted their allegiances from a strong President to an assertive Congress, and a new era of conflict, litigation, and gridlock appeared.

The environmental record of the Obama administration illustrates the possibilities and limits of policy making within our Constitutional constraints. Executive appointments, particularly at the EPA, brought pro-environmentalists back to the table with positive results such as getting CO_2 regulated as a greenhouse gas. Once again, the President and his staff (aided by climate science and recurring weather disasters) put climate change on the public agenda. The movement toward alternative energy sources made slow progress, and the construction of the Keystone pipeline was temporarily blocked. Obama's 2015 Climate Action Plan used the EPA's authority to control CO_2 as a pollutant to write regulations restricting the output of greenhouse gases, but given the gerrymandering of districts in the House of Representatives, the frequent use of Senatorial "holds" and the power of conservative judges in many Federal courts, the future of this Action Plan is questionable. The only bipartisan agreement likely to succeed is when Republicans and Democrats share the values of the core elite consensus; that is, the values of continuing exponential economic growth as reflected in policies such as technological innovation, free trade, and globalization. It should be noted that at the same time the Climate Action Plan was proposed, Obama and Republicans in Congress pushed through the Trans-Pacific Free Trade Agreement that promotes economic growth, gives the President "fast track" authority, and allows corporations to sue before special panels if national policies harm their corporate bottom lines. In other words, what environmental protection that occurs in the near future promises to be well within the confines of the mainstream environmentalism, the elite consensus, and the political process embedded in liberal policy making. The possibility of a more deeply rooted approach based upon a more radical understanding of the crises and in conformity with the fundamental principles of ecology does not seem to be immediately likely.

Given all of this, if the question is asked—Can we successfully remedy the underlying causes of our environmental crisis and produce a real, workable solution to our social, economic, and ecological ills within the framework of the American Constitution?—the answer would have to be "No." If the more modest question was asked—Will incremental, fragmented statutes such as the Clean Air Act, the Clean Water Act, the Endangered Species Act, or the Resource Conservation and Recovery Act be enough to at least stop impending crises from getting worse?—again the response would have to be "No." At this point, a cynic, or a concerned citizen, would be justified in asking—If these pieces of mainstream environmentalism cannot correct, or stop, ongoing environmental deterioration in the form of exponential economic growth, geometrically increasing human populations, energy supplies that are peaking as demand becomes greater, or global climate change threatening new, unpredictable, and irreversible calamities, then why do we have them? The answer to this question is unambiguous. Mainstream environmental protection is absolutely essential because it will force the problems to get worse at a slower rate. It is a sad commentary on the current state of our environment that marginally retarding the process of energy depletion, population explosion, or species extinction must be viewed in a positive light, but the truth is that the conservation of remaining open spaces, the regulation of toxic and hazardous substances, conservation and restoration biology, and the other efforts of mainstream environmentalism are crucial in the short run, while America looks inward and draws upon its cultural heritage for the values, ethics, and political perspectives needed to move toward an authentically just and sustainable culture. Mainstream environmentalism is necessary because it is "practical" in the near term, and because of this, it holds forth the promise of buying us a little more time to muster the collective will to make the more difficult, but more important, changes needed for a truly workable resolution of our multiple crises.

In the not too distant future, something like radical environmentalism must become the guiding vision of American politics. It may be clever and politically astute to avoid the word "radical" and, instead, use some less incendiary expression such as "cutting-edge environmentalism" or "outside-the-box environmentalism" but, in any case, the fundamental assumptions around which our political behavior is built must be critically analyzed, the consequences of our various environmental legacies must be exposed, and alternatives to destructive ecological practices must be explored. The national character and political will of America is rooted in the cultural perspectives articulated during the Founding and evolving through our history. Understanding the environmental legacies within American political thought will demonstrate this cultural will is pointed in conflicting directions but still richly diverse enough to provide an avenue for sustainable environmentalism within our most basic political values. These political ideas are the American "constitution" in the second sense of that term. Exploring these environmental legacies—interpreting this second, more fundamental American constitution—will occupy the second part of this study.

NOTES

1. Ron Chernow, Hamilton (New York: The Penguin Press, 2004). Walter Isaacson in Benjamin Franklin (New York: Simon and Schuster, 2003). Joseph J. Ellis, Founding Brothers: The Revolutionary Generation (New York: Alfred A. Knopf, 2001).
2. Robert A. Dahl, How Democratic is the American Constitution?, 2nd ed. (New Haven: Yale University Press, 2003); Sanford Levenson, Our Undemocratic Constitution (New York and Oxford: Oxford University Press, 2006); Daniel Lazare, The Frozen Republic: How the Constitution is Paralyzing Democracy (New York: Harcourt Brace and Company, 1996). Progressive historians at the turn of the twentieth century actively challenged the economic justice and democratic commitment of the Constitution. See Charles A. Beard, An Economic Interpretation of the Constitution of the United States, originally published in 1913 with a new author's introduction in 1935 (New York: The Free Press, 1965), and J. Allen Smith, The Spirit of American Government (Cambridge, MA: The Belknap Press of Harvard University, 1965), originally published in 1907.
3. In his classes, Oberlin professor David Orr conducts interesting mock trials built upon these assumptions. See David W. Orr, Down to the Wire: Confronting Climate Collapse (New York and Oxford: Oxford University Press, 2009), 138–145.
4. Barry Commoner, The Closing Circle: Nature, Man, and Technology (New York: Alfred A. Knopf/Bantam Books, 1972).

5. Garrett Hardin, Living Within Limits: Ecology, Economics, and Population Taboos (Oxford and New York: Oxford University Press, 1993).
6. Commoner, The Closing Circle, p. 29. Hardin, Living Within Limits, p.199.
7. The explanation of ecological cycles can be found in any basic environmental science textbook; for example, William P. Cunningham and Barbara Woodworth Saigo, Environmental Science: A Global Concern, 6th ed. (Boston: McGraw Hill, 2001), 67–68.
8. Commoner, The Closing Circle, 29.
9. Hardin, Living Within Limits, 200.
10. Ibid., 201.
11. Commoner, The Closing Circle, 37.
12. On the precautionary principle, see Dryzek, The Politics of the Earth, 139; and Cunningham and Saigo, Environmental Science, 227–228; and Carolyn Raffensperger and Joel Tichner, Protecting Public Health and the Environment: Implementing the Precautionary Principle (Covelo, CA: Island Press, 1999).
13. Borrowed from Herman E. Daly and John B. Cobb, Jr., For the Common Good: Redirecting the Economy Toward Community, the Environment, and a Sustainable Future (Boston: Beacon Press, 1989), 306.
14. Garrett Hardin, "The Tragedy of the Commons," Science, Vol. 162, December 1968; see also Hardin, Living Within Limits, Chapter 21.
15. Adam Smith, An Inquiry into the Nature and Causes of the Wealth of Nations (New York: The Modern Library, 1985), Smith's comments on "the invisible hand" are at 225.
16. Hardin, "The Tragedy of the Commons," 1243.
17. Gifford Pinchot, Breaking New Ground (New York: Harcourt, Brace and Company, 1947), 326.
18. Various authors have come forward to argue against the tragedy of the commons. Susan Jane Buck Cox, "No Tragedy of the Commons," Environmental Ethics, Vol.7, No. 2 (Spring 1985), 49–61; and Gary Snyder, The Practice of the Wild (San Francisco: North Point Press, 1990).
19. Carolyn Merchant, Radical Ecology: The Search for a Livable World (New York and London: Routledge, 1992), 89–91: and William R. Catton, Jr., Overshoot: The Ecological Basis of Revolutionary Change (Urbana and Chicago: University of Illinois Press, 1980), 58 and 277.
20. See, for example, Francis Moore Lappe, Diet For A Small Planet (New York: Ballantine Books, 1991).
21. Examples include Edward O. Wilson, Biophilia (Cambridge, MA: Harvard University Press, 1984). See also Paul Shepard, The Tender Carnivore and the Sacred Game (New York: Charles Scribner's Sons, 1973); Thinking Animals (New York: The Viking Press, 1978); and Nature and Madness (San Francisco: Sierra Club Books, 1982).

22. E.E. Shattschneider, The Semi-Sovereign People, cited in Bachrach and Baratz, "The Two Faces of Power," 149.
23. Among scholars who have noted this bias, see Zachary A. Smith, The Environmental Policy Paradox, 3rd ed. (Upper Saddle River, NJ: Prentice Hall, Inc., 2000), 54–55; and Walter A. Rosenbaum, Environmental Politics and Policy, 91–92. The late C.B. Macpherson labeled this combination "possessive individualism" to summarize liberalism's commitment both to the individual and to private property. C.B. Macpherson, The Political Theory of Possessive Individualism (London: Oxford University Press, 1962).
24. Classic studies in pluralism included David Truman, The Governmental Process (New York: Alfred A. Knopf, 1951); Robert A. Dahl and Charles E. Lindblom, Politics, Economics, and Welfare (New York: Harper Torchbooks, 1953); and Robert A. Dahl, Who Governs? (New Haven: Yale University Press, 1961).
25. C. Wright Mills, The Power Elite (Oxford: Oxford University Press, 1959): G. William Domhoff, Who Rules America? (Englewood Cliffs, NJ: Prentice Hall, Inc., originally published 1967, updated 2009); Thomas Dye and Harmon Ziegler, The Irony of Democracy, Millennial Edition (New York: Harcourt Brace, 2000).
26. On the "authoritarian working-class personality," see Seymour Martin Lipset, Political Man: The Social Basis of Politics, 2nd ed, (Baltimore: The Johns Hopkins Press, 1981), and Dye and Ziegler, op.cit., and Peter Bachrach, The Theory of Democratic Elitism (New York: Little, Brown Co., 1967).
27. Peter Bachrach and Morton S. Baratz, "The Two Faces of Power," American Political Science Review (December 1962), vol. 56, 947–952.
28. Standard references include Walter A. Rosenbaum, Environmental Politics and Policy, 3rd ed. (Washington, D.C.: CQ Press, 1995), 84–87; Karen O'Connor and Larry J. Sabato, The Essentials of American Government, 3rd ed. (Boston: Allyn and Bacon, 1998), 379–389. Other writers have accepted Bachrach's thesis while rejecting the policy cycle model, see Charles E. Lindblom and Edward J. Woodhouse, The Policy-Making Process, 3rd ed. (Englewood Cliffs, NJ: Prentice Hall Publishing Co., 1993), 11.
29. Steven L. Yaffee, "Why Environmental Policy Nightmares Recur," Conservation Biology (April 1997), vol. 11, no. 2, 328–337.
30. C.A. Bowers, The Culture of Denial: Why the Environmental Movement Needs a Strategy for Reforming Universities and Public Schools (Albany: State University of New York Press, 1997); Timothy W. Luke, Capitalism, Democracy, and Ecology: Departing from Marx (Urbana and Chicago: University of Illinois Press, 1999); Dave Foreman, Confessions of an Eco-Warrrior (New York: Harmony Books, 1991).

31. Yaffee, "Why Environmental Policy Nightmares Recur," 333–335.
32. Frederick R. Anderson, Daniel R. Mandelker, and A. Dan Tarlock, Environmental Protection: Law and Policy (Boston: Little, Brown and Company, 1990), Chapter IV. Robert W. Adler, Jessica C. Landman, and Diane M Cameron, The Clean Water Act 20 Years Later (Washington, D.C.: Island Press, 1993).
33. Michael E. Kraft, "Congress and Environmental Policy," in James P. Lester, ed., Environmental Politics and Policy: Theories and Evidence 2nd ed. (Durham and London: Duke University Press, 1995).
34. For a case study of these issues as seen in the Clean Air Act, see Richard E. Cohen, Washington At Work: Backrooms and Clean Air, 2nd ed., (Boston: Allyn and Bacon, 1995); Gary C. Bryner, Blue Skies, Green Politics, 2nd ed., (Washington: D.C.: CQ Press, 1995).
35. Barry G. Rabe, "Power to the States: The Promise and Pitfalls of Decentralization," pp. 31–52, and Richard N.L. Andrews, "Risk-based Decision Making," pp.208–230; both in Norman J. Vig and Michael E. Kraft, Environmental Policy in the 1990s, 3rd ed. (Washington, D.C.: CQ Press, 1997).
36. Rabe, "Power to the States: The Promise and Pitfalls of Decentralization;" David J. Brower, David R. Godschalk, and Douglas R. Porter, eds. Understanding Growth Management: Critical Issue and a Research Agenda (Chapel Hill: The Urban Land Institute & Center for Urban and Regional Studies, 1989).
37. Michael E. Kraft, "Congress and Environmental Policy," p. 197; and "Environmental Policy in Congress: Revolution, Reform, or Gridlock?" in Norman J. Vig and Michael E. Kraft, Environmental Policy in the 1990s, 3rd ed., 120–121.
38. Zachary A. Smith, The Environmental Policy Paradox, 53.
39. Robert V. Bartlett, "Evaluating Environmental Policy Success and Failure," in Norman J. Vig and Michael E. Kraft, Environmental Policy in the 1990s, 2nd ed., (Washington, D.C.: CQ Press, 1994), 167–187.
40. Ibid., 174 and 182.
41. Ibid., 178, internal quotation from Richard I Hofferbert, The Reach and Grasp of Policy Analysis: Comparative Views of the Craft (Tuscaloosa: University of Alabama Press, 1990), 156.
42. .Yaffee, "Why Environmental Policy Nightmares Recur," 336.
43. William Ophuls and A. Stephen Boyan, Jr., Ecology and the Politics of Scarcity Revisited: The Unraveling of the American Dream (New York: W.H. Freeman and Company, 1992), 246. See also Zachary Smith, The Environmental Policy Paradox, 51–52; Walter Rosenbaum, Environmental Politics and Policy, pp.90–91; and Charles E. Lindblom and Edward J. Woodhouse, The Policy-Making Process, 32.

44. Ibid., 245–246.
45. Sandra K. Hinchman and Lewis Hinchman, "'Deep Ecology' and the Revival of Natural Right," Western Political Quarterly, Vol. 43, Issue 2, 1989, 201–228.
46. Ophuls and Boyan, Ecology and the Politics of Scarcity, 246.
47. Robert Lineberry, Government in America, 2nd ed., (Boston: Little, Brown and Company, 1983). See also Michael E. Kraft, "Environmental Policy in Congress: Revolution, Reform, or Gridlock?" op. cit.
48. See John Dryzek, The Politics of the Earth, Chapter 5.
49. See John Dryzek, The Politics of the Earth, Chapter 4.
50. During the early Progressive era and in the administration of Theodore Roosevelt, conservationists passed legislation such as the Forest Management Act (1897), the River and Harbor Act (1899), the Lacey Act (1900), and the Antiquities Act (1906) that addressed early concerns over natural resources and pollution through the techniques of government land acquisition or direct regulation of economic activity. Again in the 1930's, the second Roosevelt government pushed through legislation creating the Tennessee Valley Authority (1933), the Taylor Grazing Act (1934), and the Soil Conservation Act (1935).

PART II

Cultural Legacies: America's Unwritten Constitution

CHAPTER 5

Political Theory and the American Founding: The Tension Between Logic and History

On June 29, 1787, Connecticut delegate William Samuel Johnson began that day's debate at the Constitutional Convention with the frustrated declaration that, "The controversy must be endless whilst Gentlemen differ in the grounds of their arguments ..." The specific dispute under consideration was the representational makeup of the proposed Senate, but Johnson's reference to "the grounds of their arguments" indicates his recognition that the cause of this disagreement, like the cause of many others, lay in the delegates' fundamental differences in political philosophy. Although Delawarean John Dickinson asserted that the representatives were "all united in their objects ..., [and] equally united in the means of attaining them," more prevalent was Governor Edmund Randolph's belief that "the currents of opinion are various."[1] Political theory, what was referred to as "principles," "maxims," or "models," constantly underpinned the more superficial discussions on institutional structures. Most of the 55 men who, at one time or another, participated in the Philadelphia convention hoped that eventual compromise would be possible, but this wish did not blind them to the reality that ideological diversity existed within their ranks, and that even greater disputes awaited the public exposure of their proposed Constitution. Negotiations would entail combining philosophical principles that were logically contradictory in order to achieve final agreement on a Constitution that could be ratified. Intellectual inconsistency would be the price paid for historical success.

Theoretical conflict typified much of the political discourse in the eighteenth century. Older values were being replaced and yet, as new values emerged, the prior ideals were not totally rejected. This produced a confusing and often incongruous blend as new positions were put forward at the same time that verbal homage was paid to the older framework of ideas. For contemporary scholars of the period, this creates the difficult task of trying to separate the genuine commitment to an older tradition from the disingenuous patronizing of a political position that is being attacked. For example, attitudes on the role of governmental power, the origin and authority of moral "first principles," the source of economic wealth, the optimal size of the political community, the nature of the human society being ruled, and the basic definition of such key terms as "the common good," "the people," and "republicanism" were becoming increasingly unsettled. These individual ingredients of a political theory became logically and historically arranged in unique patterns that, today, open the way for the adoption of different strategies when analyzing the Founding period. On the one hand, these various components can be logically connected into complete, consistent, and rational visions of the world that are distinguishable from competing visions, ideologies, and traditions. But while these components can be *logically* organized around certain groupings, further inquiry often reveals that *historically* they have been mixed and rationalized in unpredictable and strange ways.[2]

A clear, comprehensive, and yet comprehensible overview of the theoretical debates in eighteenth-century America can be achieved by focusing on three sets of languages and logic that repeated themselves in the political debates of the time. Thanks to the work of scholars such as J.G.A. Pocock, Bernard Bailyn, and Michael Lienesch, a clearly discernible republican or civic humanist tradition has been isolated.[3] In what follows, this civic humanism will be surveyed with specific reference to the various intellectual components that comprise its vision of politics. Much of the philosophical controversy in the last quarter of the eighteenth century involved attacks that were made on this civic humanism and alternatives that were proposed to the civic humanist political order. These attacks and alternatives are often grouped today under the name of "liberalism," but liberalism itself was far from being a consistent, unified view of politics. The theoretical literature and the political debates of the late eighteenth century reveal two versions of liberalism that were put forward as candidates to replace civic humanism. One was a social contract, rights-based liberalism associated with the English politics of John Locke, and the other was an interest-based, Scottish strain of thought that rejected much of the Lockean model, yet also repudiated civic humanism in a way that still entitles it to be called "liberalism." To appreciate the full complexity of the intellectual setting, it must also be noted that both Lockean liberalism and Scottish liberalism retained significant, albeit different and contradictory, linkages to the older civic tradition. Some of these linkages have led the Scottish tradition to be termed the "moral sentiments" school, but little scholarly doubt remains regarding the enormous impact that this heritage, and its leading exponent, David Hume, had on the American Founders. For those who are not initiated into this debate, the goal of this chapter is to survey the civic humanist, Lockean liberal, and Humean liberal idioms and to analyze how the various components of each tradition distinguished and unified them. For those who may be more familiar with this period and the academic controversies that surround it, the thesis of this overview is that a methodological commitment to an analytical approach reveals internally consistent philosophies that conflict with, and yet are blended with, other ideologies. These mixtures of compatibility and contradiction are disclosed through a developmental, historical approach to the actual formulation of the American polity that might best be illustrated through case studies showing how the most prominent Founders handled the important topics of human rights and economic theory. This chapter sets the stage for a discussion of specific American theorists by drawing the broad outlines of the categories and languages they used. Rooted in each of these ideologies is an image of humanity's role in Nature and its relationship to the environment that surrounds us. Following a discussion of the specifically political aspects of their thought, the focus will shift to their respective environmental legacies.

Civic Humanism and the Search for Community

The Founders of the American political system were as much heirs as they were innovators. Scholarship over the past 40 years has isolated and analyzed the components of the intellectual tradition upon which many Founders claimed to build. In the broadest terms, the tradition is called "classical republican," "republican," or "civic humanism," but within this general classification there exist several historically distinguishable branches of thought, and both the names and sources associated with these subdivisions vary. From an historical point of view, the most immediate source of civic humanist thought came from the radical Whig, tory, or "country" opposition to the British ministry of Sir Robert Walpole. On the British side of the Atlantic, this opposition theory was most clearly found in John Trenchard and Thomas Gordon's <u>Cato's Letters</u>, the writings of Henry St. John, Viscount Bolingbroke, and James Burgh's <u>Political Disquisitions</u>.[4] In America, the ideology of the country opposition can be traced to many of the publicly printed sermons of colonial clergy, to the numerous items found in the "pamphlet wars" preceding the War for Independence, and to portions of the political thought of specific writers such as Thomas Jefferson and George Mason.[5]

Predating and presaging the country opposition writing in British thought was the work of the seventeenth-century commonwealth men or republican writers. The central thinker here is James Harrington, whose <u>The Commonwealth of Oceana</u> (1656), along with his shorter pieces such as <u>The Rota: A Model of a Free State</u> (1660) and <u>A System of Politics</u> (1661) remain the leading voice of British civic humanism.[6] And yet, while Harrington made important and original contributions to civic humanism (most notably, the focus on landed property as the principle source of "balance" in a commonwealth), many of the overarching themes and concepts found in his work can be traced back to the republican thought of the Italian renaissance, and here, the most significant name is that of Niccolo Machiavelli. Machiavelli's republicanism, as found in <u>The Discourses</u> and borrowing on the earlier Roman and Greek thought of Cicero and Aristotle, represents the clearest attempt to piece together the various components of a logically consistent theory of politics, time, and history. There are three perceptual cornerstones that create the basis for virtually every version of civic humanism. They are first, the centrality of the political community and the forces which cause it to adhere or disintegrate; second, the inversely correlated relationship between power and liberty in any political society; and third, the essentially entropic or cyclical nature of human political history.

If it were possible to distill civic humanism down to its most fundamental essence, its basic axiom would be that humans are inherently social/political creatures who must order their lives in conformity to the moral good (usually denoted "the public interest") of the community in which they live. It is here that the tradition's roots in the philosophies of Aristotle and Machiavelli can be most clearly seen. Civic humanism sets as its ideal the life of the ethical citizen: someone who is actively involved in the political life of the community. Such a person renounces private or material concerns and promotes the general welfare through participation, civic virtue, and the armed defense of the *polis*. This republican tradition involves the "revival of the ancient notions of political *virtus*, of the *zoon politikon* whose nature was to rule, to act, to make decisions; and ... the ideology of the *vita activa*, operating in a communal climate where men were indeed called to assemble and make decisions ..."[7]

At the center of this vision lies the concept of the political community. It has been characterized as "an organic sense of structured differences, an essentially Platonic experience."[8] No mere mechanical "institution," the community for the civic humanists must be seen as both all-encompassing and living. The community was comprised of those virtuous citizens who worked to further the public interest. There could, of course, be people living *in* the community who were not *of* the community (since citizenship was far from universal), but basically the terms "citizen," "community," and "public interest" were mutually defining; a community was a group of citizens who consciously shared a common end. So defined, the public interest necessarily assumed a large degree of social consensus, a culturally homogeneous population, and a relative smallness in terms of geography and numbers that would make the consensus and homogeneity possible.[9]

Looming ever-present in the civic tradition is the fear of the one force that could destroy the community and pervert the virtuous pursuit of the common good. That force was corruption. Like a specter, corruption haunted the polity. It was easy to define: corruption was the loss of civic virtue. Public corruption existed whenever officials pursued private ends or became influenced by persons/groups pursuing their own ends. Private corruption consisted of placing one's self-interest ahead of the public interest. It meant citizens pursuing their personal values instead of being guided by the common good of the community. For the civic humanist, history provided numerous omens that signaled the impending onslaught of corruption. Usually pictured as the product of a secret but ubiquitous conspiracy, corruption's existence was indicated by a litany of evils that ran from standing armies and public debt to luxury (a sign that people were preoccupied with private, material gain), "faction" (the growth of private interest groups), "innovation" (the breakdown of the traditions that held the community together), commerce, and the growth of cities, to the selling out of the community to the placemen and stockjobbers who held office, not by virtue but through an unscrupulous system of political patronage.[10]

Perhaps most disconcerting for this vision, corruption seemed to be carried along by the very forces of time and history. Time brought change and change almost inevitably entailed corruption. Because of the fundamental entropic forces at work in human affairs, most civic humanists looked toward the future with something akin to foreboding. This produced the "dread of modernity" and the "politics of nostalgia" that commentators have found permeating these writings.[11] It should not, however, be inferred from this summary that no significant distinctions exist within this tradition, but rather that the conversation among civic humanists was usually conducted within narrow parameters. For example, classical republicans such as Aristotle, Cicero, and Machiavelli tended to view the fight against corruption in moral terms. Only civic virtue, a *ridurre ai principii*, or a public commitment to idealistic first principles could retard the spread of materialistic, self-interested corruption. English followers of Harrington, on the other hand, often saw the situation in a more economic context. For them, corruption was most often attended by a reciprocal loss of economic and political independence. Since it was taken as given that "power followed property," the surest means of preventing the corruption of the republic was to maintain the social conditions which reinforced and stimulated the existence of a landed, independent, armed, and self-sufficient "yeomanry" or "gentry." Thus, the Machiavellian strand envisioned the fight against corruption in an idealistic, moral context; while Harringtonians emphasized a more materialistic defense founded on land and economic autonomy, but both were united around a philosophy that saw politics as a battle between the republic and corruption.[12] The site of this struggle was history, and the weapons used were the forces of power and liberty.

Particularly in the seventeenth- and eighteenth-century versions of Anglo-American civic humanism, one finds the often repeated theme of a direct, inverse relationship between political power and individual liberty. A type of zero-sum correlation pervades the political community where gains in power can only be achieved and explained by losses in liberty. Furthermore, this relationship between power and liberty is not static: it is inherent in the nature of power that it seeks to enlarge itself at the price of liberty. Hence, the language of "encroachment" and "trespass" which invariably accompanies civic humanist discussions of power and the concomitant use of terms such as "jealousy" and "vigilance" regarding liberty.[13] It also follows that civic humanist discussions of politics are filled with references to "checks and balances" on those who exercise power and to attempts at institutionalizing "mixed government" and the restraint of governmental authority.

There is both a logic and a tension involved in all of this. On the one hand, the separate theoretical components such as community, virtue, citizenry, and corruption appear to be linked together in a consistent whole. On the other hand, an ambivalence regarding power, community, and the government lies just beneath the philosophical surface. Concentrated, unchecked, illegitimate economic and political power is to be restricted and resisted, but if authority is exercised in coherence with the "general welfare" or "the common good," then civic humanists are willing to extend it almost indefinitely. "Agrarian laws" to redistribute and limit economic wealth, "sumptuary laws" to restrict ostentatious displays of "luxury," established religions that weld the secular and spiritual sides of human personality, and a deep suspicion of virtually any form of diversity or any aspect of private life are all part of this tradition. The fear of power and the wish for unity coexist with a love of liberty and an uneasiness regarding who to define in or out of the social community. Strains of the republican theme can be heard today when Americans speak of a patriotic duty to the common good, of moral obligations to the their local communities, or the inner need to sacrifice the temporary pursuit of their self-interest for the larger and longer objectives of the public interest. On a less noble side, republican ideology also finds expression in what Richard Hofstadter called "the paranoid style" in American culture.[14] It is here where every logical inconsistency is seen as a lie and every untoward event is evidence of a malicious conspiracy. The wish to encourage civic virtue and the fear of encroaching corruption produce sometimes strange and frightening contrasts in America's psychic culture.

Small wonder, then, that civic humanists nervously find themselves comparing their own societies to earlier historical examples and measuring the relative degrees of corruption between their communities and the lost republics of antiquity. Comparative politics and historical anxiety are both made necessary by this perception of the world. The need to locate their social position on the slide through corruption to the collapse into anarchy or despotism leads civic humanists to see one of two variations on the theme of historical deterioration. Under one perspective, perhaps best represented in the writings of Machiavelli, the initial founding of the republic is pictured as the moment of supreme dedication to idealistic principles and, therefore, the point in history where the republic is the most virtuous. From there, practically any step that the community takes must be, almost by definition, downhill. The image is one of initial purity followed by inevitable corruption.

Several observations can be made regarding this view of history. First, understanding it helps to explain the fascination that many republican writers have for "foundings" and "founding fathers;" giving the events mythic importance and nearly deifying the men involved. But the adulation that civic humanists bestow upon the creation of the republic should not obscure the fact that, for them, this event must be taken as an actual, empirical, historical reality. The founding is neither a heuristic device nor merely a mental/rationalist construct. Instead, real men, such as Solon or Lycurgus, and actual events, such as the establishment of "the Gothic balance" or "Saxon liberty," are appealed to as examples. Deeply imbedded in the British commonwealth tradition is the idea of an "Ancient Constitution" that acts as the embodiment of established civic rights and from which the slide into corruption can be dated.[15] Not every republican writer is willing to go as far as Machiavelli and praise poverty, or as far as Andrew Fletcher and suggest that there exists an irreconcilable antagonism between virtue and civilization, but there is clearly implied the belief that the process of technological and economic "development" can lead to corruption and ruin. In the civic tradition, there are few good words spoken about the idea of "progress."[16]

Recognizing this harsh, pessimistic, entropic view permits the appreciation of a somewhat more optimistic, cyclical vision of history that some civic humanists put forth. Here, societies are seen as going through three phases or "stages of history." Initially, there is primitive, rude childishness that grows and matures into a virtuous adulthood. After this peak, the more classic pattern of history occurs, and societies slip into eventual corruption and decline. Some social growth and economic development, therefore, is possible, and desirable, within this view of republics, but still there are limits beyond which the society becomes "overly civilized," and corruption must inevitably set in. New England clergyman Samuel West summarized this view when he preached that "as in nature there is a regular progress, increase, and decline, so nations have their helpless infancy, active youth, vigorous manhood and feeble old age, followed with dissolution."[17] "Comfort" is possible, but "luxury" is still to be avoided. This is the version of civic humanism that one finds most often expressed in the period of the American Founding. It is a pastoral, small town, or agriculturally based republicanism that holds forth the promise of the virtuous citizen and the public-spirited community as the bulwark of a morally sound and socially stable polity. In either the form of New England, puritanical republicanism or Southern, agrarian republicanism, the American tradition grows from its British roots.[18] But this civic, republican ideology also produced what was, in many respects, its antithesis in the philosophies of John Locke and David Hume.

Lockean Liberalism and the Protection of Individual Rights

Using the insight that 300 years of history provides, and concentrating on the social transformations that were achieved, the contemporary scholar may understandably be drawn to focus on the differences between the whig theories of John Locke and the ideological orientation of the country opposition that surrounded much of his work, but there also exist strong theoretical linkages between Locke and his predecessors. Compare and contrast, for example, the concepts of "the ancient constitution" and "the social contract." Both are used as the source and justification for a theory of rights, the violation of which is taken as vindication of opposition, resistance, and revolution. Each represents the philosophic benchmark that delineates legitimate from illegitimate authority. At first blush, little seems to separate the two notions. But Locke's social contract is subtly different from the ancient constitution.

Locke's political writings contain no mention of "the Gothic balance" or "Saxon liberty." For a man who built much of his reputation on the epistemological arguments found in <u>An Essay Concerning Human Understanding</u>, his <u>Second Treatise</u> is almost devoid of empirical referents. This is significant. For republicans, the ancient constitution was an historical fact, and a great deal of scholarly effort was spent tracing its origins, development, and threatened demise. The reason Locke expends no energy locating the historical source of his "social contract" is that, for him, the concept is (and was never meant to be more than) an abstraction. This is speculative rationalism—not anthropology. The rights being described in the ancient constitution are the historical rights of a community, while the rights under consideration in Locke's social contract theory are "natural" or inherent rights that are discovered, not through empirical research but through logical reasoning. Viewed from one perspective, both civic humanism and Lockean liberalism are concerned with the rights of English citizens and, therefore, the differences between the two ideologies may seem minor. Seen alternatively, civic humanism claims to be built upon experience and focuses on what communities collectively *did*, while

Lockean liberalism is built on reason and is concerned with what people individually *think*. The ancient constitution purports to describe how citizens *did act* in creating and defending their historical rights. Locke's state of nature is not a description of an historical process, but is, instead, a rational, mental, heuristic construct inquiring into how contemporary members of society *would act* if political restraints on their behavior were removed.[19] This slightly different conceptualization carries with it important theoretical consequences.

Besides freeing Lockean theory from the possibility of empirical refutation, a rationalistic derivation of rights through the devices of a "state of nature" and a "social contract" makes possible a reshuffling of definitions and a reordering of theoretical priorities. Whereas Founders who looked at the world through a civic humanist perspective tended to concentrate on the concepts of community, civic virtue, and the collective morality of a group; people who accepted a Lockean perspective were more likely to think and speak in terms of the individual or the class; the emphasis on civic virtue gave way to the promotion of private virtue; the shared morality of the collectivity became less important than the diverse values of individuals; and, in general, communal idealistic cohesion became seen as less vital than personal economic ambition. Civic humanism is the language of the community seeking to preserve its integrity in the face of dissolution. Lockean liberalism is the voice of the politically excluded demanding to be let into the political process.[20]

Consider, for example, the consequences that follow from an acceptance of the argument in the famous fifth chapter of the Second Treatise. Here, Locke accomplishes multiple ideological tasks, most of which build upon his defense of labor as the source of all economic value, and his contention that the exclusive use of the property produced by human labor is the primary natural right to be defended. The distinctions between civic humanism and Lockean liberalism now become strikingly apparent. Possession of landed property, the Harringtonian concept of "estate," can no longer be pictured as the basis for inclusion in the political community. Instead, if labor is the source of value, then one can have "property" without owning land, and the definition of "citizen" must be broadened beyond the country yeomen or gentry to include the burgeoning urban bourgeoisie. Furthermore, shifting the source of wealth and the basis of citizenship from land to labor means that the relationship between power and liberty need no longer be seen as a zero-sum game. If power is based on property, and if property is land, then the total amount of power is fixed (the total quantity of the earth is finite), and the only means of acquiring more power must be by taking it from someone else, i.e., depriving them of some of their liberty. This, at least, was how classical republicans viewed the situation. But if property is labor, then, in theory, everyone can labor, everyone can have property, and increases in power need not entail decreases in liberty.[21] Of course, additional philosophic arguments must be added before this Lockean notion of infinite economic growth becomes completely rationalized. Undisturbed, undominated wilderness must now be seen as "waste," and the spoilage and sufficiency rules that previously restricted accumulation must be transcended through the advent of money and capital, but fundamentally, Lockean liberalism can be seen as the shattering of the prior philosophical limitations on power, property, and communal membership. Only later in the historical process does the liberating quality of this ideology disappear when the excluded group is admitted into the polity and then seeks to use their new political influence to bar the next set of petitioners. This is radicalism, but it is *bourgeois* radicalism.[22]

Not surprisingly, with this shift toward an individualistic, materialistic, and acquisitive concept of citizenship comes an alteration in the notion of virtue. The republican tradition carried with it an idea of virtue that implied public action, political involvement, self-sacrifice, and an intense patriotism that was usually expressed in martial, masculine terms. Virtue was *civic* virtue, and it meant the primacy of the public and the idealistic. With Lockean liberalism there begins to emerge a concept of virtue that is primarily private and materialistic. Welding the new commercial mentality with a Calvinist view of human nature, the virtuous person is now seen as someone who is industrious, hardworking, prudent, serious, and productive. The opposite of virtue is no longer corruption or self-interest, but laziness.[23] The "covenant" which seeks to affirm the shared commitment to a united faith is transformed into the "contract" which requires diverse individuals to perform certain duties in spite of their differences. In short, *zoon politikon* is replaced by *Homo economicus*.

At this point, it is necessary to return to and reemphasize the main theme of this chapter, which is the relationship between the logic of philosophy and the history of events. The American Founding can be viewed both as an attempt to put into practice theoretical "principles," "maxims," or "models" and as a reaction to certain historical phenomenon. A full picture requires historical inquiry into the occurrences of the time and a logical analysis of the perspectives through which they were viewed.

The clarity provided by logic is often muddied by the confused statements of the historical actors. After all, the men who fought the American Revolution, wrote and debated the Constitution, and helped establish the fledgling government were not consumed by the desire for philosophical consistency. Neither were they completely shallow pragmatists who had no concern for the ideological underpinnings or long-range consequences of their actions. If many members of the founding generation adopted the perspective of Lockean liberalism, it was because they found this perspective compelling and useful.

The debate over taxation during the 1760s in both Great Britain and America brought to the foreground the issue of representation and the economic source of wealth. If movable property, i.e., property created by labor, was to be taxed, then, it was argued, the basis of representation had to be extended beyond land or "freehold." Lockean liberalism made it easier to see the issue in this new light and to argue for an alternative concept of representation. In America, during the Critical Period, many states passed legislation that restricted the collection of debts, promoted paper money, and legitimized ex post facto laws.[24] Here again, Lockean liberalism made it possible to view the protection of private property as the principle component in the definition of "justice." It also supported the suspicion that it was the majority, the very "community" so praised by civic humanists, that represented the predominant threat to justice. Besides, once independence from Great Britain had been declared, it was difficult to defend any right as being a part of the "ancient rights of Englishmen," since the ties that buttressed that argument had been severed. If rights were to be defended, their source would have to be seen as resting in reason and not history.

In all of these instances, the components and perspective of Lockean liberalism are heard. They surfaced during the Constitutional Convention in the speech by Charles Pinckney contrasting the political agendas of the "landed" and the "manufacturing" interests (although Pinckney favored the former) and in the address by Luther Martin where he discusses human equality in a state of nature and directly cites Locke. Anti-federalists such as "John DeWitt" also appealed to Lockean liberalism in their use of rationally based natural law theory, a social contract philosophy of government, and a call for a Bill of Rights. And in the political thought of James Madison, one often sees the impact of Lockean liberalism in the fear of "oppressive majorities" that trample on "justice," which is invariably defined in terms of private property rights.[25] Historically, Lockean liberalism did not supplant civic humanism so much as it mingled with it. One finds it used at times when republican ideology does not adequately address the issues under consideration or does not fit the purposes of the particular Founder. The fact that civic humanism and Lockean liberalism are often *logically* contradictory does not prevent them from being *historically* blended. With the mixture of Lockean liberalism and the older civic idiom, the republican tradition became "an ideology in flux," but the situation was rendered even more complex with the addition of another, non-Lockean version of liberalism.[26]

HUMEAN LIBERALISM AND THE CLASH OF PRIVATE INTERESTS

With the background already provided, one way to picture Humean liberalism is to see it as an attack on both civic humanism and Lockean liberalism. David Hume provides the path to a third way of perceiving politics by undermining both Locke's social contract and republicanism's ancient constitution. One of the major theses of his influential History of England is that the "ancient constitution" could not be violated because, in a strict sense, it never existed. Absolute monarchy was as much in the British tradition as republicanism, and if the issue was "encroachment" on "ancient rights," then Parliament and "the people" were as guilty as any king of "trespass." In this manner, Hume reinterpreted English history and countered the radical Whig versions that were being put forth by republicans such as Paul de Rapin-Thoyras and Catharine Macaulay.[27]

In his essays, particularly in "Of The Original Contract," Hume shifts the focus of his attack away from the republican tradition and toward the social contract theory of John Locke. While admitting that the idea may be valid in a very symbolic and simplistic sense, Hume denies that a "social contract" ever existed as an empirical, historical fact. There was never any expressed consent asked for or given, and the idea of "tacit consent" is an absurdity since most citizens have no choice but to stay in the country of their birth and follow its laws. Habit, and not reason or consent, is the source of the obedience that is paid to government, according to Hume. Furthermore, it is argued that this is as it should be. Hume goes on to assert that teaching a theory based on consent, and leading to a concept of inalienable rights, is as dangerous as it is false. It promotes popular disturbances, undermines authority, and calls into question the stability of private property, which if traced back to its source, usually originates in "fraud and injustice."[28] With characteristic bluntness, Hume ridicules the position of social contract liberalism, "Were you to preach, in most parts of the world, that political connexions are founded altogether on voluntary consent or a mutual promise, the magistrate would soon imprison you, as seditious, for loosening the ties of obedience; if your friends did not before shut you up as delirious, for advancing such absurdities."[29] This essay, in which Hume specifically cites Locke as the target of his attack, points toward the distinction between Humean liberalism and both Lockean liberalism as well as civic humanism.

Like John Locke, David Hume was one of the early proponents of empiricist epistemology, but unlike Locke, Hume consistently applied his empiricism to all of his work. The Scot used his study of history to demonstrate that the republicans were wrong in their claim of an ancient constitution, and he used both history and logic to prove (at least to his satisfaction) that Locke was misguided in his political thought. In contemporary phraseology, the "good old days" never existed in experience or reason. The consequences that follow from this rejection are notable. If the past is no longer seen as a remote time of community, freedom, and equality, then moving away from it is no longer pictured as a slide into corruption. The nostalgia for a lost harmony is replaced by a wish for future progress. This faith in progress, expressed in Humean liberalism through essays such as "Of Commerce" and "Of Refinement in the Arts," demonstrates why Hume deserves to be called a "liberal." Gone is the fear of "luxury," a term that is now used to connote "the good life" and the spur to industrial production. Other points show the differences between Humean and Lockean liberalism. For example, with the rejection of the ancient constitution and the social contract, there is also gone the philosophical underpinnings of a theory of human rights.

Within Humean liberalism, one senses a discomfort with the entire idea of human "rights." There is a brief discussion of the topic in the essay, "Of the First Principles of Government." Here, Hume divides the concept into the "right of power," and the "right of property," but asserts that the right of power is simply an opinion formed over time that those who exercise power have a legitimate title to that authority. Regarding the right to property, Hume merely notes that some authors have made it the foundation of their system, but he offers no analysis of it himself. In this essay, "rights of the individual" are never mentioned.[30] Given the general thrust of Humean liberalism, this omission should not be surprising. After all, rights are not ostensible phenomena that one can point to nor are they physical experiences. Their existence is not subject to empirical proof. And since claims negating the existence of any particular right never entail a logical contradiction, the existence of rights is not subject to rational demonstration. Building a theory of politics on the supposition that civic or individual rights exist appears a weak foundation for a stable political order, and may explain why many of the American Founding Fathers apparently preferred to ignore any discussion of a bill of rights when framing the American Constitution. True, Hume and his close friend Adam Smith both believed that custom or habit could be used to support some *evolving* sense of "right" or "natural law" that could constrain lawmakers, and they believed in an innate conscience or "moral sentiment" that could act as a guide to human action and maintain some place for virtue in human conduct, but this represents a significantly altered concept of positive law, and the moral sentiments they discuss only function privately or with close associates. As societies grow more complex and individuals tend to form groups or "factions," moral sentiments and reason become less reliable guides. Human conduct becomes driven by passion and interest, Hume argues, and it is only by use of these that governments can control the governed.[31]

Humean liberalism may be conceptualized as the rejection of both civic virtue and a respect for individual rights as counselors to human action. As Hume succinctly put his dismissal of virtue, "Political writers have established it as a maxim, that, in contriving any system of government, and fixing the several checks and controuls of the constitution, every man ought to be supposed a *knave*, and to have no other end, in all his actions, than private interest." His repudiation of a rationally based theory of rights is stated with equal brevity, "Reason is, and ought only to be the slave of the passions, and can never pretend to any other office than to serve and obey them ... 'Tis not contrary to reason to prefer the destruction of the whole world to the scratching of my finger."[32] For Humean liberalism, the issue is neither the establishment of a virtuous community nor the protection of individual rights but the creation a stable political system in a social setting characterized by the narrow pursuit of individual self-interest where passionate individuals use the power of their private interest groups to oppress those who disagree with them or who block their path to further power. In this sense, Humean liberalism, as distinguished from classical and contractual philosophy, represents the clearest expression of an emerging system of modern political thought.

Hume believed our species is driven by our emotions, our passions, and our habitual (and almost incurable) tendency to competitively prefer our own interests to those of others. In this, there is almost no distinction between humans and other types of animals. The differences that do exist stem from the fact that human passions can be divided into what Hume called the "calm" and the "violent" passions. When these distinct forms of passions become attached to public figures, belief systems, or material objects, "interests" are created. Mirroring the calm and violent passions, human emotions lead to public and private interests. The problem for Hume, and for all of humanity, is the historically verifiable fact (at least verifiable to Hume's satisfaction), that our species invariably pursues violent passions and short-range private interests to the neglect of our calm passions and the long-range, public interests. Humans attach and group themselves around their shared, short-range, narrow concerns, and these "interest groups" or "factions" struggle against each other in an aggressive dual for the rewards of power, prestige, or plunder that was the prize of victory. Contrary to republicans, there was little communal cooperation in Hume's society since belligerent quarrelling was an ineradicable component of human behavior, and, contrary to the faith of the Lockean liberal Enlightenment, this enmity could not be reasoned or educated out of the human breast. For Humean liberals, what classical republicans called "corruption" was simply human nature, and what Lockean liberals called "natural rights" were merely private interests that individuals passionately wanted to pursue.[33]

Since neither the majority nor the minority could be relied upon to ameliorate this factional conflict, and since factional violence was the cause of nearly all civil unrest, it became necessary to find some instrument that would take political decision-making out of the hands of any and all social actors. It was here that David Hume found the potency and potential of economic, social, and political institutions as an alternative that would deprive humans of their ability to do mischief, calm or at least frustrate their violent passions and produce what no group actually worked for—a stable, enduring, rule-guided social structure. Social institutions—in particular economic markets, political constitutions, and judicial court systems—had to provide enough satisfaction to the self-interest of consumers (it was no longer accurate to call them "citizens") to establish some emotional attachment to the system, and then use this attachment as a goad or prod to steer humans in one direction or another. Humean liberalism can be seen as a precursor to the Utilitarian view that pain and pleasure were the prime motivators of action and an even earlier proponent of behavior modification as a technique of social control. Hume articulated the basic canon of interest group liberal thought. Humans were essentially self-seeking individuals whose only concept of rational behavior was the passionate pursuit of their own preferences. For Hume, "By this interest we must govern him, and, by means of it, make him, notwithstanding his insatiable avarice and ambition, cooperate to public good. Without this ... we shall in vain boast of the advantages of any constitution, and shall find, in the end, that we have no security for our liberties or possessions, except the good-will of our rulers; that is, we shall have no security at all."[34]

In addition to a more realistic, albeit a more negative view regarding "the natural depravity of mankind," Hume's empirical study of history pointed toward a paradox in human behavior.[35] He concluded that over time, human actions were consistent enough to permit regular, verifiable statements regarding human nature. There exists enough uniformity in our passions and interests, and enough means have been tried to achieve our ends, that it is possible to construct a science of politics. Like civic humanists, Hume believed that there was a public interest or a common good for all mankind, but unlike his predecessors, he did not believe that this common good was the result of a consensus, nor was it to be achieved by the active participation of public-spirited citizens. Herein lay the enigma. Although a public interest did, in theory, exist, actual individuals were so absorbed in the pursuit of their private self-interests that the achievement of the general welfare could never be the product of conscious action.

This leads to what is essentially a dual system of politics. On one level, interests and passions are subjective, idiosyncratic, and ethically equal. This is where we should be skeptical regarding what is best, where diversity and liberty should reign, and where we need checks, balances, and moderation. But at another level, interests are empirically testable through the study of history and capable of being hierarchically ranked. Hume never argued that diversity of interests or ethical relativism should be pushed to such an extreme that we consider the absurd possibility (for Hume) that anarchy is as good as stable and enduring government. This second level represents the calm pursuit of the public interest. This is the world in which the "experimental method" may be introduced into "moral subjects," in which "politics may be reduced to a science," and in which founding fathers may construct forms of government, forgetting their own private economic concerns and thinking exclusively of the common good for all humanity. The key to the success of the founding fathers is the construction of constitutions that contain what might be called the mechanism of unintended consequences.[36]

Humean liberalism places its faith in constitutions that are premised on the assumption that all individuals, including those who staff the government, follow their narrow self-interests. The object is not to make citizens virtuous or to instill in them a respect for the rights of others and a concern for the common good. Instead, the goal is to institutionalize the clash of private interests to make conflict economic and private rather than political and public. If the size of the nation is large enough, and if sufficient checks are built into the process, then no interest group will ever be able to command a majority and exercise their will, and the endless conflict will lead to the long-term stability of the system. Thus, although no one in the process consciously works for the common good, as defined by political stability and economic growth, this is the inevitable upshot. In retrospect, Humean liberalism can be seen as the beginning of the liberal faith in rule-driven bureaucracy as a substitute for an ethical citizenry.

It is difficult to underestimate the impact that Humean liberalism had on the American Founders. The linkages are both personal and philosophic. Benjamin Franklin was Hume's publishing agent in the colonies and had traveled to Scotland to meet him. Benjamin Rush had trained in Scotland and had gone there to recruit John Witherspoon as the new president and professor for Nassau Hall (Princeton College). Although he despised the atheistic portion of Hume's thought, Witherspoon heavily influenced his students with Hume and the other proponents of Scottish philosophy. Nine of Witherspoon's students, including James Madison, were in Philadelphia in the summer of 1787. There, they joined with Franklin, Hamilton, and the Scottish immigrant James Wilson to write a constitution. The imprint of Humean liberalism can be found throughout that document in the claims that the virtue of the citizens and the respect for rights in America are insufficient to assure political stability, in the assertion that only the conflict of private interests represents a realistic foundation for politics, and in the often stated premise that an extended republic will permit self-interested private persons to minimally participate in politics without causing disruption to the society. Gone is the civic humanist faith in active, virtuous, patriotic citizens combining in small, independent communities. "Republicanism" is redefined to mean merely representative government, and "citizenship" is reduced to the simple act of voting, where otherwise private persons go to the polls, and then go home.[37]

The Historical Tangle—Founders on Human Rights and Economic Theory

Although it is possible to isolate distinct, internally consistent visions of the political landscape that logically contradict competing perspectives, when the historical record is consulted, we find elements of these ideologies mixed in distinctive patterns. After all, neither Jefferson, Madison, Hamilton, nor any of the other Founders ever called themselves, or thought of themselves as, a "civic humanist," "Lockean liberal," or "Humean liberal." At least in the case of Jefferson, it is possible to find elements of all three traditions at work. Madison may have been a consistent liberal, but analysis can locate both Lockean and Humean strains. Hamilton, it appears, never philosophically wandered far from Hume. Republicanism, rights-based liberalism, and interest-group liberalism represent the core of American political thought, but elements in those perspectives can be separated out and relocated in alternative visions. This opens the possibility of greater flexibility in our cultural thinking, albeit purchased at the price of logical consistency in our national character.

One clear example of this is found in arguments surrounding differing conceptualizations of human rights. Rights are entitlement that are, or should be, legally and ethically binding on one group when asserted by another group. Once we understand how deeply embedded the concept of rights is in both the republican and Lockean traditions, and how uncomfortable individuals coming from the Humean tradition are with any attempt to use "rights" as part of a political agenda, we can appreciate that political debates in America are generously laced with often-conflicting notions of rights. Jefferson's position on rights is the most comprehensive and yet the most confusing. There is scarcely any definition or source of rights that Jefferson does not accept and advocate. If there are logical contradictions in these theories of rights, Jefferson seems able to transcend or ignore them. In his first major treatise on politics, the work that would bring him national recognition and lead to his assignment on the committee to write the Declaration of Independence, Thomas Jefferson explored the various conceptualizations of rights as they pertained to the relationship between King Georgia III and his American colonies. <u>A Summary View of the Rights of British America</u> is an example of political rhetoric built around the classical republican or civic humanist conceptualization of historical or traditional rights.[38] Jefferson's premise is that Britain and America are parts of the same historical and cultural community.

A Summary View contains many references to "Saxon ancestors," "the establishment ... of the British constitution at the glorious Revolution on its free and antient principles," and the "Feudalism" introduced into Britain (but not America) by "Norman lawyers." Jefferson asserts absolute knowledge of these entitlements because "History informs us" and "experience confirms" to all British subjects the existence of "the rights derived to them from their ancestors." This is the assertion of traditional rights, discovered by means of history, and legitimized through the joint membership in an ongoing community. It is Jefferson coming close to pure civic humanist argument, pleading for the Americans that "It is neither our wish nor our interest to separate from" Great Britain; urging the king "to establish fraternal love and harmony thro' the whole empire."[39]

The terminology and the thought patterns change, however, when Jefferson's Declaration of Independence is considered. Republican arguments built upon the assumption of shared membership in an historical community do not work very well when your intention is to "dissolve the political bands which have connected" one part of the empire with another. When the decision to pursue independence was made, republican logic had to give way to the historical realities and a shift to Lockean liberalism became necessary.[40] This shift was only partial and it was only temporary, but Jefferson's adoption of Lockean natural rights in the Declaration seems obvious.[41] Most recognizable are the well-known references to "the laws of nature and nature's God," the "inalienable rights," and the "self-evident" truth that all men are "endowed by their creator" with the ability to form governments based on "the consent of the governed" in order to "secure these rights." Even stronger links exist between Jefferson's use of expressions such as "governments long established should not be changed for light and transient causes," the apparent fact that "mankind are more disposed to suffer while evils are sufferable" and the final right to revolution triggered when "a long train of abuses" "evinces a design to reduce them under absolute despotism."[42] The most apparent disjunction between Jefferson's political theory and Lockean liberalism is the Virginian's refusal to include a right to property among the natural rights that all humans possess. At least as far back as the writings of St. Thomas Aquinas, property was considered an addition to the *jus naturae* and not part of it. Of course, what was added could also be subtracted, so that under this much older conceptualization, private property could be made subject to a larger, social perception of the common good.[43] Private property might still exist, but it was a convenience and not inalienable. It was a useful—some would contend an extremely useful—social institution but its possession was a civil right and not a natural right.[44] For Jefferson, there were three types or classifications of rights; traditional or historical rights that fit within his classical republican vision of politics, natural rights deduced from a Lockean liberal logic, and civil or legal rights which were socially constructed and recognized by the prevailing legislative and judicial institutions of a society. Strong communities protected historical rights, logical argument protected natural rights, and civil rights were defined and defended by a formal Bill of Rights enshrined in fundamental documents.

Thomas Jefferson's conviction that a Bill of Rights was an indispensable ingredient in the American Constitution leads to the consideration of the differences between his political theory and that of his close friend, James Madison. Few issues better illustrate the dichotomy between theoretical consistency and historical adaptation than James Madison's stance on human rights. Madison manages to combine a Lockean faith in individual rights with a Humean skepticism in rights as a solid foundation for a stable polity. Madison incorporates both elements into one consistent, liberal, and archetypically American view of the world.[45] The strongly Lockean portion of Madison's thought is found in his fervent defense of property rights. The clearest expression of this is in his 1792 *National Gazette* essay, titled "Property."[46] Madison opens his argument by giving two possible definitions of the term "property." Under his first conceptualization, Madison defines property as "that dominion which one man claims and exercises over the external things of the world, in exclusion of every other individual." This is the Lockean liberal notion that property can consist of any external object with which a person has mixed his or her labor. But Madison takes the concept of individual private property even beyond this Lockean view of land, capital, or other forms of material possessions.

After his initial definition, Madison goes on to expound what he calls the "larger and juster meaning" of the word. Under this more expansive definition, "property" can be taken to mean "everything to which a man may attach a value and have a right, and *which leaves to every one else the like advantage.*"[47] Now, Madison views property as including many "things" beyond material objects. Ideas can be property under this extended definition. Madison is saying that value is sufficient to accord the status of "property" to these ideas, beliefs, or opinions. In asserting that individuals can have property in, and hence rights to, entities that are essentially interests, Madison is merging Hume and Locke in ways that the simple explication of their logic would seem to preclude. If anything, Madison out-Lockes Locke by taking the concept beyond material goods. In a wonderfully simple summation of his philosophy, Madison declares in the "Property" essay, "In a word, as a man is said to have a right to his property, he may be equally said to have a property in his rights."[48] For Madison, almost anything can become a right because virtually anything can become property. Here is the quandary in which Madison finds himself. Under rights-based liberalism, property and rights are mutually reinforcing ideas. Since protecting individual rights is the very purpose for which government exists, it follows that only those governments who strictly guard private property rights can be consider "just." In a perfectly just world—a nation governed by Madison's "empire of reason"—rational humans would recognize and respect the importance of rights and property. Unfortunately, for this Framer and fourth President, this world of angels is not the one we inhabit. For Madison, Locke's natural rights will almost always fall prey to Hume's depravity of human passions. Although natural rights unquestionably exist for Madison, he has almost no faith that a bill of rights will be of much use in protecting those rights.

In a famous exchange of letters between Madison and Jefferson in October 1788 and March of 1789, their differing attitudes on a bill of rights are revealed. Madison opens the correspondence by bringing Jefferson (in Paris as the American ambassador) up to date on the proceedings of the Federal Constitutional Convention, the ratification process then underway, and the fact that many supporters and detractors are pushing for the addition of a bill of rights.[49] Although he avers that his "own opinion has always been in favor of a bill of rights," Madison concedes, "I have never thought the omission a material defect ...." The reason Madison is lukewarm over the suggestion for a bill of rights is his belief that such declarations are mere "paper barriers against the power of the community" and almost "too weak to be worthy of attention."[50] Madison felt strong communities were more likely to oppress, than defend, individual rights and liberty. He tells Jefferson that a bill of rights is a mere "parchment barrier;" simple "declarations on paper;" and that these "restrictions, however strongly marked on paper, will never be regarded when opposed to the decided sense of the public, and after repeated violations, in extraordinary cases will lose even their ordinary efficacy."[51] Tyrants, mobs, and, in particular, interested majorities will ignore and abridge any set of rights whose enforcement depends only on civic virtue, human reason, or pronouncements made on official documents.

The protection of Lockean natural rights must be secured, therefore, not by the use of republican virtue or Lockean reason, but paradoxically, by Humean institutional arrangements that make it politically unlikely any concerted majority will form or that give intensely motivated minorities the bureaucratic means to block policies aimed against their salient interests.[52] Thomas Jefferson, it should be noted, was only mildly impressed with his friend's logic regarding a bill of rights. In his response, Jefferson gives a point-by-point rebuttal to Madison's misgivings, affirms his belief that "the good in this instance vastly overweighs the evil," and concludes, "I am much pleased with the prospect that a declaration of rights will be added ...."[53]

No such ambiguity presents itself in Alexander Hamilton's position on human rights and a bill of rights. Hamilton's argument in Federalist 84 can be seen as an example of his rejection of both the classical republican notion of historical rights and the Lockean liberal belief in natural rights. Hamilton's attitude is suggested when he says with an insinuation of sarcasm that "popular rights" belong in that class of "aphorisms" "which would sound much better in a treatise of ethics than in a constitution of government." The historical relevance of the next-to-last Federalist paper is that it represents Hamilton's counterargument to the Anti-federalist position that without a Bill of Rights, the proposed Constitution was fatally flawed and should be rejected. There is something disingenuous about Hamilton's assertion that a bill of rights is both unnecessary and unwise. He claims that since the powers of the proposed government will be strictly limited to those enumerated in the Constitution, there is no need for further limitations. "[A] minute detail of particular rights is certainly far less applicable to a Constitution like that under consideration, which is merely intended to regulate the general political interests of the nation, than to a constitution which has the regulation of every species of personal and private concerns." He then states, "I go further and affirm that bills of rights ... are not only unnecessary in the proposed Constitution but would even be dangerous. They would contain various exceptions to powers which are not granted; and, on this very account, would afford a colorable pretext to claim more than were granted. For why declare that things shall not be done which there is no power to do?" Yet given the Federalists' penchant for a broad construction of Constitutional powers, their willingness to stretch the "necessary and proper" clause to its outermost limits, and Hamilton's prior statements in Federalist 23 that "government ought to be clothed with all the powers requisite to complete execution of its trust ..." and "there can be no limitation of that authority ...," the fear being expressed in Number 84 appears to stem from a recognition that a bill of rights *would* limit government power in ways that Hamilton found unacceptable. This was, after all, the same man who wrote in Federalist 36, "I acknowledge my aversion to every project that is calculated to disarm the government of a single weapon, which in any possible contingency might be usefully employed for the general defense and security."[54] Within Hamilton's political philosophy, a bill of rights was not so much unnecessary or unwise as it was unwanted. To the extent that rights derive from reason and interests from emotions, Hamilton clearly sided with a politics of interests over a politics of rights.[55] Shifting the focus from rights to economic theory further illustrates the philosophical tensions during the American Founding.

Historian Forrest McDonald draws our attention to the fact that, "[T]he making of the Constitution was the expression of the prevailing popular ideology of the socially desirable or normal relationship between government and economy." The writing and ratifying of the Constitution took place inside a larger cultural context within which economics is a part—a very important part—of a contest driven by competing visions of human nature, politics, the major threats confronting the society, and the image of "the good life" or goal culture toward which America should move. The belief that economic activity should be based on inalienable private property rights and the pursuit of individual self-interest was only in its nascent stages. The idea that wealth should be distributed according to trading in private markets where everyone sought to maximize their individual gains was a controversial topic. Although these components of "commercial society" (the term "capitalism" would only be adopted later) fit nicely with Lockean notions of property rights and Humean concepts of interest-group politics, republican theory staunchly held such motivations represented the essence of societal corruption.[56]

The emerging capitalist ideology contrasted sharply with the economic perspective that we now call "mercantilism." Adherents of mercantilist economics confronted supporters of the emerging "commercial society" on a number of issues. Although both groups espoused the private, individual ownership of property, mercantilists believed that property and wealth should be regulated and managed for the public good or the common welfare of the community. Only capitalists believed in the individual pursuit of private gain through unfettered markets. "Mercantilism was geared to the economic needs of the community," observes McDonald.

The fundamental premise underlying this economic system was "the concept of governmental regulation of the economy with a view to the welfare of the sovereign community as a whole rather than that of its individual members."[57] Private ownership of property and public regulation of the economy are two characteristics of mercantilism. Since mercantilists focused on the interests of the community and not on individual interests, and since they advocated public, or political, management of the economy, mercantilists had to address the question of how government management would take place and the proper level of government best suited to exercise this control. This produced further ideological division. It was possible, for example, to be in favor of governmental regulation but also believe that regulation was best implemented at the local level (Thomas Jefferson supported this position). Conversely, there were spokesmen (Alexander Hamilton was one of these) who supported public management of the economy but wanted that power exercised by a national government with broad powers to supervise interstate commerce. Many eighteenth-century Americans found it easy "to contemplate a governmental system with broad powers to regulate economic life," but although "Americans were accustomed to fairly extensive governmental interference in their economic lives," they felt that power was properly exercised only at the local level. "It was not governmental power as such that Americans feared," McDonald observes, "but the centralization of power, especially in a government far removed from local supervision and control."[58]

Today, the broader, more generic term for this national mercantilism is "corporatism." Corporatists believe in private property, but they do not believe private property is an inalienable natural right which owners can use any way they see fit. An appreciation of corporatism also sheds light on contemporary debates since many of these disputes revolve around the limits of private property rights and the possible extension of either national or local governmental authority in the name of posterity and the common good. In spite of the fact that corporatism was deeply embedded in the philosophical debates of the eighteenth century and is tremendously important to an understanding of modern-day politics, the term "corporatism" is not normally found in either historical or contemporary records. Today, a host of synonyms including "neo-mercantilism," "managed capitalism," "managed competition," "concertation" (in Japan), "codetermination" (in Germany), "national economic policy," "the reindustrialization of America," "state capitalism," "the corporate state," "national industrial policy," "corporate liberalism," and even "crony capitalism" are used to indicate this particular mix of individual property and public control.[59]

Another distinction in economic perspectives needs to be pointed out. In opposition to both mercantilists and capitalists, supporters of French physiocracy as enunciated by economic theorists such as Francois Quesnay, the Marquis de Mirabeau, Pierre Samuel du Pont de Nemours, and notably, Thomas Jefferson, contended that it was land and agriculture, rather than manufacturing and money that represented the true source of wealth. For physiocrats, finance and trading were "sterile" occupations that merely moved resources around the economy, generating no real increase in wealth. It was only agriculture that produced something that was not there before, and public policy, therefore, needed to recognize and support this superior economic (and moral) position of the landed proprietors against the intrusions of the base, and corrupt, urban speculators.[60] It was not the difference between economic *interests* so much as the distinction in economic *theory* that separated men such as John Taylor of Caroline and Robert Morris. The issue was not so much who would gain and who would lose, but rather the much more fundamental question of what "winning" and "losing" actually meant. Were bankers and stockbrokers agents of economic development and wealth creation or were they little more than gamblers and profiteers, swindling the hard-earned incomes of others through complicated, bizarre, and corrupt financial transactions? Jefferson's acute hatred for men such as Robert Walpole in Britain and his association of Hamilton's financial scheme with the detested 1712 "bubble" during Walpole's ministry meant much more than merely having different economic interests. Jefferson (and from a diametrically different view, Hamilton) saw this as a conflict over the heart, soul, and future of America. In a very real, historical sense, both men were correct. A comprehensive interpretation of the Constitution would have to include an analysis of the competing perspectives of mercantilism, capitalism, and physiocracy, and although Alexander Hamilton lost his fight to have a Bill of Rights excluded from the Constitution, he was a winner in the battle over economic theory. Even there, however, his victory was not complete. Disagreement over the extent of Federal regulation of the economy would be never ending, and the physiocratic mistrust of financial markets would remain part of America's economic culture.

Where does this leave us? What can contemporary citizens concerned with challenges facing this nation take away from an understanding of the conflicting narratives during our Founding? When political, economic, or ecological crises arise, people look for explanations why these troublesome events are occurring, and they seek solutions that seem to promise resolution of the situation. It is here where established patterns of perception manifest themselves and where an understanding of the ideological diversity of a nation helps in appreciating the available policy options. If America is a country of republicans, Lockeans, and Humeans who blend their thought patterns in different and inconsistent ways, then these traditions will provide both the analysis and the agendas for conflict resolution. When the economic component of social challenges are under consideration, the powerful influence of private property rights, individual interests, and the faith in markets needs to be appreciated. We also need to recognize, however, that embedded in our culture are alternative economic perspectives and values. During the Founding period, mercantilism and physiocracy were two of those alternatives.

It is in this larger context of political and economic perspectives that our present ecological challenges can be understood more deeply. For example, if our surface and ground water is determined to be polluted, does this represent a prima facie rationale for governmental regulation in pursuit of the public interest or are such laws intrusive restrictions on individual rights and private property? Does the influx of large numbers of illegal and unassimilated aliens represent a challenge to the cultural unity and ecological carrying capacity of America, or are these newcomers to be welcomed as contributors to the labor force and the diverse social richness of this land? Republicans and liberals will view these issues differently. Lockeans and Humeans will not adopt the same language or the same logic in making their case. Furthermore, the contradictory amalgam of thought in America clarifies how some people can take a republican position on pollution and the public interest while adopting a liberal stance on issues of population and immigration. If present-day citizens came to appreciate the logical conflict and the theoretical joining that comprise their cultural history, some of the acerbic, incommensurable irrationality that often accompanies these clashes might be appreciated, if not completely ameliorated. American politics has always been fashioned around the possibilities and limitations imposed by the vast, beautiful, and bountiful natural environment of this continent. American political thought has always had at least an implicit recognition of Nature and the natural setting within its more expressly humanistic, economic, and political viewpoints. Given the seriousness of our current ecological problems, the time has come to spell out those historical, philosophical, and cultural legacies and to make their logic and consequences open and explicit.

Notes

1. Johnson in <u>The Records of the Federal Convention of 1787</u>, ed., Max Farrand (New Haven and London: Yale University Press, 1966), Volume I, 461; hereafter cited as Farrand. For Dickinson's comment, see Benjamin Rush to Richard Price in Farrand, III, 33; see also Edmund Randolph to Beverly Randolph in Farrand, III, 36.
2. Scholars trained in the academic discipline of history, such as Daniel Rogers, Gordon Wood, and Lance Banning, criticize those who "repackage" intellectual history "along those linear lines of influence that [have] long given political theory a bad name...," Daniel T. Rogers, "Republicanism: the Career of a Concept," <u>The Journal of American History</u>, June, 1992, 17. Rogers never explains among <u>whom</u> political theory has "a bad name." Presumably, it is historians. Lance Banning distinguishes "analytical" and "developmental" approaches (while making clear his preference for the latter strategy) that parallel my "logical" and "historical" categories. For these comments and the second portion of the quotation, see Lance Banning, <u>The Jeffersonian Persuasion: The Evolution of a Party Ideology</u> (Ithaca: Cornell University Press, 1978), 50n, 63, and 63n. Gordon Wood claims that "leading scholars have anachronistically invented too many 'paradigms' and forced too much material into them...," in "The Virtues and the Interests," <u>The New Republic</u>, February 1991, 34.
3. J.G.A. Pocock, <u>The Machiavellian Moment</u>; Bernard Bailyn, <u>The Ideological Origins of the American Revolution</u>; and Michael Lienesch, <u>The New Order of the Ages: Time, the Constitution, and the Making of Modern American Political Thought</u> (Princeton: Princeton University Press, 1988).
4. John Trenchard and Thomas Gordon, <u>Cato's Letters; or, Essays on Liberty, Civil and Religious, and Other Important Subjects</u> (4 vols. in 2; 3rd edition carefully corrected; London 1733; facsimile edition New York: Russell and Russell, 1969). Analysis which places this work in the Lockean tradition is Ronald Hamowy, "<u>Cato's Letters</u>, John Locke, and the Republican Paradigm," <u>History of Political Thought</u>, Vol. XI, No. 2, (Summer 1990). The best account of Bolingbroke's thought is Isaac Kramnick, <u>Bolingbroke and His Circle: The Politics of Nostalgia in the Age of Walpole</u> (Ithaca: Cornell University Press, 1968). See also Caroline Robbins, <u>The Eighteenth Century Commonwealth man</u> (Cambridge, Mass: Harvard University Press, 1961).
5. John M. Murrin, "The Great Inversion, Or Court Versus Country: A Comparison of the Revolution Settlements in England (1688–1721) and America (1776–1816)" in <u>Three British Revolutions: 1641, 1688, 1776</u>, ed. J.G.A. Pocock (Princeton: Princeton University Press, 1980). See also Bernard Bailyn, <u>The Ideological Origins of the American Revolution</u> (Cambridge, Mass.: Harvard University Press, 1967), especially Chapter 2; and Lance Banning, <u>The Jeffersonian Persuasion</u>, especially Chapter 2.

6. The critical source here is Pocock, The Machiavellian Moment, especially Part Two. Of course, the intellectual line of development was neither straight nor completely clear, and attempts to picture it as such have led Daniel Rogers to parody the entire intellectual process: "behind the [American] revolutionary generation, the English country writers; behind the country party, James Harrington; behind Harrington, Niccolo Machiavelli and the discourse of civic humanism—all way stations on an intellectual route from the Renaissance to the Revolution that bypassed Locke altogether." Daniel T. Rogers, "Republicanism: the Career of a Concept," The Journal of American History, June, 1992, 17.
7. J.G.A. Pocock, The Machiavellian Moment, p. 335.
8. Louis Hartz, The Liberal Tradition in America (New York: Harcourt Brace Javanovich, 1955), p. 55.
9. .For further development of the concept of "the public interest," see Gordon S. Wood, The Creation of the American Republic, 1776-1787 (New York: W.W. Norton and Company, 1969), especially Chapter 2, section 3. See also Franklin A. Kalinowski, "David Hume and James Madison on Defining 'The Public Interest'" in Virtue, Corruption, and Self-Interest, ed. Richard K. Matthews (Allentown: Lehigh University Press, 1994).
10. For a discussion of the conspiratorial mind set which pervaded the thinking of the revolutionary era, see Bernard Bailyn, The Ideological Origins of the American Revolution, p. 144 ff. See also J.G.A. Pocock, "Virtue and Commerce in the Eighteenth Century," Journal of Interdisciplinary History, no. 3 (Summer, 1972).
11. J.G.A. Pocock, The Machiavellian Moment, p. 509; and Isaac Kramnick Bolingbroke and His Circle. For a summary, see Richard K. Matthews, "Liberalism, Civic Humanism, and the American Political Tradition: Understanding Genesis," especially pp. 1130-1133.
12. These themes are most fully developed in Pocock, The Machiavellian Moment.
13. The best original source for this discussion is Trenchard and Gordon's Cato's Letters. The argument and its impact on America are analyzed by Bernard Bailyn in The Ideological Origins of the American Revolution, Chapter III; and Gordon Wood, The Creation of the American Republic, Chapter 1.
14. Richard J. Hofstadter, The Paranoid Style in American Politics, and Other Essays (New York: Alfred A. Knopf, Inc., 1965).
15. The Norman conquest is typically seen as the event that upset the balance of the ancient constitution. See J.G.A. Pocock, The Ancient Constitution and the Feudal Law: English Historical Thought in the Seventeenth Century (New York: W.W. Norton and Company, 1967); and J.G.A. Pocock, The Machiavellian Moment, Part Three.
16. J.G.A. Pocock, The Machiavellian Moment, pp. 426-432.

17. This quotation and a good discussion of the cyclical vision of history are found in Drew R. McCoy, The Elusive Republic: Political Economy in Jeffersonian America (New York: W.W. Norton and Company, 1980), p. 170. For further reference to Samuel West, see Gordon Wood, The Creation of the American Republic, p. 415.
18. In addition to the sources previously cited, see Leo Marx, The Machine in the Garden: Technology and the Pastoral Ideal in America (London: Oxford University Press, 1964). On puritanical and agrarian republicanism, see Forrest McDonald, Novus Ordo Seclorum (Lawrence: University Press of Kansas, 1985), pp. 70–77.
19. Here, I follow the argument of C.B. Macpherson, The Political Theory of Possessive Individualism: Hobbes to Locke (London and Oxford: Oxford University Press, 1962), pp. 19–29.
20. This point is made by Isaac Kramnick in "Republican Revisionism Revisited," in Republicanism and Bourgeois Radicalism, p. 172.
21. Consider the normatively negative connotation that is implied in the definition of "power" as "the fear of man" put forth by Trenchard and Gordon in Cato's Letters, Vol. II, Letter 60, p. 227, and compare it with Alexander Hamilton's more benign definition of "power" as "the ability or faculty of doing a thing" in Federalist 33, p. 223.
22. John Locke, "An Essay Concerning the True Original, Extent, and End of Civil Government," in Two Treatises of Government, edited with and an introduction by Peter Laslett (New York: New American Library, 1960), para. 42, p. 339; and para. 46, p. 342. See also C.B. Macpherson, Possessive Individualism, Chapter 5. Among the many critiques of Macpherson is Alan Ryan, "Locke and the Dictatorship of the Bourgeoisie," Political Studies 13 (1965), pp. 219–230. For further reflections on "bourgeois radicalism," see Isaac Kramnick, "Liberalism, the Middle Class, and Republican Revisionism," in Republicanism and Bourgeois Radicalism.
23. This redefinition of "virtue" is discussed by many scholars. See Drew McCoy, The Elusive Republic, pp. 69–75; John Dunn, The Political Thought of John Locke (Cambridge: Cambridge University Press, 1969), especially Chapter 18; Joyce Appleby, Capitalism and a New Social Order (New York: New York University Press, 1984), pp. 14–16; Isaac Kramnick, "The Great National Discussion," in Republicanism and Bourgeois Radicalism, p. 278; Thomas Pangle, The Spirit of Modern Republicanism: The Moral Vision of the American Founders and the Philosophy of Locke (Chicago: University of Chicago Press, 1988), especially Part Two; and John P. Diggins, The Lost Soul of American Politics (Chicago: University of Chicago Press, 1984), especially Chapters 1 and 3.
24. On the 1760s tax crisis, see Isaac Kramnick, "Republican Revisionism Revisited," in Republicanism and Bourgeois Radicalism, p. 171. For a

discussion of America during the Critical Period, see Gordon Wood, The Creation of the American Republic, Chapter X.
25. Pinckney's speech is in Farrand, Vol I, pp. 397–404. Martin uses and cites Locke in Farrand, Vol. I, pp. 437–438. See also Massachusetts anti-federalist "John DeWitt" Essay II (October 27, 1787) in The Anti-Federalist Papers and the Constitutional Convention Debates, ed. Ralph Ketcham (New York: New American Library, 1986), p. 195. For Madison's position, see "Vices of the Political System of the U. States," in The Writings of James Madison, ed., Gaillard Hunt (New York: G.P. Putnam's Sons, 1901), Vol. II, p. 361–369; his speech to the Constitutional Convention on June 6, 1787, in Farrand, Vol. I, pp. 134–136; and, of course, Federalist 10. An excellent analysis of Madison's liberalism is Richard K. Matthews, "If Men Were Angels:" James Madison and the Empire of Reason (Lawrence: University Press of Kansas, 1995).
26. Drew R. McCoy, The Elusive Republic, 48.
27. Bernard Bailyn, The Ideological Origins of the American Revolution, 41–42. For a discussion of the role of monarchy in these debates, see Gordon S. Wood, The Radicalism of the American Revolution (New York: Alfred A. Knopf, 1992).
28. David Hume, "Of the Original Contract," in Essays Moral, Political, and Literary, ed., Eugene F. Miller (Indianapolis: Liberty Classics, 1985), 465–487. Hereafter cited as Hume's Essays.
29. David Hume, "Of the Original Contract," in Hume's Essays, 470.
30. David Hume, "Of the First Principles of Government," in Hume's Essays, 32. In his edition of these essays, Eugene Miller states that the "noted author" Hume refers to in the brief discussion regarding the "right to property" is "probably" James Harrington, ibid., 32n; but in a different edition containing this essay, editor Charles W. Hendel says that the reference is to Locke's Second Treatise. See David Hume's Political Essays, ed. Charles W. Hendel (Indianapolis: The Bobbs-Merril Company, Inc., 1953), 25n. Given the fact that Hume argues for labor instead of land as the source or value, I believe that the reference should be to Locke. For Hume on labor, see David Hume, "An Enquiry Concerning the Principles of Morals," in Hume's Moral and Political Philosophy, ed. Henry D. Aiken, (New York: Hafner Press, 1948), 281.
31. The literature on these topics is extensive. For Hume's theory of an evolving natural jurisprudence, see Duncan Forbes, Hume's Philosophical Politics (Cambridge: Cambridge University Press, 1975); Knud Haakonsen, The Science of the Legislature (Cambridge: Cambridge University Press, 1981); and J.G.A. Pocock, "Cambridge paradigms and Scotch philosophers: a study of the relations between the civic humanist and the civil jurisprudential interpretation of eighteenth-century social thought," in Wealth and Virtue: The Shaping of Political Economy in the

Scottish Enlightenment, ed. Istvan Hont and Michael Ignatieff (Cambridge: Cambridge University Press, 1983).
32. David Hume, "Of the Independency of Parliament," in Hume's Essays, 42, and David Hume, A Treatise of Human Nature, ed. L.A. Selby-Bigge (Oxford: The Clarendon Press, 1888), 415 and 416.
33. For an extended version of this argument, see Franklin A. Kalinowski, "David Hume on the Philosophic Underpinnings of Interest Group Politics," Polity, Vol. XXV, No. 3 (Spring 1993).
34. David Hume, "Of the Independence of Parliament," in Hume's Essays, 42.
35. David Hume, "That Politics May Be Reduced to a Science," in Hume's Essays, p. 24.
36. See Franklin A. Kalinowski, "David Hume on the Philosophic Underpinnings of Interest Group Politics," 374.
37. There is a substantial literature on the connection between Humean liberalism and the American founding. The essay that began it all was Douglass Adair, "'That Politics May Be Reduced To A Science': David Hume, James Madison, and the Tenth Federalist," in Fame and the Founding Fathers, ed. Caroline Robbins (New York: W.W. Norton and Company, 1974). The connection was popularized by Garry Wills, Inventing America: Jefferson's Declaration of Independence (New York: Random House Vintage Books, 1978), and Garry Wills, Explaining America: The Federalist (Garden City, New York: Doubleday and Company, 1981). See also David Epstein, The Political Theory of the Federalist (Chicago: University of Chicago Press, 1984); Morton White, Philosophy, The Federalist, and the Constitution (New York and Oxford: Oxford University Press, 1987); Gordon S. Wood, The Radicalism of the American Revolution; Edward J. Erler, "The Problem of the Public Good in The Federalist," Polity 13 (1981); Edmund S. Morgan, "Safety in Numbers: Madison, Hume, and the Tenth Federalist," Huntington Library Quarterly 1986; and Theodore Draper, "Hume and Madison: The Secrets of Federalist Paper No. 10," Encounter 58 (February 1982).
38. Thomas Jefferson, A Summary View of the Rights of British America. Page references are to Thomas Jefferson, Jefferson Writings, selected with notes by Merrill D. Peterson, (New York: The Library of America, 1984).
39. Thomas Jefferson, A Summary View of the Rights of British America, quotations are at 4, 7, 16, 18, and 21.
40. Ronald L. Hatzenbuehler, "I Tremble for My Country:" Thomas Jefferson and the Virginia Gentry (Gainsville, FL: University Press of Florida, 2006) sees more consistency than divergence between the Summary View and the Declaration of Independence.
41. Garry Wills, in his popular and influential Inventing America: Jefferson's Declaration of Independence, (New York: Vintage Books, 1979), downplays this connection, going so far as to suggest, "There is no indication Jefferson read the *Second Treatise* carefully or with profit. Indeed, there is

no conclusive proof he read it at all...." (174). Even if we acknowledge that Carl Becker may have overstretched the case for Jefferson as a Lockean liberal (in Becker's 1922 The Declaration of Independence), Wills seems to go too far in the other direction.
42. Thomas Jefferson, A Declaration by the Representatives of the United States of America in General Congress Assembled, in Thomas Jefferson, Jefferson Writings, selected with notes by Merrill D. Peterson, 19. The parallels with Locke's choice of words in the Second Treatise are too close to ignore. In the chapters titled "Of Tyranny" and "Of the Dissolution of Government," Locke says "People are not so easily got out of their old forms, as some are apt to suggest," and notes the people "are more apt to suffer than right themselves by Resistance." Locke goes on with an analogy that Jefferson directly borrows, "But if a long train of Abuses, Prevarications, and Artifices, all tending the same way, make the design visible to the People, and they cannot but feel, what they lie under, and see, whither they are going; tis not to be wonder'd that they should then rouze themselves, and endeavor to put the rule into such hands, which may secure to them the ends for which Government was first erected...." John Locke, The Second Treatise of Government, sections 223, 230, and 225. The edition used is John Locke, Two Treatises of Government, edited with introduction by Peter Laslett (New York: New American Library, 1960), 462, 466, and 463. Jefferson acknowledged in a letter to James Madison that "Richard H. Lee charged it as copied from Locke's treatise on Government," but Jefferson defended himself by saying, "I know only that I turned to neither book nor pamphlet while writing it." To Lee himself, Jefferson wrote regarding the Declaration, "Neither aiming at originality of principles or sentiment, nor yet copied from any particular and previous writing, it was intended to be an expression of the American mind...." Comments by Pickering and letter to Madison cited in Fawn Brodie, Thomas Jefferson: An Intimate History (New York: W.W. Norton and Company, 1974), 121. Jefferson to Henry Lee, May 8, 1825. Cited in Adrienne Koch, The Philosophy of Thomas Jefferson (Gloucester, MA: Peter Smith, 1957), 140. While acknowledging the Lockean elements in Jefferson's thought, Hatzenbuehler makes an interesting case for the additional influence of Jefferson's Virginian colleague, George Mason. See Ronald L. Hatzenbuehler, "I Tremble for My Country:" Thomas Jefferson and the Virginia Gentry, especially Chapters 2 and 3.
43. For an excellent discussion of these issues, with particular reference to their application to environmental concerns, see the "Beyond Property" section in John Rodman, "The Liberation of Nature?," Inquiry 20, 83–145.
44. Renown Jefferson biographer, Dumas Malone asked rhetorically in 1948, "Was there any significance to his omission of the word 'property.' Which

had been used by John Locke?..." and quickly concludes, "It is exceedingly doubtful that his contemporaries thought there was." The Cold War setting in which Malone wrote is immediately apparent as he hastens to assure readers that Jefferson "did not anticipate communism." Dumas Malone, Jefferson and His Time, Volume One, Jefferson the Virginian (Boston: Little, Brown, and Company, 1948), 227.
45. Richard K. Matthews, If Men Were Angels: James Madison and the Heartless Empire of Reason. See also Richard K. Matthews, "James Madison's Political Theory: Hostage to Democratic Fortune," The Review of Politics, Vol. 67, No. 1, Winter 2005, 49–67.
46. James Madison, The Papers of James Madison, ed. William T. Hutchinson et al., 17 volumes (Chicago and Charlottesville: University of Chicago Press and University of Virginia Press, 1962–), 14:266. Also found in James Madison, The Mind of the Founder: Sources of the Political Thought of James Madison, ed. with introduction and commentary by Marvin Meyers (Hanover, NH: University Press of New England, 1981), 186–188.
47. James Madison, "Property" in Meyers, Mind of the Founder, 186, emphasis in original.
48. James Madison, "Property" in Meyers, Mind of the Founder, 186.
49. James Madison to Thomas Jefferson, October 17, 1788, in Meyers, Mind of the Founder, 156.
50. See Ibid. 156 and James Madison "Speech in the House of Representatives," June 8, 1789," in Meyers, Mind of the Founder, 169.
51. Madison to Jefferson, October 17, 1788, in Meyers, Mind of the Founder, 157 and 159.
52. James Madison "Speech in the House of Representatives," June 8, 1789 in Meyers, Mind of the Founder, 170. David Epstein demonstrates the combination of Lockean rights and Humean interests in Madison's first and most essential essay, Federalist Number 10.
53. In a challenge to Madison's fear of popular majorities, Jefferson intimates that the Constitution may not afford all the protection Madison hopes. "The tyranny of the legislatures is the most formidable dread at present, and will be for long years. That [tyranny] of the executive will come in its time but it will be at a remote period." Thomas Jefferson to James Madison, March 15, 1789, in Thomas Jefferson, Jefferson Writings, selected with notes by Merrill D. Peterson, 943 and 945.
54. Hamilton, Federalist 84, 476; Federalist 28, 183; and Federalist 36, 240–241.
55. Forrest McDonald observes, "The 'age of reason,' in Great Britain, did not deify reason, nor did it even postulate that man is a rational creature. Quite the contrary: Hume and many of his contemporaries maintained that all human action springs from desire for pleasure and aversion to pain...."

Forrest McDonald, Alexander Hamilton: A Biography, (New York: W.W. Norton and Company, 1982), 36.
56. The very best analysis of these evolving economic theories is Karl Polanyi, The Great Transformation: The Political and Economic Origins of Our Time (Boston: Beacon Press, 1944).
57. Ibid., 409–410. The definitive study of mercantilism remains Eli F. Heckscher, Mercantilism, 2 volumes, (London: George Allen and Unwin Ltd., New York: The MacMillan Company, 1955).
58. The quotations are from McDonald, We The People, 410–411. See also Forrest McDonald, E Pluribus Unum: The Formation of the American Republic, 1776–1790 (Indianapolis, IN: Liberty Fund, 1979), and Forrest McDonald, Novus Ordo Seclorum: The Intellectual Origins of the Constitution (Lawrence, KS: University Press of Kansas, 1985).
59. The literature on corporatism is significant. See Martin Stanlind, What is Political Economy? (New Haven: Yale University Press, 1985), especially Chapter 4, "The Primacy of Politics;" Alan Cawson, Corporatism and Political Theory (Oxford: Basil Blackwell, 1986); and Phillippe C. Schmitter, "Still A Century of Corporatism?" Review of Politics, January 1974, Vol. 36, No. 1, 85–131. Alan Greenspan, Chairman of the Federal Reserve Board, used the expression "crony capitalism" to describe the Japanese economic system during testimony before the Senate Banking Committee, February 28, 1998. "Corporate liberalism" is from James Weinstein, The Corporate Ideal in the Liberal State: 1900–1918 (Boston: Beacon Press, 1968). Herman Daly and John Cobb discuss the German system of codetermination in For the Common Good, 300–301. Although Eldon Eisenach disparages the use of the term because of "the absence of a state with enough authority and coherence to pull it off," no other term seems adequate to confer meaning to this form of political economy, and many environmentalists argue for the creation of just this form of state. See Eisenach, The Lost Promise of Progressivism (Lawrence, KS: University Press of Kansas, 1994), 246. No doubt it was Benito Mussolini's portrayal of Fascist Italy as a "corporate state" that contributed to the unpopularity of the term and the reluctance of present-day politicians to use that label to describe their economic theory.
60. For an introduction to the economic theory of physiocracy, see Ronald L. Meek, The Economics of Physiocracy: Essays and Translations (Fairfield, NJ: Augustus M. Kelley, 1993); James J. McClain, The Economic Writings of DuPont de Nemours (Newark, DE: University of Delaware Press, 1977); and Dumas Malone, editor, Correspondence between Thomas Jefferson and Pierre Samuel DuPont de Nemours (Boston and New York: Houghton Mifflin Company, 1930).

CHAPTER 6

The Environmental Legacy of Thomas Jefferson: Cultivating the Rooted Citizen

Thomas Jefferson saw the world differently than many of his fellow Founders. They knew that and he knew that. In spite of occasional pronouncements on his part emphasizing the similarities in their ideological positions, such as in his First Inaugural Address when he declared, "We are all Republicans. We are all Federalists," Jefferson's vision of where America should head in the future differs dramatically from the perspective of Benjamin Franklin, John Adams, Alexander Hamilton, or even that of his close friend, James Madison.[1] It is important to recognize and to appreciate these differences. If we are looking for alternatives to the predominant theoretical assumptions of American politics, and if we believe that the alternatives most likely to achieve the success of widespread acceptance are ones already firmly rooted in the American tradition, then recovering something of the philosophic diversity that existed among the Founders may move us toward a recognition of options grounded in our philosophic past. This search for theoretical diversity and alternative perspectives within American political thought has not always been the prevalent approach among twentieth-century historians and political scientists. Studies such as Carl Becker's The Declaration of Independence (1958) and Joyce Appleby's Capitalism and the New Social Order (1983) make Jefferson part of the "bland uniformity" that has as its center the assumptions of Lockean individualistic liberalism.[2] There are reasons for seeing this connection between Jefferson and Locke. There are equal, or stronger, reasons, however, for rejecting this connection as incomplete and shallow.

The obvious and crucial importance of Nature in the writings of Jefferson can easily lead someone to associate this fixation on Nature with the philosophical and heuristic uses that Hobbes, Locke, and others made of the concept of a "state of nature" in their theoretical reasoning. Certainly, there are abundant references to natural rights in Jefferson's writings, and there are moments (particularly in the Declaration of Independence) where he seems to draw on the logic of the social contract to justify his theory of revolution, but the narrow association as a Lockean misses some essential elements in Jefferson's philosophy that reject the mechanistic, individualist, free-market capitalist, and representative political underpinnings of liberal thought. The current upsurge in Jefferson's popularity among environmentalists is inexplicable unless we come to understand how many contemporary theorists draw from Jefferson's work a vision that is neither Lockean nor in step with the thought of the majority of the other Founders. Jefferson as a classical republican, country opposition communitarian, and physiocratic advocate of populist direct democracy paints a very different portrait of this American icon.

What follows is an analysis of Jefferson's work with a particular eye toward those elements that seem predominant, are logically interconnected, and lead to a philosophical outlook consistent with many aspects of modern environmental theory. Although he was far removed from issues of resource scarcity, toxic materials, or the loss of biodiversity, there are elements in Jefferson's thought that add together to form a coherent perspective on the relationship between humans and Nature. Since any investigation of Jefferson's thought would be incomplete without an exploration of his ideas and actions on the questions of slavery and race relations, this chapter will conclude with some reflections on Jefferson, slavery, and America's capacity for addressing profound national ethical crises.

JEFFERSON AND NATURE

To say that Jefferson was interested in Nature would be a gross understatement.

Thomas Jefferson was obsessed with the topic his entire life. Charles A. Miller observes, "the words 'nature' and 'natural' appear with great frequency, in countless contexts, and with many meanings" throughout Jefferson's writings. Like many other scholars, Miller looks at Jefferson's book collection at Monticello and finds, "His library of sixty-five hundred volumes, the best map to a man's intellect, is filled with nature ... This is the consequence both of a great interest in the sciences and of having amassed an unrivaled collection of Americana, which at the time had to do, inevitably, largely with nature."[3] No clearer indication of Jefferson's fascination with Nature exists than the structure and reasoning found in his only major, extended piece of writing, Notes on the State of Virginia. Published in 1787 as a response to a number of questions posed by Francois, marquis de Barbe-Marbois, Jefferson's Notes significantly rearranged the ordering of Marbois' questions in a manner that demonstrates the differences between the Frenchman's and the American's priorities. Marbois, who at the time was a member of the French legation to the United States, posed a series of questions, or "queries," which began by inquiring into the governmental structure of the fledgling republic and moved on to issues involving America's commercial capabilities and mineral wealth. Jefferson, in his responses, both changed the order of presentation and inserted new topics that Marbois had not requested. For example, Jefferson begins by describing the boundaries of Virginia and quickly moves, not to government, but to a detailed description of the rivers, mountains, plants, animals, and climate of the state. For Jefferson, it is Nature that is foundational.[4] From his perspective, humanity's relationship to its environment preconditions and serves as the basis for both the instinctive underpinnings of human behavior and for the cultural institutions that humans create. If Jefferson had written at the turn of the twentieth century, instead of at the turn of the nineteenth, he might have been charged with being a left-wing Social Darwinist. This philosophical pyramid, based on wild Nature, and ascending through cultivated Nature, through instinctively conditioned human ethics, to man-made economic and political institutions is clearly illustrated in the organization and arguments found in Notes on the State of Virginia. Discarding Marbois' anthropocentric arrangement of chapters, Jefferson assembled his chapters, or "queries," in an order that reflected his philosophical priorities.

Notes on the State of Virginia may be grouped into four phases of explication that begins with Queries I through VII. In this section, Jefferson focuses almost entirely on nonhuman Nature as he describes Virginia's boundaries, rivers, mountains, cascades, minerals, plants, animals, and climate. There are parts of Jefferson's analysis that have a prescient quality, as when he suggests that human interference with Nature—in his case the large conversion of wilderness into agricultural production—may be having an effect on Virginia's climate. He describes the winter of 1780 when the York River froze thick enough to support people, and the Chesapeake Bay had ice to the depth of seven inches at Annapolis. Jefferson puts forth his belief that, "A change in our climate however is taking place very sensibly. Both heats and colds have become much more moderate within the memory even of the middle-aged. Snows are less frequent and less deep .... The rivers, which then seldom failed to freeze over in the course of the winter, scarcely ever do so now."[5] For Thomas Jefferson, these geological, ecological, and climatic facts serve as both the foundation of human conduct and as a measure of the impact our species is having on the planet.

Jefferson's <u>Notes</u> may be divided next into those chapters that move the inquiry into the study of people inhabiting the state. Consisting of Queries VIII through XII, the discussion shifts from Nature to human nature as the author takes on the issues of population and ethnology. Again, topics that would be familiar to contemporary environmental politics enter Jefferson's discourse as he reflects on the differing impact of the growth of human numbers from native births and immigration. He broaches the issue of cultural carrying capacity as he ponders the number of people Virginia "can clothe and feed, without a material change in the quality of their diet." Setting this number at around four-and-a-half million, he makes clear his preference that this should be attained by the growth of the native stock. Jefferson sounds a note very similar to contemporary neo-Malthusian Populists as he puts forward his fear that unbridled immigration will move America toward a population that is "a heterogeneous, incoherent, distracted mass." He rhetorically asks, "Is it not safer to wait with patience 27 years and three months longer, for the attainment of any degree of population desired, or expected? May not our government be more homogeneous, more peaceable, more durable?" This rebuff of liberal diversity while making the connection between homogeneous ethnic unity and a stable polity is an early signal of Jefferson's predilection to picture the world through the classical republican perspective.

When the subject becomes America's indigenous peoples, however, Jefferson's tone shifts. Toward these "Indians," he displays a sympathetic, albeit condescending, admiration. As Jefferson saw it, Indians lived closer to Nature, with very little in the way of human artifice. Their societies might be considered examples of anarchist communities since they "never submitted themselves to any laws, any coercive power, any shadow of government." In place of human created governments and laws, Indians relied upon their instinctive moral sense (one of Jefferson's earliest uses of this expression) to regulate their behavior. He describes the practice and its consequences. "Their only controuls are their manners, and that moral sense, which, like the sense of tasting and feeling, in every man makes a part of his nature. An offence against these is punished by contempt, by exclusion from society, or where the case is serious, as that of murder, by the individual whom it concerns." Jefferson comes most close to acknowledging his anarchist preferences in his openly stated opinion that a society governed with no laws is preferable to one with a multitude of governmentally sanctioned regulations. Reviewing the Indian social order, he says, "Imperfect as this species of coercion may seem, crimes are very rare among them: insomuch that were it made a question, whether no law, as among the savage Americans, or too much law, as among the civilized Europeans, submits man to the greatest evil, one who has seen both conditions of existence would pronounce it to be the last: and that the sheep are happier of themselves, than under the care of the wolves."[6]

In Queries XIII through XX, Notes on the State of Virginia becomes overtly cultural and political. This is the world human ingenuity, creativity, weakness, venality, and depravity have created through the exercise of free will. The principles of Nature may be followed, or they may be ignored, or they may be consciously suppressed, and it is in these sections where Jefferson traces what humans have done with the liberty given to them. In true republican fashion, he repeats the theme of the struggle between virtue and corruption. There is a subtle, but unmistakable, change in tone. Pathos emerges in the analysis and the quotations become more memorable while Jefferson fluctuates between progressive hope and a morose resignation. If these eight chapters were to be grouped into a single subsection, their title might well be "the slide into corruption." When humans fail to heed the lessons of Nature, according to Jefferson, the loss of virtue and the onset of depravity inevitably follow.

The section begins with Jefferson finally turning to the issue first broached by Marbois; the discussion of the Virginia Constitution. Disregarding his earlier nod to anarchism, Jefferson now builds his thinking on the assumption that humans will inevitably slide into self-interested corruption unless restrained by the force of good laws. At the same time, law itself will become the tool of corrupt men unless the process is restricted and limited by a carefully constructed constitution. Jefferson declares, "Mankind soon learn to make interested uses of every right and power which they possess, or may assume." Even the potentially virtuous Americans are not exempt from this decline. He warns his countrymen,

"They should look forward to a time, and that not a distant one, when corruption in this, as in the country from which we derive our origin, will have seized the heads of government, and be spread by them through the body of the people ..." Sadly, since "Human nature is the same on every side of the Atlantic," Jefferson pleads, "The time to guard against corruption and tyranny, is before they shall have gotten hold on us. It is better to keep the wolf out of the fold, than to trust to drawing his teeth and talon after he shall have entered."[7] This is the pathos that persistently emerges in all of Jefferson's writings. On the one hand, there is the hope that humans can learn from Nature, and human nature, can improve, progress, and move toward stable communal virtue if not outright perfectibility. Yet, this optimism is always tempered by the recognition that humans are the species most susceptible to the forces of immorality, domination, oppression, and corruption.

Nowhere is this tragic ambivalence more clearly manifested than in Jefferson's discussion of slavery and race, which emerges full blown in Query XIV. The chapter starts out quietly enough with the Virginian describing the court system and some of the laws of his native state. In an almost dry recitation, Jefferson lists the jurisdictions and procedures of the various courts and goes into a description of general laws pertaining to topics such as the care of the poor, rules for naturalization, limits to the amount of interest that can be charged for a loan (usury), and the provision for public works projects. He discusses laws for the transference of property and summarizes a plan put before the Virginia general assembly to revise the general statutes after the successful conclusion of the Revolution. One of those suggested changes, made as an amendment to the overarching report on revision, was a bill for the emancipation of all slaves born after the passing of the act, their training and education, and their resettlement, or colonization, "to such place as the circumstances of the time should render most proper ...."[8]

More careful analysis will be directed toward Jefferson's attitude on slavery, race, and ethical decision-making later in this chapter, but the outline of his position is pointedly made here in his <u>Notes on the State of Virginia</u>. Stated with an air of tortured objectivity, Jefferson asserts his belief that slavery is wrong and must end, while also suggesting with almost equal certainty that the white and black races will never be able to live together peacefully in a democratic, integrated society. He rhetorically posits, "It will probably be asked, Why not retain and incorporate the blacks into the state, and thus save the expense of supplying, by importation of white settlers, the vacancies they will leave?" and then states his belief that while the legal institution of slavery can, and must, be abolished, the psychological bias of racism admits of no such judicial remedy. "Deep rooted prejudices entertained by the whites; ten thousand recollections, by the blacks, of the injuries they have sustained; new provocations; the real distinctions which nature has made; and many other circumstances, will divide us into parties, and produce convulsions which will probably never end but in the extermination of the one or the other race."[9] The next several pages contain Jefferson's chronicle of the similarities and differences between the white, black, and red races. To contemporary readers, these paragraphs are not comfortable reading, but to Americans seeking a deep understanding of our nation and its capacity to respond to ethical crises, an analysis of this logic is essential. It is so essential, in fact, that it can only occur after a more complete explication of Jefferson's philosophy has been made. For now, the important thing to recognize is that, for Thomas Jefferson, the institution of chattel slavery is the epitome of human social corruption.

After abruptly terminating the discussion with a call to abolish slavery and physically separate the two races, Query XIV tries to end on an optimistic note with a plea for free public education for Virginia's citizens. Even here, however, Jefferson's <u>Notes</u> do not stray from their central theme. According to the author, the overarching purpose of his scheme for general instruction and the creation of an open meritocracy is the production of an informed, ethically sensitive citizenry who can detect and deter corruption before it becomes widespread. Among the purposes of Jefferson's educational reforms, "none is more important, none more legitimate, than that of rendering the people the safe, as they are the ultimate, guardians of their own liberty." Direct democracy, and not representative government, is Jefferson's means for sheltering America from the worst effects of corruption. "Every government degenerates when trusted to the rulers of the people alone," he declares. "The people themselves therefore are its only safe depositories. And to render even them safe their minds must be improved to a certain degree. This indeed is not all that is necessary, though it be essentially necessary."[10] As subsequent queries demonstrate, Jefferson considers almost every possible means for recognizing and retarding the spread of corruption. Every means, that is, except the destruction of the aristocratic, slave-owning culture of which he was a member.

Queries XV through XVII continue the republican battle between virtue, corruption, Nature, and culture. At times, Jefferson's exposition may strike the modern ear as humorous while still providing insight into his environmental theory. In Query XV, for instance, he considers the impact of his philosophy on architecture. Jefferson observes and decries what he sees as a new architectural fashion of using wood instead of stone as the primary material for home construction. If citizenship requires a strong attachment to a specific geographical space, and if our physical dwellings represent our most immediate connection to the place where we live, then the permanency of our buildings might provide a clue to the permanency of our communal roots. The building materials and construction techniques we select from Nature are windows into the inner souls of the occupants. "A country whose buildings are of wood," he argues, "can never increase in its improvements to any considerable degree. Their duration is highly estimated at 50 years. Every half century then our country becomes a tabula rasa whereon we have to set out anew, as in the first moment of seating it. Whereas when buildings are of durable materials, every new edifice is an actual and permanent acquisition to the state, adding to its value as well as to its ornament."[11] Building materials may seem like a trivial issue in the grander scheme of political philosophy, but for Jefferson, everything, right down to the walls within which we live, contributes either to virtue or to corruption.

Jefferson repeats this republican reasoning in a steady drumbeat of examples. He notes the importance of tolerance and diversity in religious conscience in the production of communal harmony and contrasts this with the sordid pursuit of economic wealth. Virtue derives from reason, ethics, and a culture of participatory, active citizens. Corruption results from passivity, apathy, and the pursuit of individual gain. "Besides," he states, "the spirit of the times may alter, will alter. Our rulers will become corrupt, our people careless .... From the conclusion of this war we shall be going downhill." The people, he fears, "will forget themselves, but in the sole faculty of making money, and will never think of uniting to effect a due respect for their rights." This entropic view of America's future leads Jefferson to one of the Notes more famous sad summaries, "Indeed, I tremble for my country when I reflect that God is just ...."[12]

The optimistic, hopeful side of Jefferson's philosophy emerges in his brief Query XIX. Consisting of only two paragraphs, the chapter on "Manufactures" neatly sums up Jefferson's vision for a virtuous and free America. At its foundation are the economic assumptions of the physiocratic theory of wealth cemented with the moral principles of republican political philosophy. Expressed in memorable language, the Virginia plantation owner voiced his conviction that agriculture and virtue are inextricably intertwined.

"Those who labour in the earth are the chosen people of God, if every he had a chosen people, whose breasts he has made his peculiar deposit for substantial and genuine virtue .... Corruption of morals in the mass of cultivators is a phenomenon of which no age nor nation has furnished an example."[13] Here as elsewhere, Jefferson does not miss the opportunity to contrast this ethical farmer with his alienated and corrupt opposite. Merchants, manufacturers, and, most of all, bankers and financiers do not look to "their own soil and industry," but are beholden, instead, on "the casualties and caprices of customers." For Jefferson, the upshot of this capitalist, commercial society, is preordained. "Dependence begets subservience and venality, suffocates the germ of virtue, and prepares fit tools for the designs of ambition." Commercial society represents the final contradiction between rural republicanism built around farms and small towns, and finance capital centered in densely populated urban centers. "The mobs of great cities," Jefferson forcefully asserts, "add just so much to the support of pure government, as sores do to the strength on the human body."[14]

Query XIX shines light onto a fuller view of Jefferson's philosophy, but the deeper tragedy inherent in his life and thought is missed if analysis does not immediately turn to the subsequent chapter in Notes on the State of Virginia. Its ostensible topic is "the commercial productions particular to the state," but keeping in mind everything revealed thus far about Jefferson, Query XX exposes a man trapped between a rich, consistent, and, in many ways, elegant philosophy and a cultural lifestyle diametrically at odds with his own pronouncements. The gulf between theory and practice is revealed. From the first sentence of the chapter, Jefferson admits that the chief product of Virginia agriculture is not wheat, meat, flax, cotton, or any other product of the soil useful for the practical and healthy life of the state's citizens, but rather, tobacco. Up to this point, everything Jefferson has written in his Notes has involved the struggle between virtue and corruption as they emerge from human's interaction with Nature. Now, toward the end of his efforts, Jefferson grudgingly admits that when human corruption and Nature meet, the botanical result is tobacco. "Those who labour in the earth" may not be all that virtuous if the product of their efforts is what Thomas Jefferson's fellow Virginian (and tobacco planter) Edmund Randolph called "that baneful weed."[15]

Like his comments on human slavery, Jefferson has not one good word to say about the cultivation of tobacco. The growing and harvesting of tobacco destroys the biological health of the soil, while the selling and trading in tobacco revenues involves a system of debt and consumption that undermines the ethical independence of the planter.[16] Jefferson does not shy from acknowledging that the raising of tobacco is "a culture productive of infinite wretchedness." Part of that "wretchedness" is related to the institution of slavery. The connection between tobacco and the soil is built on the fact that growing the crop "requires still more indispensably an uncommon fertility of the soil" that cannot last more than three of four years, with the result that "the earth is rapidly impoverished."[17] At this point, new, fertile land must be acquired (usually through debt) and cleared (always by slave labor). Like virtually all of his fellow Virginia tobacco planters, Jefferson was addicted and trapped in a system from which he was apparently incapable of escaping.[18] The irony is this system bore a close similarity to the cycle of luxury, debt, and dependence that Jefferson so vigorously condemned as the centerpiece of Alexander Hamilton's financial scheme.

If Thomas Jefferson could have looked back on his writings from the perspective of the end of the nineteenth century, he would have seen that the lavish lifestyle of the Virginia planters created debt and a need for cash. The need for a cash crop drove the extensive planting of tobacco. The cultivation of tobacco depleted the soil and caused expanding land ownership and the frequent conversion of wild landscapes to agricultural production. These labor-intensive field transfers were used to justify the institution of chattel slavery and the westward expansion of the American empire.[19] Conflict over westward expansion and slavery precipitated the American Civil War, and the Civil War ended with the violent destruction of the lavish Southern lifestyle. On some level, Thomas Jefferson saw this. He saw the corruption. He understood the connection between slavery, tobacco, debt, and dependence. He also clearly envisioned the alternative social order built on healthy food crops (wheat), free yeoman farmers, and independent virtuous citizens. But the need to advocate an internal cultural revolution as dramatic as the political revolution he supported against the British escaped his philosophical and ethical capacity. The man who the high Federalists accused of being a wild-eyed extremist was not nearly as radical as the nature of his times required. In spite of this, there is a tight, coherent pattern in Jefferson's philosophical legacy that meshes ecological perspective, political theory, and social organization. <u>Notes on the State of Virginia</u> opens the door for a broader study of his philosophy and establishes the centrality of his environmental outlook for an understanding of his contribution to American political thought.[20]

Thomas Jefferson and the Primacy of Ethics

Understanding the thought of Thomas Jefferson and its relevance to contemporary issues of environmental politics requires that we see and appreciate the absolute centrality of his notion of community and the close connection he makes between this concept and his ethical theory. The other components of his thought flow logically and inexorably from the fact that Jefferson believes humans are essentially social creatures who develop their complete capacities only as interacting members of communities. Unlike liberal thinkers who picture humans as self-interested individuals who associate primarily for economic purposes, Jefferson's perspective sees humans as communal creatures who are born with a moral sense that serves the purpose of holding their societies together. Among the many expressions of this belief, perhaps the best is found in Jefferson's 1787 letter to his nephew, Peter Carr. Here, Jefferson asserts his conviction that humans are inherently social and links this assumption with its close corollary: humans are instinctively moral. "He who made us would have been a pitiful bungler," Jefferson irreverently states, "if he had made the rules of our moral conduct a matter of science. For one man of science, there are thousands who are not. What would have become of them? Man was destined for society. His morality therefore was to be formed to this object."[21]

The beginning point for building an understanding of Jeffersonian politics is the recognition that human beings belong to the grouping of animals that we call gregarious. Closely tied to this idea of sociality is the fact that Jefferson makes ethics the basis for human solidarity. The emphasis in Jefferson's thought, therefore, is on the community and not the individual; the glue that bonds the community is morality and not economic interest. While Jefferson often defends the rights of the individual, that individualism is foreground against this contrasting background of inherent, organic, ethical community. It follows, then, that humans are more accurately portrayed as *Homo civitus* than as *Homo economicus*.[22] The ethical life consists of conduct that strengthens and perpetuates the community. In turn, the community consists of a collection of people who are actively seeking to sustain themselves. In this sense, ethics and community are mutually defining and mutually reinforcing concepts, and, as the above quotation suggests, Jefferson saw both community and morality as being instilled instinctively. When Jefferson deprecates the role of science in the shaping of moral conduct, he is acknowledging his belief that ethics are not a matter of education. That is to say, moral standards are not simply inscribed onto a *tabula rasa* by an outside agent of socialization. The assumption within Jeffersonian philosophy that all humans enter the world with essentially the same ability to make moral judgments is one part of what he meant by the assertion that "all men are created equal."

This connection between community and morality was not a passing fancy for Jefferson: it represents the core assumption of his philosophy. He enunciated it in the 1787 letter to Carr and, 27 years later, in a letter to Thomas Law, Jefferson is still reiterating this same philosophic position. He even repeats the "God as bungler" image he used long before. "The Creator would indeed have been a bungling artist, had he intended man for a social animal, without planting in him social dispositions ...," Jefferson asserts. "I sincerely, then, believe with you in the general existence of a moral instinct. I think it the brightest gem with which the human character is studded, and the want of it as more degrading than the most hideous of the bodily deformities."[23] Jefferson's conviction that there exists an instinctive basis for ethics establishes a link between his thought and that branch of the eighteenth-century Scottish Enlightenment known as the theory of moral sentiments. Proponents of this school include Francis Hutcheson, Henry Home (Lord Kames), Thomas Reid, Adam Smith, and even Jefferson's archenemy, David Hume. Jefferson owned and recommended works by all of these writers.[24] Since so much of Jefferson's political thought builds upon an understanding and acceptance of the theory of moral sentiments, it might be useful to summarize the essential elements of this doctrine.

The defining characteristic of moral sentiments philosophy is its location of the human capacity for morals in sentience and feelings, rather than reason, calculation, or logic. The primacy of emotions over reason results in ethics being seen as most properly a matter of the heart and not of the head. Although Thomas Jefferson is often portrayed as devout follower of the Enlightenment with its commitment to science, reason, and the ascendancy of logic over superstition (and there certainly are significant portions of his thought that demonstrate this Enlightenment strain in Jefferson), in his ethical system, Jefferson is adamant in asserting the importance of sentiment and feelings over logic and reason. In his previously cited letter to Carr, for example, Jefferson states that attending lectures on moral philosophy is "lost time" since morality cannot be learned in a classroom. Jefferson's most endearing testimony on the issue, however, comes in his famous letter of October 12, 1786, to Maria Cosway:

the "Dialogue between my Head and my Heart," Here, he has his Heart tell his Head, "When nature assigned us the same habitation, she gave us over it a divided empire. To you she allotted the field of science; to me that of morals … Morals were too essential to the happiness of man to be risked on the incertain combinations of the head. She laid their foundation therefore in sentiment, not in science."[25]

When we witness instances of cruelty, dishonesty, or neglect, it is not our rational thought processes that are offended. According to Jefferson, a moral person does not have to engage in an economic analysis of costs and benefits or a utilitarian calculus of pain and pleasure (what he calls "a miserable arithmetic") when confronted by a suffering veteran along a road or a needy woman begging charity. For the ethical person, the response is emotional and suprarational. It is this downplaying of reason and the substitution of feelings and emotions that most clearly distinguishes the theory of moral sentiments from rationalistic ethics. Jefferson's Heart (his moral sense) summarizes his position for his Head (the rational, empirical scientist). "When the circle is to be squared, or the orbit of a comet to be traced; when the arch of greatest strength, or the solid of least resistance is to be investigated, take up the problem; it is yours; nature has given me no cognizance of it. In like manner, in denying to you the feelings of sympathy, of benevolence, of gratitude, of justice, of love, of friendship, she has excluded you from their controul. To these she has adapted the mechanism of the heart."[26]

Throughout his writings, Jefferson repeats this basic formula: morals are based on sentiments, which are feelings ultimately reducible to impressions of pleasure and pain. To say that a particular action is "good" or "virtuous," therefore, is to say that it excites a pleasant feeling within us. The opposite is true regarding those actions we label "bad" or "wicked." There is a slight difference here between Jefferson's position and that of the Scottish philosophers. Hutcheson, Reid, Hume, and Smith simply state that morally worthy acts excite within us a pleasant feeling—they often call it "approbation"—while immoral acts create sensations or sentiments of pain. In Jefferson's thought, the source of these moral sentiments is specified. He believed that humans possess a moral *sense* which was the origin of the sentiments we feel. Again in his 1787 letter to Carr, Jefferson states that each human is, "endowed with a sense of right and wrong … This sense is as much a part of his nature as the sense of hearing, seeing, feeling; it is the true foundation of morality … The moral sense, or conscience, is as much a part of man as his leg or arm. It is given to all human beings in stronger or weaker degree, as force of members is given them in a greater or less degree. It may be strengthened by exercise, as may any particular limb of the body. This sense is submitted indeed in some degree to the guidance of reason; but it is a small stock which is required for this: even a less one than what we call common sense."[27]

The essential elements of Jefferson's ethical theory can be found in this quotation. There is the belief that humans possess a moral sense or conscience. This conscience is found within us at birth, that is to say, its existence is instinctive and is not the result of outside education or socialization. Since the moral sense is biologically imprinted in all humans, it is universal to the species. It follows then, that the moral sense establishes both an individual and a social aspect to ethics. Within each particular person, the moral sense signals if a proposed action is virtuous or vicious. When others view the acts that we perform, *their* moral sense is stimulated, and they are able to judge our actions. In a letter to his daughter, Martha, Jefferson emphasized the individual relationship between a person and their conscience. "If ever you are about to say anything amiss or to do anything wrong, consider beforehand," the father warns. "You will feel something within you which will tell you it is wrong and ought not to be said or done: this is your conscience, and be sure to obey it. Our maker has given us all, this faithful internal Monitor …"[28]

When advising his nephew, Jefferson focused instead on the social aspect of the moral sense. "Whenever you are to do a thing, though it can never be known but to yourself, ask yourself how you would act were all the world looking at you, and act accordingly."[29] An individual is not, therefore, the sole judge of his or her moral conduct. What may be called the spectator theory of morality takes the process beyond the individual to include all the members of the person's community. In fact, when the proposed action may have a global effect, such as an armed revolution against an established monarchy, the moral sense of the entire human race may need to be consulted. This linking of the moral sense to the ethical impressions of a wider community is why "the opinions of mankind" require "a decent respect," and why (moral) "facts" may be "submitted to a candid world."[30] The moral spectator may be considered dis*interested*, but that is not to say that the spectator is dis*passionate*.

Jefferson's advice to his nephew also establishes what will become a critical connection between his moral theory and his politics. Jefferson asserts that although the moral sense is embedded in each of us, it is capable of being "strengthened by exercise." The ability to make ethical judgments exists in all humans as part of their conscience, but the skill in forming and following moral precepts can be improved with regular use. This line of reasoning also carries with it the inference of its negation: the moral sense can atrophy, wither, and decay from a lack of employment. Making moral judgments is in some ways similar to the ability to sing (a favorite analogy in the Scottish Enlightenment); every human is born with the capacity to produce sound, but only those who regularly practice will perfect the talent. It is an unfortunate but obvious fact that for many of us, our inherent ethical potential languishes from a lack of attention. Jefferson offered his nephew examples of both moral reinforcement and moral atrophy. In a 1785 letter, the uncle stresses the role of habituation. "Encourage all your virtuous dispositions, and exercise them whenever an opportunity arises; being assured that they will gain strength by exercise, as a limb of the body does, and that exercise will make them habitual."[31] Two years later, Jefferson offered his nephew an example of the moral sense being made dull or numb. "State a moral case to a ploughman and a professor. The former will decide it as well, and often better than the latter, because he has not been led astray by artificial rules."[32] Merely because someone is educated, civilized, or refined in social manners does not assure that they will behave ethically (this is the model of the professor). Indeed, Jefferson suggests (significantly for future environmentalism) that someone who has simple manners, lives plainly, and occupies their life in a close relationship to the Earth (the ploughman) is most likely to exercise their moral sense properly.

Jefferson's moral theory may be summarized as follows: ethics cannot be based upon reason. Instead (1) morality is founded upon feelings or sentiments which are (2) generated within each individual by their instinctive moral sense and (3) are made social by reflecting on how our behavior will stimulate the moral sentiments of others. The ability to judge and to behave morally (4) can be improved by the regular exercise of our moral sense. Conversely, (5) immoral behavior is likely to follow if our moral sense is not exercised and, instead, is allowed to become atrophied and dull. For Jefferson, then, the ethical life is tantamount to the social exercise of our instinctive moral conscience.

At this point, it is possible to draw a picture of the individual who exhibits all of these characteristics. The social exercise of our instinctive moral conscience requires someone whose conduct represents the daily application of their moral sense, whose independence assures all that their ethical behavior comes from within, who demonstrates this ethic by being an active participant in the social and political life of their community, who is, therefore, deeply rooted in his or her community, and whose public conduct serves as a model and stimulus for the moral senses of others. At the center of Jefferson's vision is the person who is both a moral agent and an active member of the community. This is the picture of the virtuous citizen. It may not be an exaggeration to assert that for Jefferson, the entire purpose of politics, indeed the entire purpose of *life*, consists of making, improving, and fortifying the conditions under which the virtuous citizen can develop and thrive. Perhaps the most critical of these conditions is the active participation in the affairs of the community. One cannot be moral, or one cannot remain moral for long, unless he or she exercises their moral sense in association with others. It should now be evident that an intellectual chasm exists between the political thought of Thomas Jefferson and the ideas of many eighteenth- and early nineteenth-century liberals. Although David Hume and Adam Smith also constructed their ethics on the theory of moral sentiments, when they turned their attention to politics, they argued that the moral sentiments could not extend far enough (Hume) or could not hold up against the intense counter pressure from economic self-interest (Smith) and, therefore, a political substitute for ethics would have to be found.[33] For these liberals and others, such as James Madison, the substitute was the orderly, rule-guided clash of private interests. Society became seen merely as a collection of economically motivated individuals. The role of the government was to act as an amoral "umpire" in an essentially materialistic contest where morals and ethics became denigrated to little more than idiosyncratic and relativistic "values." The underlying assumption was that competition would force the participants to compromise and ameliorate their private interests in such a way that the common good would result.[34] The unexamined faith was that an "invisible hand" would guide this process or that a group of founders could build a system where beneficial social results would be the unintended consequence of ethically meaningless activity.[35] If incapable of a fully moral life, *Homo economicus* might at least be gently coerced into leading an existence that was politically stable. There exists little evidence in the writings of Thomas Jefferson to support the idea that he was ever attracted to this philosophy. On the contrary, there is much testimony that Jefferson was intellectually repulsed by many of the assumptions and by much of the logic of liberalism.

No clearer example of the differences between liberal and Jeffersonian approaches to politics exists than the distinct governmental structures each philosophy advocates and the widely divergent citizen roles that underpin each set of structures. Their differing definitions regarding what constitutes a "republic" perfectly illustrate this dichotomy. For liberalism, republican government becomes equated with *representative* democracy. Recall James Madison's definition from Federalist Number 10: "A republic, by which I mean a government in which the scheme of representation takes place." According to this view, the function of the citizen is to observe the political process, but to participate minimally. The job of the "citizen" (now in quotation marks) is merely to observer the political campaigns of others, vote, and dutifully pay taxes. Not only is the voter *permitted* to be passive politically, but also submissive acquiescence to the rule of others is *encouraged*. The desirability of this passivity follows from ethical doctrines that either view the moral sentiments as existing but inadequate for larger political purposes or that picture the individual as a moral *tabula rasa* who needs only to accept the socialization of values impressed upon them by others. Since Jefferson so clearly rejected both of these concepts of citizenship, it follows that his view of politics would also differ dramatically.

For Jefferson, politics should be less concerned with representative government than with the moral improvement of the citizens. It even extends beyond the formal rule of law. In an 1816 letter to John Taylor, he explores some of the contending denotations of the term "republic." He praises Taylor for having "successfully and completely pulverized Mr. Adams' system of orders, and his opening the mantle of republicanism to every government of laws …." For Jefferson, republican government is more than the rule of law, and it is more than the simple act of voting. Contrast Jefferson's definition of a republic with Madison's. Jefferson offers his opinion to Taylor. "Were I to assign to this term a precise and definite idea, I would say, simply, it means a government by its citizens is mass, acting directly and personally, according to rules established by the majority." Jefferson goes on to acknowledge "that every other government is more or less republican, in proportion as it has in its composition more or less of this ingredient of the direct action of the citizens." Direct action, participatory democracy, and immediate citizen involvement are the defining characteristics of a "republic" for Jefferson. In other words, a republic is a place where virtuous citizens outnumber the corrupt individuals. He sees that, "Such a government is evidently restrained to very narrow limits of space and population."[36] It is in the political life of the community that the citizen is afforded the greatest opportunity to exercise their moral instincts. Political activity should be structured, therefore, in such a way as to offer the greatest occasion for the maximum number of citizens to join in the debate, formulation, and implementation of collective decisions. He recommends ward republics as the political arrangement best suited for achieving these goals.[37]

The belief that genuine democracy requires citizens to be organized in units small enough to guarantee active participation and close contact with their fellow citizens establishes a connection between Jefferson's thought and the philosophies of Montesquieu and others in the classical republican tradition. According to Jefferson's theoretical structure of power, political activity and moral education would begin in local units of government smaller than counties. These ward republics would represent the basic component of a four-tiered system extending to counties, states, and, eventually, the federal government. In an 1816 letter to Joseph Cabell, Jefferson establishes two key justifications for creating ward republics: they would encourage maximum citizen participation (which, significantly, he contrasts with mere voting), and they would incline each citizen toward the patriotic defense of democracy. "Where every man is a sharer in the direction of his ward-republic, or of some of the higher ones, and feels that he is a participator in the government of affairs, not merely at an election one day in the year, but every day," Jefferson writes, "when there shall not be a man in the state who will not be a member of some one of its councils, great or small, he will let the heart be torn out of his body sooner than his power be wrested from him by a Caesar or a Bonaparte."[38] This is far removed from passive voters expressing individual values in a representative democracy built upon an extended republic. In Jefferson's scheme of ward republics, the goal is the creation of a *participatory* democracy.

At all levels of government, Jefferson evaluated institutions according to how well they contributed to the ethical "spirit" of the people. In a very important letter to Sam Kercheval, Jefferson reviews the ethical impact of the Virginia state constitution and finds it wanting. The moral sense of the people has persisted in spite of, and not because of, their governing institutions. "Where then is our republicanism to be found?" Jefferson asks. "Not in our constitution certainly, but merely in the spirit of our people … Owing to this spirit, and to nothing in the form of our constitution, all things have gone well. But this fact, so triumphantly misquoted by the enemies of reformation, is not the fruit of our constitution, but has prevailed in spite of it." In the same letter, Jefferson discusses ward republics as an organizational structure, and argues they are superior because they call the citizens to action, which arouses the "strongest feelings" in the "exercise" of free government. "Divide the counties into wards of such size as that every citizen can attend, when called on, and act in person ...," he urges Kercheval. "[M]aking every citizen an acting member of the government, and in the offices nearest and most interesting to him, will attach him by his strongest feelings to the independence of his country and its republican constitution ... These wards, called townships in New England, are the vital principle of their governments, and have proved themselves the wisest invention ever devised by the wit of man for the perfect exercise of self-government, and for its preservation."[39]

The critical issue then becomes the need to find some tangible means by which citizens can become actively attached to their ward republics, or local communities. This leads to an analysis of the role and limits of private property, and specifically, to a discussion of the function of land within Jeffersonian philosophy. Here, we find an apparent paradox. On one hand, Jefferson sees owning and working the Earth as the best means for developing virtuous citizens who are attached to their local communities. At the same time, however, he seeks to place restrictions on the idea of private property which he insists is *not* a natural, or inalienable, right in any liberal or Lockean sense. This paradox between the centrality and the limits of the idea of property is resolved by focusing on Jefferson's long-held belief that "the Earth belongs to the living." The final connection between Jefferson's thought and modern environmentalism can be established by investigating this connection between morality, community, and land. Of course, the fate of the Earth—"this corporeal globe, and everything upon it" as Jefferson calls it in the Kercheval letter—is just what moves contemporary environmentalists to action.[40]

Given that land and the opportunity to live an economically self-sufficient life while becoming rooted to one's community are so important in the moral development of the individual, one might expect Jefferson to argue for the existence of a natural or an inherent right to own property. The belief that the right to private property, or "estate," is the most fundamental of all rights possessed by the individual was certainly a widespread conviction during Jefferson's time. But Jefferson explicitly and repeatedly rejects the idea that there is an inalienable right to private property. It is tremendously significant that Jefferson changed Locke's famous trilogy of natural rights—life, liberty, and estate—to life, liberty, and the pursuit of happiness. There exists no clearer evidence of the disparity between Jeffersonian republicanism and Lockean liberalism than Jefferson's unwillingness to consider property a natural, inalienable right.

One fascinating insight into this aspect of Jefferson's thought is found in a letter to James Madison dated September 6, 1789. Here, Jefferson announces and explains his idea that "the earth belongs to the living in usufruct."[41] Contemporary scholars use this letter to illustrate Jefferson's belief that long-term debt violates the rights of those expected to pay (one generation cannot put succeeding generations into debt), and his even more radical belief that all contracts, constitutions, and laws should expire every 19 years.[42] Of particular interest to an ecological interpretation of Jefferson, however, are the implications that follow from the idea that we own land in "usufruct," or trust, and not by natural "right." During their lifetime, Jefferson argues, people can own land, manage it, and acquire the proceeds from it "as they please." But if the ownership of the land was considered a right to which the individual had an absolute claim, then the individual might burden the land with debt and thereby "eat up the usufruct of the lands for several generations to come, and then the lands would belong to the dead, and not to the living, which would be the reverse of our principle." Instead, Jefferson argues, "The portion [of the Earth] occupied by an individual ceases to be his when he himself ceases to be, and reverts to the society."[43] Here, Jefferson's thought comes full circle and returns to its starting point, for this principle asserts the primacy of the community over the individual and of morals over economics. Individuals are permitted to own and care for a part of the Earth in order to exercise and demonstrate their moral sense. Ownership of land also serves the function of connecting the individual to their community and fixing the ward republic in which they would act as citizens. With the death of the individual (or even more radically; every 20 years!), the community has the opportunity to observe how they behaved toward this trust. Jefferson suggests that the society may then express *its* moral sentiments by assigning the land to the wife, child, creditor, or legatee of the deceased, "not by any natural right, but by a law of the society of which they are members, and to which they are subject. Then no man can by *natural right* oblige the lands he occupied, or the persons who succeed him in that occupation ..."[44] Each person, each community, and each generation must publicly demonstrate their moral worth by how they treat the Earth.

Jefferson was certainly not opposed to private property; in fact, protecting private property was an important function of his government. But he believed that property must be managed for the common good of the community. It is easy to view "the earth belongs in usufruct to the living" as establishing humans as stewards of the Earth, but when we consider

Jefferson's philosophy as a whole and keep in view the crucial role played by the moral senses, we see something other than humans acting in guardianship over a passive and presumably incompetent Nature. In Jefferson's thought, virtually nothing is passive. Perhaps a better image than stewardship is the idea of metaphorical mirroring: how we treat the Earth is a reflection of our own moral sense.[45] It displays our moral development or our moral degeneracy. It connects us to each other and either reflects our commitment to the health of our community or our egoistic pursuit of our material self-interests. Finally, our interaction with the Earth reveals our relationship with posterity. It is what we will give them and how they will know us. When future generations see what we have done to the Earth, *their* moral sentiments will be stimulated. They will praise us or curse us and in that final judgment will exist the meaning of our lives.

Jefferson's Legacy and Contemporary Environmentalism

Although Thomas Jefferson's writings were composed over two centuries ago, the values he enunciated remain part of American culture. The vision of social life and the perspective on our relationship with Nature he articulated still play a very important role in conditioning our perception of environmental politics. For many inhabitants of this nation, Jefferson's ideals comprise a significant element in their conception of what it means to be an American.

Over the past 40 years, certain environmentalists have elevated the theoretical writings of Thomas Jefferson to near icon status. As one might expect, authors who focus on an agricultural perspective, such as Wendell Berry, Wes Jackson, and William Vitek, look to Jefferson as one of the central founders of American pastoralism.[46] Mainstream politicians ranging from Stewart Udall to Al Gore also use him as an inspiration for their ecological concerns. At the other end of the political spectrum, Christopher Manes, one of the most militant of the Earth First! deep ecologists, sees Jefferson as a potential source for radical environmental politics.[47] Rural reformers and environmental decentralists, including Daniel Kemmis and bioregional activist Kirkpatrick Sale, see in Jefferson a blending of a populist, participatory commitment to democratic politics and an ethical concern for the environment.[48] Beyond these examples, a large and philosophically diverse group of environmentalists has paid homage to various elements in Jefferson's thought.[49]

A Jeffersonian vision of politics merges with environmental theory when ecosystem degradation becomes viewed as a mirrored image of larger and deeper ethical concerns. When we finally come to see environmental destruction as resulting from the atrophying of our moral sense and the absence of an active, ethical citizenry, and when we see that this moral decay results from human alienation from natural processes, then we can see that any approach that reinforces the idea that humans are, or should be, passive, materialistic consumers in liberal, capitalistic societies is part of the problem, and not part of the solution. It is here where the political thought of Thomas Jefferson provides the philosophic critique and the historical context that embeds radical environmentalism in American political culture. To the extent he professes any economic concerns, he is a corporatist physiocrat and not a capitalist. His legacy is populist and agrarian and, therefore, he belongs in neither the classical nor progressive schools of liberalism.

Specifically, there are at least three major components that comprise Jefferson's legacy to contemporary radical environmentalism. First is Jefferson's moral theory, his assertion of the primacy of ethics over economics, and his belief that ethics require an active exercise rather than a passive acceptance. There is a ready kinship, for instance, between the thought of Jefferson and the thought of Aldo Leopold. Leopold's enormously important book, A Sand County Almanac, is widely regarded as the twentieth century's leading exposition of an alternative approach to the relationship between humans and Nature that is built on ethics and not economics. Leopold's "land ethic" and the moral theory that substantiates it harken directly back to the theory of moral sentiments.[50] The second ingredient in Jefferson's legacy to contemporary environmentalism is his belief that healthy, ethical communities must primarily be based in rural settings where individuals are rooted in and committed to the protection of the Earth and the nurturing of human relationships. The popularity of this bioregionalism testifies to the deep psychological need that many Americans feel for environmental relationships that are more meaningful and permanent than those found in the insipid materialism of the shopping mall and desolate anomie of the housing "development."[51] The third element that demonstrates Jefferson's ongoing legacy is the persistent strength of American Populism. The corporatist wish to link politics, economics, and ethics has often been united with the belief that this corporatism must have its basis in strong, active, local governments. Jefferson's belief that "the earth belongs to the living" can be found in a shallow version among those contemporary environmentalists who call for human stewardship over the Earth and in a deeper, richer, more meaningful sense among those radical thinkers who recognize that how we treat Nature is but a reflection of our inner moral sensitivities.[52]

Beyond these important connections between the political thought of Thomas Jefferson and contemporary environmentalism, however, the sad truth must be admitted that much of this contemporary environmental literature perpetuates the two major flaws in Jefferson's philosophy; his inability to recognize and his unwillingness to deal with the critical significance of limits to population growth and the issue of race relations in America. The need to check the size of the human population may be the more basic of these two questions, for if population was small enough, diverse ethnic or racial groups could be dispersed and empowered in their own ward republics, or they could form multiracial ward republics with each citizen actively and equally participating. Jefferson himself was not ignorant of the importance of population to his overall theory. In a letter to physiocrat Jean Baptiste Say, Jefferson notes that he has read Thomas Robert Malthus' Essay on the Principle of Population but he deprecates the impact of population in America where there is "an immense extent of uncultivated and fertile land." In an earlier letter to James Madison (who had also read Malthus), Jefferson sees that a small ratio of people to the Earth is essential to his theory, but he naively believes that the ratio will long remain beneficially modest. As Jefferson expresses himself to Madison, "I think our governments will remain virtuous for many centuries; as long as they are chiefly agricultural; and this will be as long as there shall be vacant lands in any part of America." This is a repetition of the perspective he enunciated in his First Inaugural where he describes America as "a chosen country, with room enough for our descendants to the hundredth and thousandth generation ..."[53] Alas, if America is to support a large population for this many years, it will have to rely on technology, massive institutions, and a gigantic economy. In an America with 340 million human inhabitants, or in a world with ten billion people (two often quoted demographic projections), there is no place for Jeffersonian republicanism.

It is difficult to determine if contemporary environmentalism will develop an ecological perspective that combines the best of Jeffersonian thought with a recognition of the need for population limits.[54] The political thought of Thomas Jefferson is not without its flaws. Still, it may represent the only tradition in America's cultural heritage that is capable of combining a political perspective built upon active, engaged citizens living in strong, healthy communities with an ethical concern for humans and the Earth that we all occupy. The critical factor is whether Jeffersonian theory and contemporary American environmentalism have the moral courage to push the logic of their argument through to the radical consequences and revolutionary social change necessary to confront and resolve the overarching ethical crisis of our time.

Jefferson's Other Legacy

Nations become trapped in ethical crises when their fundamental institutions and habits lead them along one historical path and their basic communal values and moral principles lead them in a different, contradictory direction. The morally obscene becomes the customary and the practical.[55] Since institutions, customs, and ethics are all part of a nation's culture, these crises can be viewed as one aspect of our traditions doing battle against other parts of our way of life: either the nation's institutions and habits must change, or the community's ethical principles must be ignored. It may be the case, however, that every nation tries to deny or gloss over these ethical crises since no nation likes to admit that its culture is internally flawed. America is certainly no exception to this pattern of denial, nor is America immune from cultural conflicts between its practices and its professed principles. This nation's historical responses to slavery and racism are examples of ethical crises entrenched in a political order based on freedom and equality coexisting within a social structure that accepts, or at least tolerates, the values embedded in slavery and racism. Today, the devastation of the environment is another such ethical crisis.

As much as any other American, Thomas Jefferson was a key figure in establishing the organizational foundations of our democracy, and, as much as any other American, he is emblematic of this nation's inability to openly and boldly resolve the moral pestilence of first slavery, and then racism, that seem so engrained and intractable within the United States. His stirring words in the Declaration of Independence, proclaiming this nation's belief that "We hold these truths to be self-evident; that all men are created equal; that they are endowed by their creator with inalienable rights; that among these are life, liberty, and the pursuit of happiness ..." stand as the exemplar of our political ethics, and yet, as Thomas Jefferson wrote these words, he was the owner of 175 human beings who worked as slaves on his Virginia plantation.[56] How can Thomas Jefferson the Virginia slave master be philosophically and ethically squared with Thomas Jefferson the author of the Declaration of Independence and America's foremost advocate for human rights?

Jefferson is important because he symbolizes both the power of American ideals and our failure to live up to the high standards of our moral values. Race relations are momentous both in their own right and, in the current context, for what they portend as the United States begins to confront its next great ethical crisis. A discussion of Jefferson and race would move from his personal behavior, to his policy on the critical social issue of his time, to values that drive to the foundation of America's views on equality, citizenship, and community. Among those contemporary commentators who address Jefferson and race these levels of analysis are displayed in arguments regarding Jefferson's relationship with his female slave, Sally Hemmings; his overall policy on emancipation and the political vigor (or lack thereof) with which he attacked the institution of slavery; and Jefferson's thoughts on the possibility of America resolving the crisis of racism by becoming an inclusive, harmonious republic.

The question whether or not Jefferson had a sexual relationship with Sally Hemmings has lingered since political hatchet man James Thomson Callender exposed the rumor during the 1802 political campaign. Since then, speculation and allegations have periodically surfaced that Jefferson kept Hemmings as his concubine from the time she accompanied his daughter to France in 1787 until his death in 1826. Since so much of the dispute hinges on the emotional bond between Jefferson and Hemmings, and since the nature of that attachment—how oppressive or loving it may have been—is unknown and unknowable, there seems to be little, beyond historical titillation, to be gained from this debate.[57] Apparently, Thomas Jefferson (and one might add Alexander Hamilton, Gouverneur Morris, Benjamin Franklin, and a host of our other Founding Fathers) were able to make a sharp distinction between sexual promiscuity and *civic* virtue.

There were, however, aspects of Thomas Jefferson's life that were indisputably public in their consequences. Throughout his long life, Jefferson never said a word in support of slavery or the slave trade. Denunciations of slavery are found everywhere in Jefferson's writings. His original draft of the Declaration excoriates King George for permitting and supporting the African slave trade, which Jefferson reviles as "this execrable commerce." Jefferson declares the king "has waged cruel war against human nature itself, violating its most sacred rights of life and liberty in the persons of a distant people who never offended him, captivating & carrying them into slavery in another hemisphere, or to incur miserable death in the transportation thither."[58] In <u>Notes on the State of Virginia</u>, Jefferson refers to slavery as "this blot in our country" and "this great political and moral evil."[59] In letters written over the course of his life, he said, "The love of justice and the love of country plead equally the cause of these people, and it is a moral reproach to us that they should have pleaded so long in vain ...." In his <u>Autobiography</u>, written 45 years after the Declaration of Independence, Jefferson was still affirming, "Nothing is more certainly written in the book of fate than that these people are to be free."[60] While condemning the institution of slavery and calling for the eventual emancipation of all slaves, Jefferson always insisted that once free, all Americans of African descent would be subject to "colonization," a policy designed to deport them to new homes in either Africa or Santo Domingo.

All of his verbal condemnation of slavery stands in stark contrast to Jefferson's actual behavior. In 1804, he refused to support Connecticut Representative James Hillhouse, who proposed an amendment to prohibit the extension of slavery into the newly acquired Louisiana Purchase. In 1808, Congress, at the urging of President Jefferson, abolished the African slave trade. This was to be the only major effort toward abolition during Jefferson's public service. When Edward Coles, private secretary to James Madison and Virginia plantation owner, wrote Jefferson with a plan to sell his plantation, free his slaves, and resettle them in Ohio, he was rebuffed by Jefferson with the claim that freed blacks were as "incapable as children of taking care of themselves."[61] Perhaps the darkest moment in Jefferson's record on slavery came with the 1819/1820 controversy over the admission of Missouri into the Union. The retired President placed his political weight and reputation against all efforts to restrict the expansion of slavery beyond the Mississippi River or above the extended Mason–Dixon Line. Rejecting attempts at a Missouri Compromise, Jefferson advocated a policy of "diffusion," believing the extension of slavery into the American west would decrease the number of large slave owners, mitigate the worst effects of the institution, and gradually open the possibility for eventual emancipation.[62] The policies of colonization and diffusion were, of course, pure delusion whose only real effects were to postpone the inevitable day of reckoning as America drifted toward civil war and to obscure the essential ethical quandary surrounding the contradiction between slavery and freedom. To support colonization and diffusion was crackpot realism of the most destructive kind, and yet, the question remains; how and why did Jefferson not see the immediacy of abolishing this institution he recognized as a "moral and political depravity"?[63] Why did the man who made human rights and the moral sense the centerpieces of his political philosophy not act more vigorously to support those rights in the greatest ethical struggle of his time?

To merely say that Thomas Jefferson was hypocritical is an insufficient response. It is insufficient because Jefferson's thinking is too complex to permit a simplistic response to such a critical question, and it is inadequate because it fails to move us toward an appreciation of how America deals with ethical crises. There are parallels here with our ecological crisis. In academic and scholarly writings, the environment is often addressed as an ethical issue—our responsibility to future generations, the harmony of different species in the ecosystem, our moral obligation not to do irreparable harm to biological processes—yet in our policies and actions, humanity's relationship to the planet is very often reduced to an economic or political matter—the need to measure and internalize the full costs of our actions, the balancing of various economic interests, or the construction of compromises among conflicting political groups. Jefferson's legacy of waffling, dodging, and eventual reactionary intransigence on slavery is matched today by Prometheans, free marketeers, and many mainstream environmentalists who resist taking the radical steps required to resolve the contradictions between our ethical concern for the Earth and the cultural practices undermining ecological health. Jefferson's behavior on race relations and abolition exhibits his lack of individual will to act according to his principles, and it also demonstrates how an American President can construct public policy premised on completely delusional beliefs. Beyond these personal failings, the much more significant concern is Jefferson's thoughts on the capacity of our national character to resolve ethical crises, and on the sources of racial animosity that, he felt, made the chances for eventual racial harmony very slim. Understanding this part of Jefferson's legacy must build on an appreciation of his views on human nature, human equality, and obstructions to the formation of a just community.

Jefferson had a tripartite concept of humans. The various members of our species could be compared and contrasted according to their physical attributes, their intellectual capacities, or, as we have seen, by the development of their moral sense. Seeing all humans as an amalgamation of body, mind, and ethics permits a more nuanced exploration of Jefferson's tense and tortured position on race and ethics.[64] The pieces of Jefferson's logic fit together, if not pleasantly, then at least consistently. Jefferson's perspective on each of these characteristics explains how he can believe the races are morally equal and yet socially incapable of living together.

Regarding the physical and intellectual attributes of the white and black races, Jefferson acknowledged blacks to be equal in strength, endurance, and courage, and although he had misgivings regarding the intellectual capacities of blacks, he was sensitive enough to admit that life within the institution of slavery hardly gave Africans the opportunity to develop and demonstrate their mental potential. The critical factor was not the physical or even the intellectual comparison of the two races. People of dissimilar muscular power or intellectual development can still live together in peaceful unity. Physical strength and mental prowess are irrelevant to meaningful discussions of equality, social harmony, and the pursuit of justice, and, therefore, if there are dissimilarities between the races that impact their moral and social interactions, these differences should be sought in the instincts and in our learned attitudes. The time has come to bring together Jefferson's theories of ethics, politics, and race.

When Thomas Jefferson wrote, "all men are created equal," he meant by this expression that all men (he certainly would have included women in this concept) were born with the same moral sense, or conscience, and the same potential for developing those sympathetic feelings of right and wrong. Since the moral sense is instinctive, its presence is unaffected by one's race or station in life. Men and women, blacks and whites are all "endowed by their creator" with "inherent and inalienable rights," but from this perspective, it also follows there is no inconsistency in Jefferson's belief in ethical equality mingling with his fear that the races could not coexist.[65] Jefferson specifically acknowledged these beliefs. In Query XIV of the Notes, Jefferson makes the distinction between learning (the head) and the moral instinct (the heart) and declares the two races equal in the latter category. "Whether further observation will or will not verify the conjecture, that nature has been less bountiful to them in the endowments of the head, I believe that in those of the heart she will be found to have done them justice." He goes on to acknowledge, "we find among them numerous instances of the most rigid integrity, and as many as among their better instructed masters, of benevolence, gratitude, and unshaken fidelity."[66]

The tragedy for Jefferson, and America, is that the acknowledgment of this moral equality does not necessarily translate into the reality of communal harmony. It is the recognition of this sad possibility that underpins Jefferson's perspective on slavery, race, ethics, and politics. Because Jefferson believed blacks and whites are ethically equal, he condemned slavery as an unqualified evil, but because he also believed the two races could never live together peacefully, he held that blacks must be colonized to another country as soon as emancipated. The situation was even more complex and tragic. Since Jefferson probably recognized the impossibility of colonization, and since all of his schemes for abolition hinged on colonization immediately following manumission, he became politically paralyzed at the fundamental first step of emancipation, spending his life arguing against slavery while doing pathetically little to end it. Toward the end of his life, Jefferson confessed this moral paralysis in an 1820 letter to John Holmes, where he provides the weak conclusion, "we have the wolf by the ears, and we can neither hold him, nor safely let him go. Justice is in one scale, and self-preservation in the other."[67] Given the ethical choice between justice and self-interest, Jefferson the citizen could not subdue Jefferson the slave owner. The larger lesson may be that it takes more than a recognition of intrinsic value in order to create a sense of harmony and social justice among races, ethnic groups, or even species. It may also take a great deal of ethical courage and the shared acknowledgment that both groups are members of the same community with a common past and a common future, but that shared history does not come from an instinctive conscience. It is learned.

Why did Jefferson think that blacks and whites could never live together in an integrated, biracial, republican community? From his writings, it is possible to distill three causes for this pessimistic conclusion. As was alluded to earlier, the first cause of social disharmony would be the long history of animosity between the races. The fear among whites that blacks would use emancipation as a trigger for violent attacks against their former oppressors, an impression reinforced by paranoia over slave revolts, and the indisputable memory among blacks of the injustices they had endured, combined in Jefferson's mind to make interracial harmony highly unlikely. The second reason Jefferson doubted the feasibility of an interracial republic was his suspicion that slavery and racism had so poisoned the minds of blacks that they could never have the true love of America that was the essential foundation for any sustainable communal commitment. In Query XVIII, he expressed his sad conviction that slavery has destroyed the "amor patriae" of all slaves and their descendants. "For if a slave can have a country in this world," Jefferson declares, "it must be any other in preference to that in which he is born to live and labour for another; in which he must lock up the faculties of his nature, contribute as far as depends on his individual endeavors to the evanishment of the human race, or entail his own miserable condition on the endless generations proceeding from him."[68] For Jefferson, blacks could never be rooted to a land that had treated them so badly.

Given what we now know about the interconnections in Jefferson's thought among the primacy of ethics, community, and the potential for learning to expand or contract the moral sense, the third explanation for discounting the possibility of biracial harmony must be viewed as the most poignant. Jefferson felt slavery had so dulled, desensitized, and extinguished the moral sense in the hearts of white Americans that they could never expand their feelings of ethical community to include blacks. The most tragic consequence of slavery and racism in America (certainly among those white elites who were slave owners) was to convert them into self-interested, narrow, egotistical despots. Jefferson recognized, "The whole commerce between master and slave is a perpetual exercise of the most boisterous passions, the most unremitting despotism on the one part, and degrading submissions on the other." The impact of this extreme social inequality as a learning experience is a terrible as it is obvious. "Our children see this," says the slave-owning parent of two young girls, "and learn to imitate it; for man is an imitative animal." An example follows. "The parent storms, the child looks on, catches the lineaments of wrath, puts on the same airs in the circle of smaller slaves, gives a loose to his worst of passions, and thus nursed, educated, and daily exercised in tyranny, cannot but be stamped by it with odious peculiarities." Slavery and racism may have provided economic benefits for white society, but the cost of this wealth creation was the destruction of our sympathies, compassion, and empathy for a large segment of our fellow Americans. The cost of slavery for the white masters was the shriveling of their hearts. A good candidate for the most agonizing sentence in all of Jefferson's multivolume writings is the one that follows these comments on the cultural and ethical consequences of the master/slave, oppressor/oppressed, white/black relationship. "The man must be a prodigy," the self-reflective Jefferson admits, "who can retain his manners and morals undepraved by such circumstances."[69]

This extraordinarily revealing sentence may be a pitiful confession by the leading spokesman for the moral sentiments tradition in America declaring his own conscience had become at least partially "depraved." Jefferson may be seen as engaging in an act of self-condemnation and acknowledging that all those volumes and volumes of Enlightenment literature in his study were worthless as an antidote to the ethical autism permeating the numbed spirit of his soul. If Jefferson is conceding that he is no "prodigy," then the parts of his philosophy fall consistently, if tragically, into place. Jeffersonian republicanism has as it basis the ethical citizen who is a member of a community that is held together by shared feelings of sympathy, compassion, and common interests. The quandary in Jefferson's political thought does not derive from his concept of equality; the crisis results from his inability to expand and extend his notion of community.

The complete scope of Jefferson's environmental legacy is now exposed. On the positive side, the thought of Thomas Jefferson contains those elements that are essential to the resolution of our ecological crisis, and it embeds those perspectives firmly within the American philosophical tradition. If this country is to move toward a harmonious, stable, and sustainable relationship between its people and its ecosystem, those humans will need to build the qualities of citizens who are rooted to a place they love, who limit their materialist consumption with a strong moral sense, and who are capable of shifting their point of view and seeing the impact of their actions on other species as well as the landscape. Most critically, future environmentalism will have to do what Thomas Jefferson was incapable of doing himself: we must expand our feelings of kinship. Jefferson's America (and ours) needs to recognize the white and black races as members of the same community. Similarly, humans must extend our moral sense beyond our own species and recognize all components of our environment as interconnected, ethically equal, and mutually sustaining parts of the same community. This will not be easy. Just as a racist would argue that the white and black races were fundamentally different, an anthropocentric humanist today will ignore the interdependencies and focus exclusively on the difference between humans and animals, between animals and plant species, and between sentient and nonsentient components of the ecosystem. The question is not whether or not the differences exist—they do. The issue is whether or not they will be used to blunt our consciousness and conscience in a way that numbs the recognition of our interdependence and shared destinies. To believe the American environment can be completely humanized and technologically controlled is as morally wrong as it is physically impossible. It is the present-day equivalent of the delusional belief that colonization or territorial expansion could cure the abomination of an American republic that tolerated chattel slavery.

The legacy of Thomas Jefferson is an admirable set of principles coupled with a personal determination that was too often timid and fainthearted. Individual will takes private courage, but more importantly, the collective will of a nation requires a shared community, a recognition of entwined destinies. Tragically, two centuries after Jefferson wrote, it is still not clear if the white and black races of America are capable of living together in a harmonious community, but it is absolutely certain that we cannot separate. Directly analogous is the fact that Americans and their environment cannot become alienated from each other. Summoning the collective will to respond to the risks that threaten both human society and ecological health begins with an acknowledgment of that entwined destiny. Today's fealty to technology and environmental "decoupling" as solutions to our predicament are as reprehensible and delusional as the trust in colonization, diffusion, and racial segregation were. Just as reliance on these unworkable racial policies served to dull Jefferson's moral sense, our contemporary devotion to technological palliatives is numbing our appreciation of the entwined destiny we share with Nature. In this sense, decoupling, technology, and the domination of Nature are not part of the solution, they are the problem.

Notes

1. "First Inaugural Address," March 4, 1801, in Andrew A. Lipscomb and Albert Ellery Bergh, eds., The Writings of Thomas Jefferson, 20 vols. (Washington, DC: Thomas Jefferson Memorial Association, 1903–1904; hereafter cited as Writings of TJ with reference to volume and page), III: 319.
2. Daniel Boorstin, The Genius of American Politics (Chicago: University of Chicago Press, 1953). Louis Hartz, The Liberal Tradition in America (New York: Harcourt Brace Jovanovich, 1955). Carl L. Becker, The Declaration of Independence: A Study in the History of Political Ideas (New York: Random House, 1958). Joyce Oldham Appleby, Capitalism and the New Social Order: The Republican Vision of the 1790's (New York and London: New York University Press, 1983). The reference to the "bland uniformity" of American thought is from the presidential address to the American Political Science Association by Judith N. Shklar, published as "Redeeming American Political Theory," American Political Science Review, vol. 85, no. 1 (March 1991), 3.
3. Charles A. Miller, Jefferson and Nature: An Interpretation, (Baltimore: The Johns Hopkins University Press, 1988), 1 and 3.
4. Thomas Jefferson, Notes on the State of Virginia, in Merrill D. Peterson, ed., Thomas Jefferson: Writings (New York: The Library of American, 1984), 123–325. For a discussion of Jefferson's and Marbois' arrangement of questions, see Ronald L. Hatzebuehler, "I Tremble for My Country": Thomas Jefferson and the Virginia Gentry (Gainesville, FL: University Press of Florida, 2006), 71–74.

5. Ibid., Query VI, 206–207.
6. Ibid., Query XI, 220.
7. Ibid., Query XIII, 246.
8. Ibid., Query, XIV, 264.
9. Ibid., 264.
10. Ibid., 274.
11. Ibid., 280.
12. Ibid., 287 and 289.
13. Ibid., 290.
14. Ibid., 290 and 291.
15. Cited in T.H. Breen, Tobacco Culture: The Mentality of the Great Tidewater Planters on the Eve of Revolution (Princeton: Princeton University Press, 1985), 200.
16. Jefferson to Dr. Benjamin Waterhouse, June 26, 1822, in Merrill D, Peterson, ed., Thomas Jefferson: Writings, 1458.
17. Thomas Jefferson, Notes on the State of Virginia, Query XX, 293.
18. See Jefferson's warnings to Lucy Ludwell Paradise, in Julian P. Boyd editor, The Papers of Thomas Jefferson, 26 vols. (Princeton, NJ: Princeton University Press, 1950–1995), 10, 305. Hereafter cited as Papers of TJ with reference to volume and page.
19. For analysis of these issues, see T.H. Breen, Tobacco Culture, and Stephen E. Ambrose, Undaunted Courage: Meriwether Lewis, Jefferson, and the Opening of the American West (New York: Simon and Schuster, 1996), especially Chapter 2.
20. The final two chapters of Notes on the State of Virginia are anticlimactic. The argument does not so much end as dwindle and eventually simply peters out in a long listing of the various statutes passed by the House of Burgesses over a number of years.
21. Jefferson to Peter Carr, August 10, 1787, in Julian P. Boyd, editor, Papers of TJ, 12: 14–15. Jefferson's revealing failure to capitalize the name of the deity is from the original.
22. An excellent review of Jefferson's political thought (from which I borrow heavily), is Richard K. Matthews, The Radical Politics of Thomas Jefferson (Lawrence: University Press of Kansas, 1984), here see p. 12.
23. Jefferson to Thomas Law, June 13, 1814, in Writings of TJ, XIV: 138–144.
24. See Jefferson's letter and list of books to Robert Skipwith, August 3, 1771, in Papers of TJ, 1: 76, where the only conspicuously absent name is Hutcheson. Beyond Scotland, the most obvious exclusion is Jean-Jacques Rousseau who also deserves to be counted part of the "moral sense school." The connection between Jefferson and the Scottish Enlightenment was given wide exposure by Garry Wills, Explaining America: Jefferson's Declaration of Independence (Garden City, N.Y.: Doubleday & Co., Inc., 1978).

25. "Dialogue Between My Head and My Heart," letter to Maria Cosway, October 12, 1786, in Papers of TJ, 10: 451.
26. Ibid.
27. Jefferson to Peter Carr, August 10, 1787, in Papers of TJ, 12: 14.
28. Jefferson to Martha Jefferson, December 11, 1783, in Papers of TJ, 6: 380.
29. Jefferson to Peter Carr, August 19, 1785, in Papers of TJ, 8: 406.
30. "Declaration of Independence," in Papers of TJ, 1: 423–424.
31. Jefferson to Peter Carr, August 19, 1785, in Papers of TJ, 8: 406.
32. Jefferson to Peter Carr, August 10, 1787, in Papers of TJ, 12: 15.
33. Adam Smith, The Theory of Moral Sentiments in Robert L. Heilbroner, ed., The Essential Adam Smith (New York and London: W.W. Norton and Company, 1986). David Hume, An Enquiry Concerning the Principles of Morals in Hume's Moral and Political Philosophy, ed., Henry D. Aiken (New York: Hafner Press, 1948).
34. Franklin A. Kalinowski, "David Hume and James Madison on Defining 'The Public Interest'," in Virtue, Corruption, and Self-Interest, ed., Richard K. Matthews (Allentown, PA: Lehigh University Press, 1994).
35. Franklin A. Kalinowski, "David Hume on the Philosophic Underpinnings of Interest Group Politics," Polity, Spring 1993.
36. Jefferson to John Taylor, May 28, 1816, in Merrill D. Peterson, ed., Thomas Jefferson: Writings, 1392.
37. The first person to recognize the radically populist implications of Jefferson's ward republics was Hannah Arendt, On Revolution (New York: Viking Press, 1963). Ward republics are also discussed in Richard K. Matthews, The Radical Politics of Thomas Jefferson; and Garrett Ward Sheldon, The Political Philosophy of Thomas Jefferson (Baltimore and London: The Johns Hopkins University Press, 1991)
38. Jefferson to Joseph C. Cabell, February 2, 1816, in Writings of TJ, XIV: 422.
39. Jefferson to Sam Kercheval, July 12, 1816, in Writings of TJ, XV: 35 and 37–38.
40. Jefferson concludes his letter to Kercheval by observing that "[T]he dead have no rights. They are nothing; and nothing cannot own something.... This corporeal globe, and everything upon it, belong to its present corporeal inhabitants, during their generation." Ibid., XV: 43.
41. Jefferson to James Madison, September 6, 1789, in Papers of TJ, 15: 392–397.
42. Herbert Sloan, "The Earth Belongs in Usufruct to the Living," in Jeffersonian Legacies, ed. Peter S. Onuf (Charlottesville, VA: University Press of Virginia, 1993). Especially insightful is Richard K. Matthews, The Radical Politics of Thomas Jefferson, chaps. 2 and 3.

43. Jefferson to Madison, September 6, 1789, in Papers of TJ, 15: 392 and 393.
44. Ibid., 15: 393.
45. The idea of metaphorical mirroring is borrowed from John Rodman, "The Liberation of Nature?" Inquiry 20; 83–131.
46. Wendell Berry, The Unsettling of America: Culture and Agriculture (San Francisco: Sierra Club Books, 1977), especially Chapter 8. Wes Jackson, Becoming Native to This Place (Lexington, KY: The University Press of Kentucky, 1994). William Vitek and Wes Jackson (ed.), Rooted in the Land: Essays on Community and Place (New Haven: Yale University Press, 1996), especially, William Vitek, "Rediscovering the Landscape." For an overview of Jefferson's place in American pastoralism, see Leo Marx, The Machine in the Garden: Technology and the Pastoral Ideal in American (New York and Oxford: Oxford University Press, 1964).
47. Stewart L. Udall, The Quiet Crisis (New York: Avon Books, 1963). Al Gore, Earth In the Balance: Ecology and the Human Spirit (Boston and New York: Houghton Mifflin Co., 1992). Christopher Manes, Green Rage: Radical Environmentalism and the Unmaking of Civilization (Boston and New York: Little, Brown and Co., 1990).
48. Daniel Kemmis, Community and the Politics of Place, (Norman, OK and London: University of Oklahoma Press, 1990). Kirkpatrick Sale, Dwellers in the Land: The Bioregional Vision (Philadelphia: New Society Publishers, 1991). Of course, multiple interpretations of Jefferson's work are possible. This has lead one scholar to proclaim Jefferson "the American sphinx," Joseph J. Ellis, American Sphinx: The Character of Thomas Jefferson (New York: Alfred A. Knopf, 1997).
49. For example, Amory Lovins, Soft Energy Paths Toward a Durable Peace (San Francisco: Friends of the Earth, 1977), p. 24; William Ophuls, Ecology and the Politics of Scarcity (San Francisco: W.H. Freeman and Co., 1977), pp. 241–242; Mark Sagoff, The Economy of the Earth (Cambridge: Cambridge University Press, 1988), pp. 135–136; Lynton Caldwell and Kristin Shrader-Frechette, Policy for Land: Law and Ethics (Lanham, MD; Rowman and Littlefield Publishers, Inc., 1993), pp. 225–231; David W. Orr, Earth in Mind (Covelo, CA: Island Press, 1994), pp. 177 and 181.
50. A fuller discussion of Leopold, his land ethic, and the connection to the theory of moral sentiments is found in J. Baird Callicott, "The Conceptual Foundations of the Land Ethic," in A Companion to A Sand County Almanac (Madison: The University of Wisconsin Press, 1987), and in J. Baird Callicott, In Defense of the Land Ethic (Albany: State University of New York Press, 1989). See also Franklin A. Kalinowski, "Aldo Leopold as Hunter and Communitarian," in Rooted in the Land: Essays on Community and Place, ed., William Vitek and Wes Jackson.

51. In addition to the sources previously cited, see Annie Dillard, <u>Pilgrim at Tinker Creek</u> (New York: Harper's Magazine Press, 1974).
52. Daniel Kemmis' <u>Community and the Politics of Place</u> speaks eloquently to these issues. As the leading spokesman for the bioregional approach to radical environmentalism, Kikpatrick Sale's <u>Human Scale</u> and <u>Dwellers in the Land</u> acknowledge their intellectual debt to Jefferson's ward republics. See John R. Rodman, "The Liberation of Nature?" as well as any of the works by the late Paul Shepard, especially <u>The Tender Carnivore and the Sacred Game</u> (New York: Charles Scribner's Sons, 1973). A good introduction to environmental ethical corporatism (although they do not use this expression) is Herman E. Daly and John B. Cobb, Jr., <u>For the Common Good: Redirecting the Economy Toward Community, the Environment, and a Sustainable Future</u> (Boston: Beacon Press, 1989).
53. Jefferson to Jean Baptiste Say, February 1, 1804, in <u>Writings of TJ</u>, XI: 2. Jefferson to James Madison, December 20, 1787, in <u>Papers of TJ</u>, 12: 442. "First Inaugural Address," in <u>Writings of TJ</u>, III: 320.
54. Christopher Manes' <u>Green Rage</u> sees and welcomes the radical potential in Jefferson's thought. William R. Catton, <u>Overshoot: The Ecological Basis for Revolutionary Change</u> (Urbana and Chicago: University of Illinois Press, 1980) both recognizes Jefferson's contribution and criticizes the founder for his failure to address the issue of limits. See also Garrett Hardin, <u>Living Within Limits: Ecology, Economics, and Population Taboos</u> (New York and Oxford: Oxford University Press, 1993); and Joel E. Cohen, <u>How Many People Can the Earth Support?</u> (New York and London: W.W. Norton and Company, Inc., 1995).
55. A brilliant, if painful, exploration of this issue is Hannah Arendt, <u>Eichmann in Jerusalem: A Report on the Banality of Evil</u> (New York: Penguin Classics, 1994). Perhaps the ultimate perversity in our conduct toward the environment is that our alienation from Nature and our damage to the ecosystem has become so commonplace; so utterly banal.
56. At various times during his life, Jefferson owned either slightly over, or slightly under, 200 African American slaves (making him the second largest slave owner in Albemarle County). Factoring in purchases, sales, births, and deaths, it has been estimated that over the course of his life, Thomas Jefferson was the owner of approximately 400 different human beings. In his will, Jefferson freed only five of these slaves. See Paul Finkelman, "Jefferson and Slavery: 'Treason Against the Hope of the World," in Peter S. Onuf (ed.) <u>Jeffersonian Legacies</u>, 181 and 218. For other accounts of Jefferson's slave holdings, see Lucia Stanton, "'Those Who Labor for My Happiness:' Thomas Jefferson and His Slaves," in Peter S. Onuf (ed.), <u>Jeffersonian Legacies</u>, 148: Hatzenbuehler, <u>"I Tremble For My Country: Thomas Jefferson and the Virginia Gentry</u>, 18: Fawn M. Brodie, <u>Thomas</u>

Jefferson: An Intimate History (New York: W.W. Norton and Company, 1974), 89 and 465: and John Chester Miller, The Wolf By the Ears: Thomas Jefferson and Slavery (Charlottesville: University of Virginia Press, 1991), 107. DNA data have demonstrated that a member of the Jefferson family fathered at least one of Sally Hemmings' children, and historical research has established Thomas Jefferson's presence at Monticello nine months before the birth of each of Sally's children. See Thomas Jefferson Foundation, Report of the Research Committee on Thomas Jefferson and Sally Hemmings, http://www.moticello.org/site/plantation-and-slavery/thomas-jefferson-and-sally-hemmings-brief-account, January 2000.

57. A good review of the literature on both sides of this issue is Scott A. French and Edward L. Ayers, "The Strange Career of Thomas Jefferson: Race and Slavery in the American Memory, 1943–1993," in Peter S. Onuf (ed.), Jeffersonian Legacies, 418–456. In 1960, noted scholar Douglass Adair defended Jefferson against Callender's "ugly tale" and "scandalous stories." Douglass Adair, "The Jefferson Scandals," in Trevor Colbourn, ed., Fame and the Founding Fathers: Essays by Douglass Adair (New York: W.W. Norton and Company, 1974); 161, 166, and 169. Fawn M. Brodie exploded the issue when she argued that the sexual relationship between Jefferson and Hemmings not only existed but also may have been based on a sincere emotion attachment. See her Thomas Jefferson: An Intimate History, (New York: W.W. Norton $ Company, 1974). Rebuttals to Brodie are Dumas Malone, Merrill Peterson, John Chester Miller, and Gordon Wood. John Chester Miller, The Wolf By The Ears: Thomas Jefferson and Slavery (Charlottesville, VA: University of Virginia Press, 1991), 207n. Gordon S. Wood, "The Trials and Tribulations of Thomas Jefferson" in Peter S. Onuf (ed.), Jeffersonian Legacies, op. cit., 398. An accessible review of the literature on the Jefferson/Hemmings affair can be in Joseph J. Ellis, American Sphinx, Appendix, 303–307.
58. Thomas Jefferson, "The Declaration of Independence," (original draft) as found in "The Autobiography," in Merrill D. Peterson, ed., Thomas Jefferson: Writings, 22.
59. Thomas Jefferson, Notes on the State of Virginia, in Merrill D. Peterson, ed., Thomas Jefferson: Writings, 214.
60. Thomas Jefferson to Edward Coles, August 25, 1814; TJ to Jared Sparks, February 4, 1824; and Autobiography, all in Merrill D. Peterson, ed., Thomas Jefferson: Writings, 1343, 1484, and 44.
61. .Ibid., 142 and 205–207. See also Paul Finkleman, "Jefferson and Slavery: 'Treason Against the Hope of the World'," in Onuf, Jeffersonian Legacies, 188; and TJ to Edward Coles, August 25, 1814 in Merrill D. Peterson, ed., Thomas Jefferson: Writings, 1343–1346.

62. See John Chester Miller, The Wolf By the Ears: Thomas Jefferson and Slavery, especially Chapters 24, 25, and 26.
63. Ibid., 251..
64. This analysis borrows heavily from The Radical Politics of Thomas Jefferson by Richard K. Matthews, especially Chapter 4, "The Nature of Man: Red, White, and Black," Matthews' analysis unifies and penetrates to the core of Jefferson's philosophy and should be the starting point for any discussion of Jefferson, race, and ethics.
65. Thomas Jefferson, "Declaration of Independence," the original draft as cited in Autobiography, Merrill D. Peterson, ed., Thomas Jefferson: Writings, 19. Given this insight, Paul Finkelman displays his misunderstanding of Jeffersonian thought when he writes, "Jefferson could assert the equality of mankind only be excluding blacks." See Paul Finkelman, "Jefferson and Slavery: 'Treason Against the Hope of the World'," in Onuf, Jeffersonian Legacies, 185.
66. Thomas Jefferson, Notes on the State of Virginia, in Merrill D. Peterson, ed., Thomas Jefferson: Writings, Query XIV, 268–269.
67. Thomas Jefferson to John Holmes, April 22, 1820, in Merrill D. Peterson, ed., Thomas Jefferson: Writings, 1434.
68. Thomas Jefferson, Notes on the State of Virginia, in Merrill D. Peterson, ed., Thomas Jefferson: Writings, Query XVIII, 288.
69. Ibid.

CHAPTER 7

The Environmental Legacy of Alexander Hamilton: Manufacturing Power from Delusion

In 1790, Great Britain and Spain nearly went to war over their possessions in North America. If war developed, it was very likely Britain would use its colonies in Canada as a launching area for attacks against the Spanish in Florida and New Orleans. The most obvious route of assault would be down the Mississippi River, but that would threaten to draw the newly independent United States into the conflict. Even at this early stage of statehood, many Americans saw the western frontier with its vast space and seemingly endless resources as the key to their future freedom and happiness. Free navigation of the Mississippi was essential to various visions of America's relationship to its environment. As the crisis between Spain and Britain deepened, President George Washington asked his cabinet members for written opinions regarding possible American policy. On September 15, 1790, Secretary of the Treasury Alexander Hamilton submitted his judgment to Washington.

Since the anticipated conflict never materialized, and since the British never asked permission to cross American territory, Hamilton's actual position is not that historically significant (in fact, he advised Washington to grant the permission if Britain requested it).

What is important, however, is that during his long and carefully reasoned paper on international law and American foreign policy, Hamilton chose a certain rhetorical image that gives a noteworthy clue regarding his perception of the relationship between America and its western frontier. In his report, Hamilton casually speculates, "Let it be supposed that

Spain retains her possessions on our right ..." The former colonel in the Revolutionary army continues the military practice of viewing the flanks of his position as left and right in the next paragraph, "In regard to the possessions of Great Britain on our left ..."[1] From where Alexander Hamilton philosophically stood, Florida was on America's right, and Canada was on our left. Hamilton's vision was not focused toward the west but rather was aimed eastward—back to Europe.

Unlike Thomas Jefferson who looked to the American frontier as a source of hope for "vacant lands" where citizens might build new communities, or James Madison, who saw the vast western lands as a place where factions might proliferate, fragment, and neutralize each other, Hamilton's attention was turned toward Europe and to those nations who treated the young republic with contempt, condescension, or pity. Hamilton wanted a mighty and powerful country that could resist the "arrogant pretensions of the Europeans." In the eleventh number of The Federalist Papers, Hamilton voiced his resentment at European attitudes toward the young American republic and outlined a course of action. "Europe, by her arms and by her negotiations, by force and by fraud, has in different degrees extended her dominion ..." around the globe. "The superiority she has long maintained," he bitterly observes, "has tempted her to plume herself as the mistress of the world, and to consider the rest of mankind as created for her benefit." Hamilton then forcefully proclaims his aspiration for our new nation. "Let Americans disdain to be the instruments of European greatness! Let the thirteen States, bound together in a strict and indissoluble Union, concur in erecting one great American system superior to the control of all transatlantic force or influence and able to dictate the terms of connection between the old and the new world!" A militarily powerful and economically rich United States would stand up to the conceited and patronizing states of Europe and "teach that assuming brother moderation."[2]

The political legacy of Alexander Hamilton lives on in an American culture that judges itself not by its virtue or even by its political stability but by the amount of power we wield, and that power, in turn, has traditionally been evaluated in comparison to the strength of the nations of Europe. In order to convert America into a global juggernaut, Hamilton believed the nation would have to be altered from a predominantly agricultural economy to one focused on manufacturing and commerce. Decentralized, agrarian democracies could never be central actors on the world stage. The problem for Hamilton was his conviction that given free choice, individuals would almost invariably prefer a simple life as yeomen to the drudgeries inherent in the existence as industrial workers. Hence, Hamilton had to channel humanity's normal inclinations and socialize them to new behavior patterns. A different economic and social system needed to be created, one based not on loosely connected self-sufficient farming communities but, instead, a national economy founded on finance capital, industrialization, trade, and continuous exponential economic growth. To accomplish this end, labor would convert Nature into property, land would be converted into capital, and farmers would be converted into workers. In the process, Hamilton hoped the passions and vanity of people could be accelerated and channeled into the endless consumption of consumer goods. An ever-increasing cycle of debt, consumption, and work would provide the engine for an ever-enlarging source of taxes, spending, and government power. A small, insignificant collection of former colonies would become the object of fear and envy for the entire world. Its leaders would become eternally renowned for their accomplishments, and their sometimes-virtuous citizens would be transformed into domesticated, sometimes-happy consumers.

Alexander Hamilton's environmental legacy lives today in the dreams of cornucopians and Website who deny the ecological limits placed on economic expansion, who advocate the unending domination of Nature through technology and industrial expansion, and who delude themselves into believing that a nation can be powerful when it is populated merely with happy consumers instead of virtuous citizens. His environmental legacy exists also in the dreams of Progressive liberals who cling to the hope that massive consumption can be made "sustainable" and compatible with environmental protection if properly managed by a powerful and efficient national government. Since many contemporary Americans who style themselves "conservatives" are sympathetic to cornucopian/Promethean environmental thought, and similarly, many self-proclaimed "liberals" support a version of "sustainable development" that embraces the oxymoronic belief in infinite growth, it should come as no surprise that Hamilton is presently experiencing a strong level of approval from scholars on both sides of the mainstream political discourse.[3] How one reacts to the political and environmental legacy of Alexander Hamilton will depend upon the values the observer brings to an investigation of his thought, but what he believed, and what he left embedded in American culture can be determined with some objectivity through a careful analysis of his work. It must be acknowledged, however, there are no neutral commentators on the life and thought of Alexander Hamilton. In death, as in his life, Hamilton's work has sparked both acrimony and panegyric.[4] In order to understand fully the philosophy of Alexander Hamilton, we may have to bore down to a level of analysis that is basically ecological in its understanding of Nature and human nature: an understanding that holds together Hamilton's political and his economic thought. At that level, Hamilton's political theory and economic policies may be seen simply as superstructure to an even more vital environmental perspective. If the subsequent interpretation is no less biased than previous studies, it does claim the grace of being deeper and more penetrating.

The Thought of Alexander Hamilton

This chapter began with an allusion to "clues" that may be found in the writings of Alexander Hamilton. The word was well chosen, for an analysis of Hamilton's thought must ultimately be based, not on any explicit, logical, theoretical argument he enunciated but rather on suggestions, patterns, and sources that can be found scattered throughout his work and that need to be patched together to form an overarching picture of his beliefs. Analyzing Hamilton's very considerable contribution to American culture resembles more an investigation of a mystery than a formulaic academic exercise. The fault for this situation lies with Hamilton himself.

Alexander Hamilton was one of the most prolific authors among a generation noted for the quantity of its written output. Although he died at the relatively young age of 47, his collected works encompass 26 volumes.[5] The object of his essays, reports, editorials, published speeches, and letters was almost always persuasion, however. The clear explication of theoretical premises was of secondary importance. He was a propagandist and a policy advocate, first and foremost, and although he no doubt studied political philosophy and used the ideas of others in shaping his vision of the world, he was not personally interested in the actual explication of ideas. He never wrote a book or an extended treatise. His skepticism regarding theory is evident in a speech he gave at the New York state ratifying convention, "In my reasoning on the subject of government, I rely more on the interests and opinions of men, than upon any speculative parchment provisions whatever ... I am therefore disposed not to dwell long on curious speculations, or pay much attention to modes and forms, but to adopt a system whose principles have been sanctioned by experience ..." And yet, when Hamilton wrote to Delaware Representative James A. Bayard encouraging him to switch his presidential vote to Thomas Jefferson in the contested election of 1800, Hamilton's decision was based on the opinion that Jefferson was the lesser evil compared to Aaron Burr, who Hamilton despised for having no core values beyond the pursuit of personal ambition. Hamilton rhetorically asked Bayard, "is it a recommendation to have *no theory?* Can that man be a systematic or able statesman who has none? I believe not. *No general principles* will hardly work much better than erroneous ones."[6] Hamilton believed in the necessity of theory, but disdained a lengthy articulation of his own principles, preferring to rely on inductive experience rather than a rationally deduced plan. He was, in that sense, a typical representative of eighteenth-century empiricism and an even more archetypical example of Americans' penchant for concealing their philosophical predispositions under the name of "practicality."[7] As Hamilton admits, he recognizes the importance of having a political theory: as Hamilton hints, finding his will take some investigating.

By the time Hamilton began his post-war political career, his attitude on Americans was firmly established and would never waiver.[8] During his adult life, his social, economic, and political thought is rooted in his profound pessimism regarding the patriotism and virtue of the American people. Hamilton's feelings can be summarized by his derisive assertion in a letter to his friend and fellow officer, John Laurens, "Our countrymen have all the folly of the ass and all the passiveness of the sheep in their compositions. They are determined not to be free and they can neither be frightened, discouraged nor persuaded to change their resolution." It is interesting to note that Hamilton's rejection of virtue is built upon the incompatibility of self-sacrifice and commercial society. Given the alternatives of America as a collection of virtuous, agrarian republics or America as a unified, capitalistic, world power, Hamilton was committed to the latter course. In 1779, he wrote Laurens, "there is no virtue in America—that commerce which presided over the birth and education of these states has fitted their inhabitants for the chain, and the only condition they sincerely desire is that it may be a golden one." Even before the end of Revolutionary War, resentment, mistrust, and cynicism became the foundation of his view of the world. In 1780, he wrote to Laurens, "I hate Congress—I hate the army—I hate the world—I hate myself."[9] How could a man who had such contempt for his fellow countrymen fight in a revolution to bring them independence? A clue may be found in the following quotation. "The same state of the passions which fits the multitude, who have not a sufficient stock of reason and knowledge to guide them, for opposition to tyranny and oppression, very naturally leads them to a contempt and disregard of all authority," Hamilton states. "In such tempestuous times, it requires the greatest skill in the political pilots to keep men steady and within proper bounds, on which account I am always more or less alarmed at every thing which is done of mere will and pleasure, without any proper authority."[10] Hamilton consistently believed men with superior experience and knowledge were separate from the unthinking "multitude" and, therefore, enjoyed a type of historical and natural "right" that the general population could neither recognize nor possess. Revolution was necessary to separate the United States from Britain and install American leaders who would then guide, discipline, and domesticate the unruly and unthinking populace of this nation. These foundational beliefs in the unconquerable passions of the common people, the necessity of a far-sighted elite, and the requirement for politics to act as a tool for political socialization place Hamilton squarely in the philosophical tradition of Scottish, interest-group liberalism. Appreciating Hamilton's attachment to this school and its central exponent, David Hume, brings clarity and consistency to all of his thought.

There was no thinker that Alexander Hamilton cited more often than David Hume. During the Constitutional Convention of 1787, Hamilton used Hume, whom he called "one of the ablest politicians," to support basing government on passions rather than virtue, and his closing comments as Publius are praise for "a writer equally solid and ingenious," followed by a footnote to Hume's essay "Of the Rise and Progress of the Arts and Sciences."[11] Elsewhere, Hamilton referred to "the cautious and accurate Hume" and "the Judicious Hume." Virtually, every commentator on Hamilton's political thought agrees that, "Hamilton's master was David Hume."[12] This is particularly significant given the fact that by the writing of the Constitution, Hume was so out of favor in America that citing him was generally prohibited by the decorum of the day. His History of England, although widely read in both Britain and the colonies, had committed heresy against Whig orthodoxy by supporting the claims of the Stuarts against the rising power of Parliament; he was an outspoken atheist; he had refused to come to the aid of his friend and publisher, Benjamin Franklin, when the American was humiliated before the British Privy Council; and although he was often sympathetic to the cause of American independence, when radical activity actually began, Hume backed off his support. Not every American was as vehement as Thomas Jefferson, who referred to Hume as, "the great apostle of toryism," and "this degenerate son of science, this traitor to his fellow men," but still, in

America at the close of the eighteenth century, Hume was a philosopher more often studied and used than named.[13] Along with possibly Jean-Jacques Rousseau, Hume was read by almost all the educated upper class in early America but quoted by almost no one. Unlike James Madison who promulgated an unacknowledged version of Hume's thought, Hamilton defied convention by his open avowal of Hume's influence. The fact that Hamilton actually identifies Hume as his mentor indicates the depth of his admiration for, and utilization of, Hume's thought.[14]

Like all adherents to interest-group liberalism, Hamilton begins with the assumption that humans are motivated almost exclusively by individual self-interest. The supremacy of interests over reason or virtue stems from the premise that in their most basic makeup, humans are sentient creatures driven by passions and a desire to seek pleasure and avoid pain. In that narrow sense, Hamilton, Hume, and others recognize a certain degree of human equality, but it was a sameness devoid of dignity. Hamilton's cynicism and scorn for his fellow humans cannot be explained, or explained away, by references to personal psychology or idiosyncratic historical conditions. It was philosophical in its roots, and to the extent that it was shared, the infection has permeated generations of interest-group liberals. All humans are fundamentally equal because all humans are degraded to a condition of emotional, unthinking sheep. In their most basic makeup, there is no difference between humans and animals: they deserve to be treated the same. The story is told that during a dinner party in New York, one of the guests was extolling the virtues of the Americans, enthusiastically predicting a happy future for the country, and declaring himself an earnest "friend of the people." Reportedly, Hamilton, frustrated and showing a good deal of emotion himself, pounded his fist on the table and yelled, "Your people, sir—your people is a great beast!" The incident reflects Hamilton's general attitude, "Experience is a continual comment on the worthlessness of the human race. I know few men estimable, fewer amiable."[15] More than simply a contemptuous pessimism or a twisted offshoot of Calvinism's belief in the "natural depravity of mankind," the certainty that self-interested emotion is the primary motivator of all conduct was used to construct an elaborate theoretical edifice.[16] Like many contemporary environmental thinkers, interest-group liberals stress the similarities between humans and other animals, but in direct opposition to environmental thought, liberals assert that all species must be dominated and domesticated. Humans, animals, and all of nature were "worthless" until they had been developed and improved. Building on this view, Hume, Adam Smith, and other liberals designed a plan, which Hamilton implemented, to redirect and reconstruct the world around them. This helps explain Hamilton's odd role in the writing and adoption of the Constitution, and it also clarifies the relationship between interest-group liberalism and the pluralist theory of power.

On the surface, Hamilton's attitude toward the Constitution seems paradoxical. No one, not even James Madison, did more to get the Constitutional Convention of 1787 called and to have the product of that Convention ratified and put into practice. Yet, during the Convention itself, Hamilton participated very little, and, indeed, he was very seldom in attendance. When Madison arranged the Annapolis Convention in September of 1786, ostensibly to discuss interstate commercial agreements, Hamilton rushed to the Maryland capital and, before all the delegates could arrive, declared the meeting a failure and offered a resolution for delegates to convene in Philadelphia for the much broader purpose of revising the Articles of Confederation. His apparent fear was that a system of interstate trade agreements, under the aegis of the Articles of Confederation, would prove successful and thereby, undercut efforts to junk the Articles in favor of a system with a much stronger central government. After the Philadelphia Convention, Hamilton worked tirelessly to have the proposed document ratified. He was the guiding force behind The Federalist Papers, and his powerful advocacy at the Poughkeepsie ratifying convention helped reverse a popular vote in New York that was 16,000 to 7000 against ratification.[17]

Yet, during the actual writing of the Constitution, Hamilton's role was minimal, and his attitude toward its various provisions seemed to fluctuate between indifference and hostility. During the first three weeks of the Convention, Hamilton participated little. On June 18, he made his only significant contribution to the discussion; a six-hour speech outlining his considerably different view of American politics and political institutions. On June 29, he left Philadelphia and did not return until September 6. On September 17, as the delegates were preparing to sign the finished document, Hamilton honestly proclaimed, "No man's ideas were more remote from the plan than his own were known to be;" and 15 years later, in a letter to his good friend Gouverneur Morris, he referred to the Constitution as "the frail and worthless fabric."[18] When the convention adjourned on September 17, 1787, Hamilton had attended less than half the meetings. His formidable effort to get the Constitution put into place and his halfhearted regard for its actual content indicate Hamilton was more concerned that *some* powerful central government be installed, rather than with the actual form of that government or the restrictions placed upon its powers.

His position on Constitutional authority, the role played by societal pluralism, and the deeper implications of Humean liberalism can be clarified by distinguishing between horizontal and vertical "axes of power." Within American Constitutional theory, along the "horizontal axis of power," power needs to be separated, checked, and balanced in order to prevent one interest from becoming the majority and oppressing competing interests. The assumption here is that the various interests share the same, low ethical stand; they are all "adverse to the rights of other citizens, or to the permanent and aggregate interests of the community" (as Madison says in Federalist 10); and the only solution, therefore, is a "policy of supplying, by opposite and rival interests, the defect of better motives" (as Madison states in Number 51).[19] This is the rationalization for a pluralist distribution of power. It rests on the assumption that the "interests" being discussed are all private interests whose adherents are motivated exclusively by the violent passions.[20]

Hamilton takes a different position on the distribution of power. His emphasis is not on the prevention of oppression but on the promotion of what he sees as the public good, and for him, this necessitates restructuring power on a "vertical axis" with the national government exercising dominance over both the state governments and over individual citizens. The emphasis is on using, not restraining, power. The talk is not about checks, ambition counteracting ambition, or separation of powers, but rather, Hamilton's focus is on authority in government, the strength to do what is necessary and proper, the equation of a powerful executive with an efficient regime, and the implication that the efficient execution of the law is an essential ingredient for the socially worthwhile execution of the law. "A feeble executive implies a feeble execution of the government," Hamilton declares in Federalist 70. "A feeble execution is but another phrase for a bad execution; and a government ill executed, whatever it may be in theory, must be, in practice, a bad government."[21] Madison's horizontal axis assumes the equivalence of short-range private interests, while Hamilton's vertical axis is premised on the social superiority of calm passions and the public interest. It also assumes that those in authority at the top of the vertical axis know what the public interest is and how to go about achieving it. Madison's horizontal axis of power builds on pluralism and assumes the ethical relativity of interests. Hamilton's focus on vertical power assumes the ethical superiority of those in elite positions of political power.

For Hamilton, as long as the Constitution allowed a powerful executive branch to pursue his vision of the common good, and a powerful Court had the prerogative to review the actions of Congress and the state governments, he was satisfied with the document. If it served the additional function of permitting one private interest to check another, Hamilton was willing to go along with the arrangement, especially if the arrangement procured the political support necessary for ratification. For him, the specific provisions of the Constitution were much less significant than the overall shape of the political and economic system being adopted. In <u>Federalist</u> 83, Hamilton expressed his fear of "the danger of encumbering the government with any constitutional provisions the propriety of which is not indisputable," and frankly declared his attitude. "The truth is that the general GENIUS of a government is all that can be substantially relied upon for permanent effects. Particular provisions, though not altogether useless, have far less virtue and efficacy than are commonly ascribed to them, and the want of them will never be with men of sound discernment a decisive objection to any plan ..."[22] Of course, when Madison later applied his system of checks and balances to the different branches of the Federal authority, the real disjunction between the competing theories would become evident, and the two men would clash politically. The underpinnings and aims of Hamilton's philosophy can now be brought into focus. It is hard to believe he ever had much of a commitment to democracy or rights-based social contract theory. To the extent he expressed any sympathy at all with these positions, it should be taken as a mere rhetorical flirtation of his youth. His real foundation rested upon the unwavering conviction that humans could never be reasoned or cajoled out of their lethargic devotion to their private, selfish interests and irrational passions. Building a political society on reason, consent, or the enlightened self-interest of the masses was, as he had expressed to Laurens, "an idle dream."[23]

One analyst has stated, "Hamilton was a romantic to the core," and although romanticism is a term not usually associated with finance ministers, in one sense this is correct.[24] Hamilton thought humans living in a "state of nature" would never rationally or voluntarily consent to leave. He was probably closer to Rousseau than to Locke in this view. Contrary to Locke's thinking, Hamilton assumed humans would lack the reason necessary to picture life in civilized, industrial societies; they would not be capable of weighing the costs and benefits of their primitive but simple lifestyle with the developed, yet complicated existence of commercial society; and there would not be an inducement strong enough to break generation upon generation of habit in order to strike out into the unknown. No matter how bleak their existence (and this is open to much dispute), humans would not create a social contract based on mutual consent and leave the state of nature. The process of becoming civilized would be one of forcing the unwilling to change to the unwanted. For Hamilton, coercion, and not consent, was the basis of all social development.

To the extent the theoretical "state of nature" could be equated with the anthropological condition of hunting and gathering societies, Hamilton was convinced humans would never voluntarily choose to leave such a world. Perhaps his experiences with native peoples in the Caribbean and America convinced him of this. Perhaps he adopted Madison's view, "It must not be inferred ... that the transition from the Hunter or even the Herdsman state to the Agricultural, is a matter of course. The first steps in this transition are attended with difficulty; and what is more with disinclination."[25] If humans were to quit their natural condition and move toward private property, agriculture, and government, they would have to be coerced into it. From the perspective of the eighteenth-century American Enlightenment, alienating humans from Nature was both necessary and difficult. The species would have to be forced against their will to become civilized, and since the process of distancing humans from Nature was what it meant to become "civilized," they would have to be trained through the application of pain and pleasure to adopt the behaviors of residents of civil society. Both ecological Nature and human nature needed to be dominated and domesticated. David Hume had written, "Almost all the governments, which exist at present, or of which there remains any record in story, have been founded originally, either on usurpation or conquest, or both, without any pretence of a fair consent, or voluntary subjection on the people."[26] Hamilton took that premise and applied it the creation and structure of the American government. He took Madison's assertion that humans do not voluntarily move from hunting societies to agricultural societies and extrapolated it to the movement from agriculture to capitalistic "commercial" society. In fact, from Hamilton's perspective, every significant change in the human condition was accomplished against the languor and resistance of the masses. Altering the behavior of an entire nation for all future generations would require a government with vast powers and an administration with distant vision.

His clearest explanation of the methods to be used came during his protracted speech at the Constitutional Convention of 1787. After the customary self-effacing introduction, Hamilton launched into a long, but carefully prepared, defense of the five "principles of civil obedience" or "supports of government" as he saw them.[27] Incorporating much of the thought of Hume, with some novel applications of his own, Hamilton defended his belief that these "great and essential principles necessary for the support of Government" are interest, opinion, habit, force, and influence. By "interest," Hamilton meant the short-range, private interests of citizens. In other words, government would have to offer its citizens some tangible or psychic reward for supporting its actions. When he used the term "opinion," Hamilton seemed to have in mind Hume's essay, "Of the First Principles of Government," where the Scot said, "as force is always on the side of the governed, the governors have nothing to support them but opinion. It is on opinion only that government is founded …"[28] In this context, the term "opinion" should be taken to mean the general feeling on the part of the masses that government serves a useful function and that they are probably better off with it than without it.

Nowhere does Hamilton reveal his allegiance to the Scottish school and David Hume more than in his focus on "habit" as an essential underpinning of the state. Hume had made habit, or custom, the benchmark for much of his philosophy. Once pain, pleasure, and experience replaced rational thought as the defining characteristics of human conduct, the distinctions between humans and animals melted, and custom became the means for manipulating the conduct of each. When dealing with sentient creatures, it was possible to vary their experiences, vary or alter the linkages they made between their past behavior and the pain or pleasure that had been inflicted upon them, and, therefore, instill within them different habits or "customary conjunctions."[29] Like domesticated animals, people could be civilized, not because they were rational but because their wild, free inclinations could be beat out of them. Hume and Hamilton replaced the image of humans as creations of God with a picture of all beings driven by passion as simply creatures of habit. For Rene Descartes' famous dictum, "Cogito Ergo Sum," they, in effect, substituted a new maxim, "I feel, therefore, I'm trainable."

These thoughts clearly captured Hamilton's attention. After listing "habit" as the third of his principles, Hamilton discussed "force" as an essential support. He divided force into "coercion of law" and "coercion of arms," but it is obvious these apparently different elements were connected in his mind. Law and "arms" (meaning the potential for the exercise of violence by the police and the military) were the means by which Hamilton hoped to instill and reinforce habits in the citizenry. During his discussion of "force," Hamilton reveals an interesting sidelight to this line of reasoning. He argues "A certain portion of military force is absolutely necessary in large communities."[30] Rather than seeing a large republic as a method for averting the use of force (as Hume and Madison hoped), Hamilton believed that a vast nation would require the creation of a permanent, standing military.

Hamilton's final "principle of civil obedience" caused him problems when he enunciated it, for in stating his belief that "influence" was a support of government, he came very close to endorsing corruption as a prop to political stability. Hume's defense of corruption, or patronage, in the British system had earned him the enduring enmity of American republicans. Now, Hamilton was apparently following Hume's lead when he said to the Convention that "a certain dispensation of those regular honors & emoluments, which produce an attachment to the Govt." was part of any enduring system. Although he was quick to point out "that he did not mean corruption," observers then, and ever since, have taken his statement to imply the wish to attach the interests of the rich to the fortunes of the state.[31]

Throughout this elaboration of his five principles, Hamilton stressed that the state governments enjoyed the support of the principle more than the national government, and that neither the Virginia nor the New Jersey Plan provided sufficient power, or "energy," to the proposed Federal government. Hamilton's conclusions must have shocked many of the delegates. He recommended that the existing states be abolished, that the substitute governments be made administrative appendages of the national government, that the President be elected for life and have an absolute veto over all laws, that Senators serve "during good behavior," that all state laws "contrary to the Constitution or laws of the United States be utterly void," and that the governors of the various states be appointed by the national government.[32] Small wonder that three days later, Connecticut representative William Samuel Johnson observed that Hamilton had "been praised by every body" and "supported by none."[33]

The major fiscal recommendation made during Hamilton's Convention speech pointed the way toward his future policy. He rhetorically asked the delegates to consider how the future government of the nation was to secure its funding. Immediately, he rejected the notion that taxes could be apportioned among population, and he likewise scoffed at the widely held belief among many physiocrats that land was the primary source of wealth. If the future Federal government was to have revenue for its military, domestic, and international purposes, Hamilton made clear his conviction there was only one source adequate to the task. America would have to become a commercial society. "Whence then is the national revenue to be drawn? From Commerce, even from exports which notwithstanding the common opinion are fit objects of moderate taxation, from excise, &c &c. These tho' not equal, are less unequal than quotas."[34] A capitalist economy, or a "commercial society" as it was known then, would provide the revenue necessary to support a strong government, and a strong government would provide the institutions of social manipulation necessary to produce the habits and values of a capitalistic culture.

In order to appreciate fully the scope of Hamilton's political and economic philosophy, it is essential to comprehend both the ends he sought and the means he designed to achieve his goals. A great amount of understanding is lost if Hamilton's financial plan is seen merely as a means of acquiring revenue. The most insightful of contemporary commentators recognize Hamilton's objective was to convert his countrymen into "free, opulent, and law-abiding citizens, whether they liked it or not ..." He aimed at, "the task of making the citizens in every regard more well-behaved, healthier, wiser, richer, and more secure." Procuring money to run the government was only part of his design; and it was not even the most significant part. "The British system was designed solely as a means of raising revenue for purposes of government ... Hamilton's system was designed to employ financial means to achieve political, economic, and social ends; and that made all the difference in the world."[35] Hamilton's task was the creation of an economic system that could be used to shape and direct human behavior; a system that would force the social movement from a predominantly agricultural society to one with a mix of agriculture and manufacturing and finally to a modern industrial state. Economic policy was his means; taming the violent passions of Americans and channeling their self-interests in ways that served the power of the government was his object.

The overarching scheme can be viewed as an attempt to have illusion conquer reality. In a letter to an unknown member of Congress, Hamilton spelled out the central financial problem facing the nation: if all the money in America was procured by the Federal government, there would still be insufficient funds to pay all the debts owed by the Continental Congress.[36]

The government could collect all the gold, silver, and paper currency in the country, and there would not be enough to pay the debts of the state and national governments. The stark reality was that the country was bankrupt, Congressional attempts to correct the situation by issuing paper money had predictably led to runaway inflation, and the value of the Continental currency was, therefore, reduced to nearly nothing. In sum, there was not enough money, and the money that did exist was virtually worthless. Some policy was needed that would increase the amount of paper money in circulation and, at the same time, keep its value stable and high. In this early sketch of his thought, Hamilton put forth two suggestions. A national bank needed to be created, and the funds issued from the bank needed to be denominated in pounds sterling. Calling the currency "pounds" instead of "dollars" would create the illusion that there was some connection between the American money and the funds of the nation we were fighting, and since British money was highly valued, perhaps this false association could raise the worth of American currency. Hamilton frankly admitted the deception involved. "The denomination of the money is altered because it will produce an useful illusion. Mankind are much led by sounds and appearances; and the currency having changed its name will seem to have changed its nature." Hamilton understood this was all delusion, but since he viewed Americans as driven by passion rather than knowledge, he felt the ruse might work. His financial system would use illusion to manipulate the masses whose opinions (unfortunately for Hamilton) must be taken into account, but whose gullibility made them tempting targets for duplicity. "A great source of error in disquisitions of this nature is the judging of events by abstract calculations, which though geometrically true are false as they relate to the concerns of beings governed more by passion and prejudice than by an enlightened sense of their interests. A degree of illusion mixes itself in all the affairs of society." Hamilton believed that in the valuation of the national currency, "Opinion will operate here also; and a thousand circumstances may promote or counteract the principle."[37] He recognized that a significant part of inflation is controlled by the confidence, or opinion, of consumers, and, therefore, manipulating that public trust was one method of controlling inflation.

In September of 1780, Hamilton wrote to James Duane, member of the Continental Congress from New York and ardent nationalist, who had solicited Hamilton's advice on the nation's problems. In this letter, Hamilton focused on the institutional changes that needed to be made.

He described the crisis simply: "the fundamental defect is a want of power in Congress" and attributed this lack of authority to three causes; "an excess of the spirit of liberty," "a diffidence in Congress of their own powers," and "a want of sufficient means at their disposal to answer the public exigencies."[38] Foreshadowing the Federalist position on the necessary and proper clause, Hamilton warned Duane against not acting merely because Congress lacked the authority to act. Legal niceties should not be permitted to stand in the way. "It may be pleaded, that Congress had never any definitive powers granted them and of course could exercise none ..." He went on to give an early expression to the theory of broad, or loose, construction of governmental authority. "Undefined powers are discretionary powers, limited only by the object for which they were given—in the present case, the independence and freedom of America."[39] If the Articles of Confederation did not bestow the power that Hamilton and Duane thought was necessary, Congress should simply assume it. After reviewing the sad condition of the military and renewing his plea for a national bank, Hamilton closed with another reference to opinion and artifice. "The manner in which a thing is done has more influence than is commonly imagined. Men are governed by opinion; this opinion is as much influenced by appearances as by realities ..."[40]

One of the clearest early expressions of Hamilton's plan and its purposes came in a revealing letter to Robert Morris dated April 30, 1781. Written only months before Hamilton won military fame at the Battle of Yorktown; the letter was Hamilton's attempt to articulate his financial plans to the newly appointed Superintendent of Finance. Although the final details would not be filled in until Hamilton became Secretary of the Treasury under the new government in 1789, the letter to Morris contained most of the fundamental details. The key to it all was to increase dramatically the amount of money in circulation and to convince foreign lenders, wealthy Americans, and the public at large that this newly printed paper money had real, tangible worth. In fact, the essence of Hamilton's financial scheme was the creation of the illusion that the value of cash was *more* real, more authentic, and more substantial than the mere objects of Nature such as agricultural commodities, homes, or land.

There is a fine connotative line between the words "illusion" and "delusion," with the first term suggesting the false perception of something that does have objective existence, while "delusion" is properly used to mean the creation of beliefs that have absolutely no foundation in the real word. Alexander Hamilton produced a system where political and economic power would be created out of nothing; out of sheer delusion. Referring to Hamilton's 1790 plan to monetize the public debt, Hamilton biographer Forrest McDonald admits, "Hamilton brought about funding, assumption, and the sinking fund, which transformed the paper into a form of capital ... Thus $30 million in liquid capital had been manufactured, as it were, out of thin air. Indeed, the new capital was of stuff even more ephemeral: all that happened was that the public, instilled with illusions and expectations, changed its opinion about the value of those pieces of paper."[41]

From Hume's basic argument that facts were merely customary perceptions, to his faith that average humans could not be educated away from their pursuit of self-interest, to the belief that opinions were the foundation of society, to the conclusion that whoever created habits created reality, Hamilton linked together an economic and political plan to manufacture power from delusion, fabricating what might be the first example of American crackpot realism. The essential components of Hamilton's plan were the creation of a huge public debt through assumption, increasing the cash in circulation by monetizing the debt, creating a national bank to issue securities and manage interest rates, and building investor confidence through the establishment of a sinking fund to pay down, but not eliminate, both the assumed debt and the subscription fees used to establish the bank. Each element was controversial, but each was critical for the functioning of the entire proposal. The elimination of any one would render the entire project unworkable, and although difficult initially to put into practice, once instituted, it would be almost beyond the realm of practicality to fundamentally alter the system. Hamilton himself acknowledged the power of institutional arrangements—particularly financial arrangements—to trap people into a system from which it was difficult, or impossible, to extricate themselves. In a 1790 newspaper article in the Gazette of the United States, he wrote, "Whoever considers the nature of our government with discernment will see that though obstacles and delays will frequently stand in the way of the adoption of good measures, yet when once adopted, they are likely to be stable and permanent. It will be far more difficult to *undo* than to *do*."[42]

At first blush, it seems the last thing the national government needed was to take over, or assume, the debts of the various state governments. It had plenty of debt of its own. Beyond the economic problems caused by debt, there was a considerable history in republican thought that associated debt with political dependence, political dependence with corruption, corruption with the decline of civic virtue, and the loss of civic virtue with the death of liberty. British opposition leaders such as Viscount Bolingbroke, and John Trenchard and Thomas Gordon in their series, <u>Cato's Letters</u>, had lambasted the ministry of Robert Walpole for just this sort of fiscal manipulation.[43] Early liberal thinkers also had misgivings regarding public encumbrances. In the final chapter of his <u>Wealth of Nations</u>, Adam Smith expressed the view that a public debt was an evil side effect of modern, commercial societies where wealthy citizens are eager to lend the government money at interest, and the government is eager to use long-term borrowing as an alternative to the more politically painful method of raising unpopular taxes. David Hume was even more emphatic. In his essay, "Of Public Credit," Hume labeled public debt "a practice which appears ruinous, beyond all controversy ..." Comparing modern commercial societies and their debts with the simple but frugal republics of antiquity, Hume declared, "the ancient maxims are, in this respect, more prudent than the modern ... If the funds of the former be greater, its necessary expenses are proportionably larger; if its resources be more numerous, they are not infinite ..." Hume concluded, "either the nation must destroy public credit, or public credit will destroy the nation."[44] The wish to avoid overextending our credit and the fear of going too deeply into debt are engrained elements of our cultural values.

But Hume also left the door open a crack on the issue of public credit and debt, acknowledging, "our national debts furnish merchants with a species of money, that is continually multiplying in their hands, and produces sure gain, besides the profits of their commerce. This must enable them to trade upon less profit. The small profit of the merchant renders the commodity cheaper, causes a greater consumption, quickens the labour of the common people, and helps to spread arts and industry throughout the whole society." What Hume was saying, in short, was that under certain circumstances, public debt stimulates economic growth. "More men, therefore, with large stocks and incomes, may naturally be supposed to continue in trade, where there are public debts; and this, it must be owned, is of some advantage to commerce, by diminishing its profits, promoting circulation, and encouraging industry."[45]

It was this idea that Hamilton seized upon. America needed more cash in circulation; the national bank would provide it by facilitating public and private borrowing; and, therefore, the bigger the debt, the more money could be pumped into the economy. In an incredible accounting switch, the public debt would be converted from a liability into an asset—debt would be redefined as credit. The line from Hume that must have really intrigued Hamilton, however, was when the Scot said that debt "causes a greater consumption" and "quickens the labour of the common people." This was the link between financial policy and social control Hamilton was looking for. The proposed government could issue public securities, the bank would accept these securities as collateral for loans of cash, there would be more money in circulation, but the value of the newly printed paper would stay high if people worked much harder and stimulated a strong demand for cash. The critical element in spectacularly increasing the workload of the masses was to convince people to use their increased wages to purchase more and more consumable products. Hamilton's system of public debt would work if it was matched with an exponentially increasing private debt, and if both could be paid down by dramatically increasing labor as Americans shifted from being farmers to being industrial workers. Hume's admonition that, in modern economies, "if its resources be more numerous, they are not infinite," would have to be ignored.

In his letter to Morris, Hamilton revealed his recognition of these connections. He began by reviewing the finances of several European countries in an attempt to estimate the maximum taxing power of a government. Citing David Hume yet again, Hamilton estimated that the total cash in circulation in Great Britain was approximately 40 million pounds sterling, and the annual domestic revenue of the British government at that time was 10 million pounds. He concluded that a fairly representative government could impose taxes each year that acquire about one-fourth of the cash in circulation.[46] Stated simply, Hamilton's plan was to determine the amount of money the government needed, put four times that amount into circulation, and impose taxes that skim the necessary funds. "When I say that one fourth part of its stock of Wealth is the revenue which a nation is capable of affording to Government," he suggests, "I must be understood in a qualified, not in an absolute sense. It would be presumptuous to fix a precise boundary to the ingenuity of financiers, or to the patience of the people ... This suffices for a Standard to us, and we may proceed to the application."[47]

The next step in Hamilton's logic was to connect circulating cash with labor potential and these with credit. He expressed to Morris his belief that "our current cash is not a competent representative of the labour and commodities of the Country" and then proposed as a solution to "erect a mass of credit that will supply the defect of monied capitals and answer all the purposes of cash ..." The linchpin that held these financial components together was a national bank "The tendency of a national bank is to increase public and private credit. The former gives power to the state for the protection of its rights and interests, and the latter facilitates and extends the operation of commerce among individuals."[48] The bank would help the nation and the people go into debt, and making payments on the outstanding balance would necessitate that the people work harder and harder. Hamilton concluded his letter to Morris by driving home the connection between government finances and social manipulation. He took Hume's digression regarding the possible benefits of debt and made it the centerpiece of his scheme. "A national debt if it is not excessive will be to us a national blessing; it will be a powerful cement of our union. It will also create a necessity for keeping up taxation to a degree which without being oppressive, will be a spur to industry ..." Then Hamilton declared the real purpose of his plan. Americans did not work hard enough. If we were to match the power of the industrial states of Europe, our people must be driven to increase their efforts. Besides, Hamilton ended; hard work is good for the masses. "We labour less now than any civilized nation of Europe, and a habit of labour in the people is as essential to the health and vigor of their minds and bodies as it is conducive to the welfare of the State."[49] If the United States was to become the powerful nation Hamilton wanted, Americans would have to acquire the habits of working harder, consuming more, and paying higher taxes. Getting them deep in debt would be the "spur" to force them along, for as McDonald has observed, once Hamilton's system was put into place, "the choice is between paying one's debts and failing to survive."[50]

Once in office, Hamilton set to work putting his political and economic theories into practice. In January 1790, the new Secretary of the Treasury issued the first of his <u>Reports to Congress on Public Credit</u>. The central features of the report were a call for the Federal government to assume the wartime debts of the states, to stabilize and increase the credit of the nation by funding this debt with regular payments from the national treasury, and to provide the revenue for this funding through tariffs on imports and a broad system of excise taxes on domestically produced wine and liquor.[51] Toward the conclusion of his Report, Hamilton once again reiterated the connections between his financial plans, public opinion, and the creation of appearances. "In nothing are appearances of greater moment, than in whatever regards credit. Opinion is the soul of it, and this is affected by appearances, as well as realities."[52]

Twelve months later, he issued his Report on a National Bank.[53] The chief objectives of the bank were monetizing of the national debt and attaching the interests of the rich to the new Federal government. In the latter case, the government would issue securities to the holders of the debt. These would be redeemable, with interest, over the course of several years. All those Americans and foreigners who held these securities would develop an immediate attachment to that government which would now be sending them regular payments. If the government lacked the funds to redeem the securities as they came due, they could simply issue more securities (and create more supporters). Hamilton knew that as the regular interest payments were made, ambitious investors would rush to purchase whatever bonds, bills, or notes the Treasury Department might issue. He stated one of the main goals of the bank was to "unite immediately the interest and influence of the monied men in its establishment and preservation ..."[54] The debt would be monetized by issuing the securities in very high denominations and redeemable only at a distant time. These securities could, however, be used as collateral at any bank, so holders need only present their government holdings and borrow cash against its future value. This would, in turn, have two important consequences; there would be an immediate, enormous influx of cash into the economy, and Americans would develop the habit of basing their personal plans on future earnings rather than past savings. While his reports on Credit and The Bank provide insight into his financial scheme, Hamilton's more entrenched assumptions concerning Americans and their environment become brilliantly clear when his 1791 Report on the Subject of Manufactures is compared with David Hume's essays, "Of Luxury," "Of Commerce," and "Of the Rise and Progress of the Arts and Sciences." These are the essays that demonstrate most clearly the type of society and the kind of humans Hume and Hamilton wanted to mold.

Hamilton begins his Report on Manufactures with a cautious attack on the agrarian economic theory of the French physiocrats. Since there is little evidence Hamilton ever specifically read the work of Francois Quesnay or any of the other physiocrats, most of his analysis seems to be a secondhand recounting of Adam Smith's criticism as it appeared in The Wealth of Nations.[55] The subtlety of his assault was necessitated by the wide popularity of the physiocratic theory that agricultural pursuits were economically more productive than manufacturing.[56] Hamilton painted his analysis as merely a call for the balancing of agricultural and manufacturing pursuits, but behind this circumspection was a more profound truth; on the critical issue of the applicability of America to agriculture, rather than commercial industrialization, Hamilton, in fact, agreed with the physiocrats. In recapitulating the argument of the physiocrats, Hamilton acknowledged, "This policy is not only recommended to the United States, by considerations which affect all nations, it is, in a manner, dictated to them by the imperious force of a very peculiar situation. The smallness of their population compared with their territory—the constant allurements to emigration from the settled to the unsettled parts of the country—the facility, with which the less independent condition of an artisan can be exchanged for the more independent condition of a farmer, these and similar causes conspire to produce, and for a length of time must continue to occasion, a scarcity of hands for manufacturing occupation, and dearness of labor generally."[57] As long as land was cheap, plentiful, and easily accessible, Hamilton recognized that humans would much prefer the more free life of agrarian yeoman to the toil of existence as workers in industrial cities.

Indeed, it was this perspective that would later cause Hamilton to express serious misgivings about Jefferson's purchase of the Louisiana Territory. While he supported the acquisition of the vital port of New Orleans, and while his position on presidential power precluded him from questioning Jefferson's Constitutional authority, Hamilton did not like the idea of adding more agricultural land to America. Although he stated that the area west of the Mississippi "was not valuable to the United States for settlement," his real fear was that it not only *was* valuable but that many citizens would rather be farmers in Nebraska and Oregon than factory hands in New Jersey or Massachusetts. In an 1803 editorial on the purchase, Hamilton wrote, "But it may be added, that should our own citizens, more enterprising than wise, become desirous of settling this country, and emigrate thither, it must ... be attended with all the injuries of a too widely dispersed population ..."[58] This, in fact, was the focus of his argument in the Report on Manufactures. His financial policies of funded debt, taxation, and monetization made it essential for America to become a commercial, capitalist, industrial nation. Without that, government could never attain the fiscal base essential for great power. But from a social point of view, Americans were not industrial workers and, for the foreseeable future, had little desire or incentive to move in that cultural direction.

A specific set of government policies was necessary, therefore, to tempt Americans into taking up industrial pursuits and instilling in them a set of values more conducive to a commercial society. Hamilton explicitly rejected the physiocrat's policy of *laissez-faire*, fearing that if government did nothing, Americans would remain small, subsistence farmers on their own land. "The spontaneous transition to new pursuits, in a community long habituated to different ones, may be expected to be attended with proportionably greater difficulty. When former occupations ceased to yield a profit adequate to the subsistence of their followers, or when there was an absolute deficiency of employment in them ... changes would ensue; but these changes would be likely to be more tardy than might consist with the interest either of individuals or of the Society ... To produce the desireable changes, as early as may be expedient, may therefore require the incitement and patronage of government."[59] Hamilton's policies included protective tariffs; "bounties" in the form of tax breaks or cash payments to industrial entrepreneurs; and the encouragement of child labor, immigration, and the introduction of women to run factory machines.[60] Americans would be moved from the farms of the West to the mines and sweatshops of the East.

There remained, however, the problem that in a free society, individuals could, if they wished, quit their jobs and seek happier employment elsewhere. Even more important than policies, a capitalistic culture required a shift in social values. Here, Hamilton pinned his hopes on his theory of human passion and turned again to the writings of Smith and Hume. In his Theory of Moral Sentiments, Adam Smith addressed the question why humans work very hard. Waxing almost poetic, Smith asked, "to what purpose is all the toil and bustle of this world? What is the end of avarice and ambition, of the pursuit of wealth, of power, and preeminence?" Interestingly, Smith rejected the idea that the goal is the satisfaction of human needs. Interest-group liberalism is built on the premise that human needs cannot be satisfied; they are "insatiable" as Hume argues in his Treatise. Furthermore, the satisfaction of human needs is not the goal since such an end, even if it could be achieved, would be socially destructive in as much as it would lead to "indolence," as Hume claims in "Of Commerce" and Hamilton concurs in his 1781 letter to Robert Morris.[61] It was Smith who struck nearest to the issue when he observed that satisfying our needs is irrelevant to the question. Humans are capable of providing for all their necessities and for many of their comforts with very little labor. Continuing his inquiry why humans work so hard, Smith asked, "Is it to supply the necessities of nature? The wages of the meanest labourer can supply them. We see that they afford him food and clothing, the comfort of a house, and of a family. If we examined his economy with rigour, we should find that he spends a great part of them upon conveniences, which may be regarded as superfluities, and that, upon extraordinary occasions, he can give something even to vanity and distinction." Adam Smith then posed and answered the critical question, "[W]hat are the advantages which we propose to that great purpose of human life which we call bettering our condition? To be observed, to be attended to, to be taken notice of with sympathy, complacency, and approbation, are all the advantages which we can propose to derive from it. It is the vanity, not the ease, or the pleasure, which interests us."[62] Here, the author of capitalism's *magnum opus* put his finger on the central value that defines the capitalist culture. Commercial society requires people to work much harder than they need to, and in order to induce them in that direction, their simple wish to be loved must be perverted into the vainglorious attempt to seek acceptance through the acquisition of "personal belongings." At the heart of commercial society is consumerism and at the heart of consumerism is the desperate hope that if only we work hard enough and buy enough merchandise, these "goods" will help make us loved, esteemed, or, at least, noticed. As capitalism slowly erodes those communities that once gave meaning to our lives, we seek solace in our material possessions. Our vanity increases in direct proportion as our lives become more vain.

The recognition of capitalism's power to shape behavior by molding the values of consumers was not original to Adam Smith. In 1714, Bernard de Mandeville had claimed, in The Fable of the Bees, that feeding private vanity would produce public benefits, and in the 1739 edition of his Treatise, David Hume opined, "the possessor has also a secondary satisfaction in riches arising from the love and esteem he acquires by them ... This secondary satisfaction or vanity becomes one of the principle recommendations of riches, and is the chief reason, why we either desire them for ourselves, or esteem them in others."[63] In the Essays that Hamilton so admired, Hume considered the relative benefits of free societies over despotisms. In "Of Civil Liberty," he argued, "there is something hurtful to commerce inherent in the very nature of absolute government and inseparable from it ... Commerce, therefore, in my opinion, is apt to decay in absolute governments, not because it is there less *secure*, but because it is less *honorable*." Hamilton agreed with this position so much that he almost plagiarized Hume in a note he sent at the Constitutional Convention, "A free government [is] to be preferred to an absolute monarchy not because of the occasional violations of *liberty* or *property* but because of the tendency of the Free Government to interest the passions of the community in its favour ..."⁶⁴ Unlike fixed, feudal societies where the luxury of the rich incites little envy in the poor (since they can never aspire for such things), people in commercial societies believe that a little more work and a little more effort can help them acquire the material objects they grow accustom to desiring.

One of the benefits of commercial society was its capacity to incite envy, avarice, ambition, and vanity. Capitalism turns individuals into greedy, shallow consumers, and this is exactly the type of society Hume wanted. "Thus men become acquainted with the pleasures of luxury and the profits of commerce ... It arouses men from their indolence and, presenting the gayer and more opulent part of the nation with objects of luxury which they never before dreamed on, raises in them a desire of a more splendid way of life than what their ancestors enjoyed ..." Hume goes on to describe human behavior in this future society, "They flock into cities; love to receive and communicate knowledge, to show their wit or their breeding, their taste in conversation or living, in clothes or furniture. Curiosity allures the wise, vanity the foolish, and pleasure both."⁶⁵ For his part, Hamilton also saw the cultural destructiveness of commercial society. To Morris, he confessed, "Great power, commerce and riches, or in other words great national prosperity, may in like manner be denominated evils; for they lead to insolence, an inordinate ambition, a vicious luxury, licentiousness of morals, and all those vices which corrupt government, enslave the people and precipitate the ruin of a nation." Nevertheless, Hamilton went on, "But no wise statesman will reject the good from an apprehension of the ill."⁶⁶

What, it might be asked, could possibly compensate a nation for the unhappiness and ethical ruin of its citizens? Hume and Hamilton's answer is that, in such a consumer society, people are willing to work so hard they produce a surplus of economic riches, and it is this surplus that can be taxed and converted into government power. A stratified, but open, society would be an incitement to vanity; vanity would be an incitement to debt; debt would be an incitement to industry; and industry would produce vast quantities of taxable wealth. In "Of Commerce," Hume stated his position with incredible frankness. "It is a violent method, and in most cases impracticable, to oblige the labourer to toil, in order to raise from the land more than what subsists himself and family. Furnish him with manufactures and commodities, and he will do it himself. Afterwards you will find it easy to seize some part of his superfluous labour, and employ it in the public service without giving him his wonted return. Being accustomed to industry, he will think this less grievous, than if, at once, you obliged him to an augmentation of labour without any reward. The case is the same with regard to the other members of society. The greater is the stock of labour of all kinds, the greater quantity may be taken from the heap, without making any sensible alteration in it." Hume conceded that, under certain conditions, civic pride or virtue might make citizens willing to sacrifice for the common good, but liberals had rejected appeals to civic virtue, and "as these principles are too disinterested and too difficult to support, it is requisite to govern men by other passions and animate them with a spirit of avarice and industry, art and luxury. The camp is, in this case, loaded with a superfluous retinue, but the provisions flow in proportionately larger."[67] The reference to "the provisions flowing in" points to the devastating environmental consequences of consumer capitalism. Nature must be turned into commodities, not only to satisfy human needs and wants, and not only to provide revenue for public expenditures, but also in order to provide all the "superfluous retinue" that human vanity can possibly be prodded into lusting after. There is abundant evidence that Hamilton understood and accepted this logic. Beneath his political and economic policies is an environmental perspective which seeks to feed human vanity, greed, and avarice; offer credit in order to purchase the material trappings of this vanity, allow people to make marginal payments on their outstanding debt (plus interest); and keep them busy working to pay on the loans that are feeding the vanity. Alexander Hamilton helped conceive and put into practice what we today call the rat race.

It seems it was always part of his vision. In Federalist 6, he observed, "the spirit of commerce, in many instances, [has] administered new incentives to the appetite ..." To Bayard, he spoke of "the strongest and most active passion of the human heart, *vanity*!" In his Report on Manufactures, he paraphrased Hume, "Even things in themselves not positively advantageous, sometimes become so, by their tendency to provoke exertion. Every new scene, which is opened to the busy nature of man to rouse and exert itself, is the addition of new energy to the general stock of effort."[68] During secret negotiations with British diplomat Major George Beckwith in 1789, Hamilton stated, "I do think we are and shall be great consumers."[69] This is a revealing sentence. Hamilton's employment of the word "consumers" might be the earliest time an American refers to his fellow Americans not as citizens, or even as people, but classifies and characterizes them merely by their ability to use up natural resources.[70] In his Report on Manufactures, Hamilton made specific reference to the environmental price tag of his plan, "It merits particular observation, that the multiplication of manufactories not only furnishes a Market for those articles, which have been accustomed to be produced in abundance, in a country; but it likewise creates a demand for such as were either unknown or produced in inconsiderable quantities. The bowels as well as the surface of the earth are ransacked for articles which were before neglected." [71]

This is the environmental legacy that underpins Hume and Hamilton's political and economic thought. It is a program of social manipulation and behavior modification where debt is the stick and vanity the carrot to induce people to work hard and consume more. The "bowels of the earth" would be "ransacked" in order to provide the "heap" of "superfluous retinue" that humans neither needed nor, in many cases, wanted. Over the years, rural simplicity would be replaced with cosmopolitan consumption; public involvement would be replaced with private aggrandizement; and a confederation of small, agrarian communities would be replaced with a unified, free society of discontented consumers pathetically attempting to satisfy the insatiable and struggling ever more frantically to shop till they drop—their only source of civic pride deriving from the flag of the most powerful nation on the planet waving from their cars as they drive to the mall. It is a system almost perfectly designed to produce the greatest unhappiness for the greatest number in the shortest period of time. It is a gigantic mechanism built to dominate humans and ransack Nature; fueled by the will to power.

Hamilton's Environmental Legacy

It would be an obvious exaggeration to make Alexander Hamilton singly responsible for the debt-ridden, high consumption, fast-paced, industrial society America has become. Numerous people and institutions contributed to creating the ecologically unsustainable, environmentally destructive set of institutions and values that we call "modernity." Hamilton's role in all of this was critical, however. In many respects, he is the archetype and the model for the values, perspective, and philosophy of the contemporary liberal-capitalist state. But Hamilton was more than simply an effective theoretician and advocate. He was also a policy maker and an institution builder. In this regard, there is no better person to act both as a symbol of the philosophic change that took place and as the chief builder of current America. It is difficult to imagine anyone who has had a greater impact than Alexander Hamilton. In our behaviors, values, and interrelationships, we daily act out Hamilton's plan for this nation.

Environmentalists have not been the first to recognize and comment on the destructiveness of the consumer society Hamilton helped create. Throughout the course of this country's history, important writers have noted the connections between the fast pace of the economy and the social unhappiness of many Americans. The young Alexis de Tocqueville visited America, witnessed the culture developing here, and wrote in 1825: "In America I saw the freest and most enlightened men, placed in the happiest circumstances which the world affords; it seemed to me as if a cloud habitually hung upon their brow, and I thought them serious and almost sad even in their pleasures ... A native of the United States clings to this world's goods as if he were certain never to die; and he is so hasty in grasping at all within his reach, that one would suppose he was constantly afraid of not living long enough to enjoy them. He clutches everything, he holds nothing fast, but soon loosens his grasp to pursue fresh gratifications ... Death at length overtakes him, but it is before he is weary of his bootless chase of that complete felicity which is for ever on the wing."[72]

During the height of commercial society's emergence in the Gilded Age, economist Thorstein Veblen published <u>The Theory of the Leisure Class</u>, which sought to describe and explain America at the close of the nineteenth century. Veblen saw around him the consumption of goods far beyond any rationally based satisfaction of human needs. It was consumption for the pure, ostentatious, vainglorious purpose of proving the ability to consume. Veblen coined the phrase "conspicuous consumption" as the ideal description of a behavior another economist has labeled "economic psychopathology." Veblen went on to theorize on the underlying motivation for such activity. American society was held together, not by an "invisible hand" and the logical pursuit of self-interest by reasonable people, but by a mad lust to demonstrate superiority in force, cunning, and the seizing of wealth without work or the production of anything of benefit.[73] Hamilton's world of illusion had become manifested in a "predatory culture" that sought destruction and consumption for their own sake.

Critics and supporters of the consumer culture continued to debate over the years with an apparent accommodation being reached in the period following World War II. Exemplified by authors such as John Kenneth Galbraith, liberals joined conservatives in praising the results of Hamilton's work. The high consumption "affluent society" was hailed by conservatives as the just reward for clever entrepreneurship and equally exalted by liberals for its ability to generate the material basis for the end of poverty if only properly distributed by a progressive national government.[74] With both conservatives and liberals capitulating to the values of consumer capitalism, the role of critic has been picked up by contemporary environmentalists who question the ecological sustainability, the mental health, and the moral worth of Hamilton's legacy.

What are the contemporary consequences that follow from the realization of Hamilton's dream? The most conspicuous outcome is that America has become, unquestionably, the most powerful nation on the Earth. Not only do we face east toward Europe (and West, South, and anywhere else we choose to turn), but now the entire world looks toward *us*. We may be feared, hated, respected, or envied, but we are never ignored. The "arrogant pretensions of the European" that so grated on Hamilton have been crushed as we have fulfilled his wish to "teach that assuming brother moderation."[75] From our vast army, to our huge navy, to our awe-inspiring nuclear arsenal, America's military might now guarantee that we can "dictate the terms of connection" between this country and any other on the Earth. This unparalleled military power derives from and reflects America's equally colossal economic strength. With a gross domestic product over $16.8 Trillion (2013), the United States has the economic base to support the tax revenues to finance the military and governmental apparatus to project our will globally. At the base of this political and economic power is the debt-ridden, high consumption industrial society that fuels the upper strata of institutions. The public debt of the Federal government is now approximately $17.7 Trillion, or $55,667 for every person in the country. Corporations owe $9.6 Trillion, while individual households account for $11.52 Trillion worth of debt. These figures add up to mean the total indebtedness in the United States in 2014 was a staggering $38.82 Trillion or twice as much as the Gross Domestic Product. Economic stability can only be maintained by continuing the exponential growth of the economy and that can be continued only with the ecological "ransacking" of the environment. Beneath this fast-paced consumer society lie the costs and consequences being paid by the ecosystem.

With roughly 317 million inhabitants, the United States represents about 4.4 % of the world's 7.1 billion people. Yet that relatively small population consumes 30 % of all the world's resources yearly, including 25 % of all energy from nonrenewable fossil fuels. In addition to our seemingly insatiable energy demands, the American consumer society reaps devastation on the environment in many other ways. It has been estimated that providing the average family with all of the material products they demand in a year necessitates four million pounds of natural resources. Americans spend more on trash bags than 90 of the world's 210 countries spend for *all* their yearly needs.[76] Noted biologist E.O. Wilson estimates that human beings now appropriate between 20 and 40 % of all the solar energy captured in organic material by plants. A large portion of this net primary product of photosynthesis is utilized by Americans. The consequences of these facts are brought into stark relief when one realizes that in order for all the humans on Earth to consume resources in the volume and at the rate of Americans would require the additional supplies from another two planets![77] Continuing such a pattern of ecological and economic exploitation is not possible. It defies fundamental laws of physics and biology.

Americans not only consume vast stores of energy and resources, we also produce large quantities of materials. This nation's 103 nuclear power plants have manufactured, thus far, over 77,000 tons of high-level, radioactive waste that will remain lethal to anyone exposed to it for the next 10,000 years. If it is true that storing that waste will cost three million dollars per plant per year, which amounts to a total expenditure of 30 trillion dollars per plant, then the expenditures for storage are far higher than the income from the electricity sold during the plant's commercial life.[78] Whether or not nuclear energy is safe and clean is almost beyond the point. If long-range costs are figured into the equation (and it is sheer delusion not to include them) nuclear power does not produce energy. It is an economic and energy sink. Yet America continues to allow these facilities to operate. Unfortunately, atomic energy with a negative economic value is not the only commodity being produced in abundance. On the average, over two-thirds of a ton of garbage is produced by every man, woman, and child in America each year. Industry generates hazardous and toxic waste at a rate of 265 metric tons of *officially* classified materials per year. The air in the United States is the recipient of some 147 million metric tons of human-generated pollution each year. This figure does not include CO_2 emissions or wind-blown soil created by poor agricultural practices. CO_2 is a critical greenhouse gas which makes a significant contribution to the problem of global warming, and here again, America accounts for approximately one-fifth of the human-generated CO_2 in the atmosphere.[79] When former Vice President Dick Cheney says that reducing America's production of greenhouse gases to 1991 levels (as called for in the Kyoto Protocol) would devastate our economy, he is probably correct.

The total costs of the Hamiltonian system extend beyond economics and ecology, however. There is a price in human mental health that is extracted by life in the rat race. The statistics for mental illness and dysfunctional human relations shed some light on life in consumer society. According to the American Foundation for Suicide Prevention, in 2011 (the most recent year for which data are available), 39,518 Americans committed suicide, making it the tenth leading cause of death in this country. Fourteen million Americans use illegal drugs, 12 million are classified as heavy drinkers, 60 million cannot stop smoking cigarettes, and another five million are compulsive gamblers. Over half of all marriages in the United States end in divorce. According to figures from the National Institute on Drug Abuse, over 10 million individuals are now classified as "problem shoppers," buying and running up debt as a pathological response to personal unhappiness and low self-esteem. James Kunstler, reporting in Orion magazine, claims that one out of every three Americans is taking some form of prescription antidepressant drug.[80] Of course, millions of others assert their individual happiness and contentment, work and buy the commodities they believe they need, and continue to pursue the values and behaviors that we now define as "normal." Alexander Hamilton's cynical conclusion that the inhabitants of the United States seek only to have their chains made from gold has, in many cases, been converted into a self-fulfilling prophecy.

Alexander Hamilton left America with a set of values and a perspective on power that remains embedded in our culture. Our institutions are constructed upon assumptions regarding the insatiable character of human desire, the infinite manipulability of the environment, and the arrogant confidence that political/economic/military power can solve any problem. Yet as Hamilton himself confessed, many of these assumptions are believed simply because we hold to the illusion they are true. The real issue is whether or not economic systems ever have to face reality, or can they operate indefinitely in a crackpot world of illusion of their own making. At present, most of the public and private institutions in America are proceeding in the belief that our economic system can continue to operate based on premises directly contradictory to the laws of physics, ecology, and what we know of human mental health. The political and economic thought of Alexander Hamilton may be the major underpinning of contemporary American society, but it is all based on illusion. More accurately, it is all based on delusion. It is all a lie.

NOTES

1. Alexander Hamilton, "Answer to Questions Proposed by the President of the United States to the Secretary of the Treasury," Hamilton to Washington, September 15, 1790, Cabinet Paper, in The Works of Alexander Hamilton, edited by Henry Cabot Lodge, 10 Volumes, (New York and London: G.P. Putnam's Sons, 1885), Volume IV, 43. Hereafter cited as (Lodge) Works of AH with reference to volume and page.
2. Alexander Hamilton, "Federalist 11," in James Madison, Alexander Hamilton, and John Jay, The Federalist Papers, edited with an introduction by Isaac Kramnick (New York and London: Penquin Books, 1987), 133–134. All future citations to The Federalist Papers refer to the Kramnick edition.
3. Liberal Progressive supporters of Hamilton include Herbert Croly, The Promise of American Life (Boston: Northeastern University Press, 1989); Charles Beard, An Economic Interpretation of the Constitution of the United States (New York and London: The Free Press, 1965); and Vernon Parrington, Main Currents in American Thought (New York: Harcourt, Brace & Co., Inc., 1927). Conservative Hamiltonians might include Richard Brookhiser, Alexander Hamilton: American (New York: The Free Press, 1999); and Karl-Friedrich Walling, Republican Empire: Alexander Hamilton on War and Free Government (Lawrence, KS: University Press of Kansas, 1999).
4. A good starting place for reviewing the secondary literature on Hamilton is Stephen F. Knott, Alexander Hamilton and the Persistence of Myth (Lawrence, KS: University Press of Kansas, 2002). Indispensable sources include Jacob E. Cooke, Alexander Hamilton: A Profile (New York: Will and Wang, 1967); Broadus Mitchell, Alexander Hamilton: Youth to Maturity, 1755–1788 (New York: Macmillan and Company, 1957); Broadus Mitchell, Alexander Hamilton: The National Adventure, 1788–1804 (New York: Thomas Y. Crowell, 1962); John C. Miller, Alexander Hamilton: Portrait in Paradox (New York: Harper and Brothers, 1959); and Clinton Rossiter, Alexander Hamilton and the Constitution (New York: Harcourt, Brace, and World, Inc., 1964). Particularly important in analyzing Hamilton's thought are Douglas Adair and Marvin Harvey, "Was Alexander Hamilton a Christian Statesman?" William and Mary Quarterly, reprinted in Fame and the Founding Fathers: Essays by Douglass Adair, edited by Trevor Colbourn (New York: W.W. Norton and Company, Inc., 1974); Gerald Stourzh, Alexander Hamilton and the Idea of Republican Government (Stanford, CA: Standford University Press, 1970); and Forrest McDonald, Alexander Hamilton: A Biography (New York: W.W. Norton and Company, 1979). Some entertainment may be

found in Gore Vidal, Burr: A Novel (New York: Random House, 1973), and Arnold A. Rogow, A Fatal Friendship: Alexander Hamilton and Aaron Burr (New York: Hill and Wang, 1998).

5. The date of Hamilton's birth, and hence his age when he was killed, is the subject of dispute. Most early biographers (and Hamilton himself after he came to America) say he was born in 1757, which would make him 47 in 1804. Ron Chernow, however, produces convincing evidence that Hamilton was born in 1755, making him 49 when he was shot by Burr. See Ron Chernow, Alexander Hamilton (New York: The Penguin Press, 2004), 16–17. The definitive collection of Hamilton's works is Harold Syrett and Jacob E. Cooke, The Papers of Alexander Hamilton, 26 volumes (New York: Columbia University Press); hereafter cited as (Syrett) Papers of AH, with reference to volume and page.
6. (Lodge) Works of AH, Vol. 1, 428. Hamilton to Bayard, January 16, 1801, (Lodge) Works of AH, Vol. 8, 583–584, emphasis in the original.
7. Jacob E. Cooke's recognition of the problems in studying Hamilton is similar. See his "Introduction" in Alexander Hamilton: A Profile. More generally, see Alexis de Tocqueville, "Philosophical method among the Americans," in Democracy in America, Volume 11, Chapter 1, (New York: Schocken Books, 1961).
8. There is a dispute among current commentators on the consistency of Hamilton's thought over his life. Vernon Parrington claims, "Throughout his career Hamilton was surprisingly consistent…" Main Currents in American Thought, while Forrest McDonald observes there are "striking differences between his writings before the age of twenty-five and the ideas he expressed in maturity …" Alexander Hamilton: A Biography, xii. John Miller shares this view of Hamilton's shifting philosophy in Alexander Hamilton: Portrait in Paradox, 15 and 99. 49–51.
9. McDonald, Alexander Hamilton, A Biography, 18–22. The quotations are from Hamilton to Laurens, September 11, 1779, and Hamilton to Laurens, September 12, 1780, in (Syrett) Papers of AH, Vol. 2, 167 and 428.
10. (Syrett) Papers of AH, Vol. 1, 176–177.
11. Hamilton's comments on June 22, cited in Max Farrand (ed.), The Records of the Federal Convention of 1787, 5 Volumes (New Haven: Yale University Press, 1987), Vol. 1, 376. Hereafter cited as (Farrand) Records with reference to volume and page. Hamilton, Federalist 85, 486.
12. Quoted in McDonald, Alexander Hamilton: A Biography, 35, and note McDonald's references to the connections between Hume and Hamilton in footnote 15, 373. See also Clinton Rossiter, Alexander Hamilton and the Constitution (New York: Harcourt, Brace, and World, Inc., 1964), 187, where Rossiter states, "he [Hamilton] was a disciple of David Hume."

Stourzh calls Hume Hamilton's "great master," <u>Alexander Hamilton and the Idea of Republican Government</u>, 24; and John Miller says, "the influence of David Hume was of paramount importance," <u>Alexander Hamilton: Portrait in Paradox</u>, 46.

13. Quoted in Henry D. Aiken "Introduction," <u>Hume's Moral and Political Philosophy</u> (New York: Hafner Press, 1948), ix. For support of Jefferson's position, see Sheldon S. Wolin, "Hume and Conservatism," in Donald W. Livingston and James T. King (eds.), <u>Hume: A Re-Evaluation</u> (New York: Fordham University Press, 1976), 239–256.
14. Since Douglass Adair's seminal essay Hume's very considerable impact in America has gained its proper place alongside the rights-based liberalism of John Locke. See Adair, "'That Politics May be Reduced to a Science': David Hume, James Madison, and the Tenth Federalist," in Trevor Colbourn (ed.), <u>Fame and the Founding Fathers</u>, 93–106.
15. Jacob Cooke recounts the diner story in his "Introduction" to <u>Alexander Hamilton: A Profile</u>, xx, and notes that it is a thirdhand recollection that may, or may not, be true. The second quotation is historically confirmed in Miller, <u>Alexander Hamilton: Portrait in Paradox</u>, 49.
16. The links between Calvinism and liberalism are well established. For an excellent overview, see Sheldon S. Wolin, <u>Politics and Vision</u> (Boston: Little, Brown and Company, 1960), Chapters 6 and 9. For a weaker version, note Garret Ward Sheldon, <u>The Political Philosophy of James Madison</u> (Baltimore: The Johns Hopkins University Press, 2001), and see the review of Sheldon by Franklin A. Kalinowski in <u>The American Political Science Review</u>, Spring 2002.
17. This analysis conforms with McDonald's in <u>Alexander Hamilton: A Biography</u>, 90–94.
18. Hamilton, comments of September 17, 1787, in (Farrand) <u>Records</u>, Vol. II, 645–646. Hamilton to Gouverneur Morris, February 27, 1802, in (Lodge) <u>Works of AH</u>, Vol. 8, 591.
19. Madison, <u>Federalist</u> 10, p. 123; and <u>Federalist</u> 51, 320.
20. McDonald, <u>Alexander Hamilton: A Biography</u>, 109–111. McDonald's opinion on the best of the <u>Federalist Papers</u> is expressed in footnote 33, 388. For McDonald's further reflections on the Constitution, see Forrest McDonald, <u>Novus Ordo Seclorum: The Intellectual Origins of the Constitution</u> (Lawrence, KS: University Press of Kansas, 1985).
21. Hamilton, <u>Federalist</u> 70, 402. McDonald thinks Hamilton's best contributions are those on the executive and the judiciary, "plus 25, 35, 36, and 84." For a different view of the relationship between Madison and Hamilton as Publius, see Garry Wills, <u>Explaining America: The Federalist</u> (Garden City, New York: Doubleday and Company, Inc., 1981).
22. Hamilton, <u>Federalist</u> 83, 469 and 472–473.

23. Hamilton to Lieutenant Colonel John Laurens, September 11, 1779, in (Syrett) Papers of AH, Vol. 2, p.167.
24. McDonald, Alexander Hamilton: A Biography, 5.
25. James Madison, "Address delivered before the Albemarle, Va., Agricultural Society, May 12, 1818," in The Complete Madison: His Basic Writings (ed.) Saul K. Padover, (New York: Harper and Brothers), 278.
26. Hume, "Of the Original Contract," in Essays, 471.
27. (Farrand) Records, Vol. 1, 282–311. Farrand used the notes found in the Library of Congress as the basis for Hamilton's outline of his June 18 speech, as augmented by a similar outline in John C. Hamilton's Life of Hamilton. James Madison, in his Notes of Debates in the Federal Convention of 1787 (New York: W.W. Norton and Company, Inc., 1966), 139, explains his account of Hamilton's speech: "The speech introducing the plan, as above taken down & written out was seen by Mr. Hamilton, who approved its correctness, with one or two verbal changes, which were made as he suggested ...."
28. Hume, "Of the First Principles of Government," in Essays, 32.
29. Hume discusses "Of the reason of animals," in the Treatise, 176–179.
30. (Farrand) Records, Vol. 1, 285. Madison's debt to Hume on the issue of size is convincingly made by Douglass Adair in "'That Politics May Be Reduced to A Science': David Hume, James Madison, and the Tenth Federalist," in Fame and the Founding Fathers. For Hume's original thoughts on the issue of size, see, "Idea of a Perfect Commonwealth," Essays, 512–529.
31. (Farrand) Records, Vol. 1, 285. In "Of the Independence of Parliament," Hume states, "The crown has so many offices at its disposal that, when assisted by the honest and disinterested part of the House, it will always command the resolutions of the whole .... We may therefore give to this influence what name we please; we may call it by the invidious appellations of *corruption* and *dependence*; but some degree and some kind of it are inseparable from the very nature of the constitution ...." Essays, 45. Jefferson recounts the story where Hamilton paraphrase Hume's support for corruption, or the British system of patronage, in his Anas, in Merrill D. Peterson, Thomas Jefferson: Writings (New York: The Library of America, 1984), 671. McDonald gives his version in Alexander Hamilton: A Biography, 214–215.
32. (Farrand) Records, Vol. 1, 291–293.
33. (Farrand) Records, Vol. 1, 363.
34. (Farrand) Records, Vol. 1, 286.
35. Henry Steele Commager as quoted in McDonald, Alexander Hamilton: A Biography, 117. The contrast of the British system with Hamilton's is McDonald's at 161.

36. Hamilton to _____, December 1779–March 1780, (Syrett) Papers of AH, Vol. 2, 236–251. John Miller speculates that the recipient was Hamilton's future father-in-law, Philip Schuyler: see Alexander Hamilton: Portrait in Paradox, 52. McDonald concurs with this judgment in Alexander Hamilton: A Biography, 375.
37. Hamilton to _____, in (Syrett) Papers of AH, 247 and 242.
38. Hamilton to James Duane, September 3, 1780, (Syrett) Papers of AH, Vol. 2, 400–418.
39. Ibid., 401. Emphasis in the original.
40. Ibid., 417.
41. McDonald, Alexander Hamilton: A Biography, 189.
42. Alexander Hamilton, untitled article, Gazette of the United States, September 1 1790; cited in Ron Chernow, Alexander Hamilton (New York: The Penguin Press, 2004), 331. Forrest MacDonald recognizes, once the system was created it "would be almost impossible to dismantle the machinery short of dismantling the whole society." McDonald, Alexander Hamilton: A Biography, 122.
43. The best introduction to the thought of Viscount Bolingbroke is Isaac Kramnick, Bolingbroke and His Circle: The Politics of Nostalgia in the Age of Walpole (Ithaca: Cornell University Press, 1968). See also, John Trenchard and Thomas Gordon, Cato's Letters; or, Essays on Liberty, Civil and Religious, and Other Important Subjects, 4 Volumes in 2; 3rd edition corrected; London 1733; facsimile edition New York: Russell and Russell, 1969.
44. Adam Smith, An Inquiry into the Nature and Causes of the Wealth of Nations (New York: Random House, 1985). Hume, "Of Public Credit" in Essays, 350–351, and 360–361.
45. Hume, "Of Public Credit," Essays, 353 and 354.
46. Hamilton to Morris, April 30, 1781, (Syrett) Papers of AH, Vol. 2, 608.
47. Ibid., 609.
48. Ibid., 611, 617, and 618.
49. Ibid., 635.
50. McDonald, Alexander Hamilton: A Biography, 227.
51. Hamilton, "Report Relative to a Provision for the Support of Public Credit," January 9, 1790, (Syrett) Papers of AH, Vol. 6, 51–168.
52. Hamilton, "Report on Credit," January 1790, 97.
53. Hamilton, "Report on a National Bank," December 14, 1790, (Syrett) Papers of AH, Vol. 7, 236–342.
54. Hamilton to Morris, April 30, 1781, (Syrett) Papers of AH, Vol. 2, 620–621.
55. Hamilton, "Report on the Subject of Manufactures," December 5, 1791, (Syrett) Papers of AH, Vol. 10, 230–340. The editors of the Hamilton

Papers confirm his reliance on Smith and lack of primary reading on the physiocrats in footnote 127, 231.
56. For a general introduction to physiocratic economics, see Ronald Meek, The Economics of Physiocracy (Cambridge, MA: Harvard University Press, 1963); M. Beer, An Inquiry Into Physiocracy (London: George Allen and Unwin LTD. 1939); and Warren J. Samuels, "The Physiocratic Theory of Economic Policy," Quarterly Journal of Economics, 76, 1962, 145–162.
57. Hamilton, "Report on Manufactures," 233.
58. Douglass Adair (ed.) "Hamilton on the Louisiana Purchase: A Newly Identified Editorial from the New-York Evening Post," in Fame and the Founding Fathers, 260–271. The quotations from Hamilton are at 276.
59. Hamilton, "Report on Manufactures," 267.
60. Ibid., 253. A few months before he issued his "Report on Manufactures," Hamilton had been involved in the creation of the Society for Establishing Useful Manufactures and had made similar proposals along with Governor Patterson of New Jersey to endow a section of that state for the establishment of an experimental industrial city (to be named after the governor). See "Prospectus of the Society for Establishing Useful Manufactures," (Syrett) Papers of AH, Vol. P, 144–153.
61. Adam Smith, The Theory of Moral Sentiments, in Robert L. Heilbroner (ed.), The Essential Adam Smith (New York: W.W. Norton and Company, 1986) 78. See also David Hume, Treatise, 491–492; Hume, "Of Commerce," Essays, 264; Hamilton to Robert Morris, Vol. 2, 615.
62. Smith, The Theory of Moral Sentiments, 79.
63. David Hume, Treatise, 365.
64. David Hume, "Of Civil Liberty," in Essays, 93 (emphasis in original). Hamilton note in (Farrand) Records, Vol. 1, 145 (emphasis also in original).
65. David Hume, "Of Commerce," 264, and "Of Refinement in the Arts," ("Of Luxury"), 271.
66. Hamilton to Morris, April 30, 1781, (Syrett) Papers of AH, Vol. 2, 617–618.
67. David Hume, "Of Commerce," in Essays, 262 and 263.
68. Hamilton, Federalist 6, 106; Hamilton to Bayard, April 1802, (Lodge) Works of AH, Vol. 8, 597.
69. Originally cited in (Syrett) Papers of AH, Vol. 5, 486; and discussed in Ron Chernow, Alexander Hamilton, 294.
70. According to the Oxford English Dictionary, the first description of human beings as "consumers" is found in De Foe's, English Tradesman, in 1745.
71. Hamilton, "Report on Manufactures," (Syrett) Papers of AH, Vol. 10, 256 and 260.

72. Alexis de Tocqueville, Democracy in America (New York: Schocken Books, 1961), Vol. 2, 161–162.
73. Thorstein Veblen, The Theory of the Leisure Class (New York: Modern Library, 1934). For commentary see Robert L. Heilbroner, The Worldly Philosophers (New York: Simon and Schuster, Inc., 1953), 218–248.
74. John K. Galbraith, The Affluent Society (Boston: Houghton Mifflin, 1998).
75. Hamilton, Federalist 11, 133–134.
76. Paul Hawken, Amory, and Hunter Lovins, Natural Capitalism, quoted in John de Graaf, David Wann, and Thomas H. Naylor, Affluenza: The All-Consuming Epidemic (San Francisco: Berrett-Koehler Publishers, Inc., 2001), 85.
77. Edward O. Wilson, The Diversity of Life, cited in Stephanie Mills, "Can't Get That Extinction Crisis Out of My Mind" 204; and David W. Orr, "The Ecology of Giving and Consuming" 143; both found in Roger Rosenblatt (ed.), Consuming Desires: Consumption, Culture, and the Pursuit of Happiness (Covelo, CA: Island Press, 1999).
78. Information and calculations from CNN report February 11, 2002.
79. Cunningham and Saigo, Environmental Science, 527, 536, 397, and 391.
80. De Graaf, Wann, and Naylor, Affluenza, 104.

CHAPTER 8

The Environmental Legacy of James Madison: Pursuing Stability in a World of Limits

Like his neighbor and friend, Thomas Jefferson, and, unlike his sometime collaborator and frequent political opponent, Alexander Hamilton, James Madison looked to the west as a source of hope for the future of American society. Jefferson felt the "vacant lands" that he had made part of the United States through the Louisiana Purchase would make possible a multitude of small, self-sufficient, "ward republics" populated by agrarian citizens who exercised and strengthened their civic virtue through direct participation in the political, economic, and social life of their communities. Hamilton feared the open and easily available farm land of the American west would lure settlers and reduce the number of laborers needed in the industrial centers of the East. The vision of James Madison differed from both of these. What Madison saw in his mind when he looked west was vast forests containing lumber to build houses, rich soil for agricultural crops, and abundant plains of grass to feed horses and cattle. But what James Madison mostly saw in the American west was space. There was room in the west for the rapidly expanding population to spread out, leave the crowded cities of the east, pursue their separate dreams, and not get in each other's way. In the West, Madison hoped, American liberty would find expression in the pursuit of personal affairs; individuals would be absorbed in their daily lives and stay out of politics; they would not cause trouble, At least for a while.

The essential factor that differentiates James Madison's political thought from the viewpoints of Jefferson and Hamilton is Madison's recognition

© The Author(s) 2016
F. Kalinowski, *America's Environmental Legacies*,
DOI 10.1057/978-1-349-94898-7_8

of, and unwavering focus on, the role of limits in human affairs. Thomas Jefferson believed humans were capable of almost unending improvement in their ethical development as their moral sense expanded and strengthened through "exercise." Alexander Hamilton thought there were virtually no problems that could not be solved with money and power, and he relied on the continuous, never-ending conquest of Nature to provide that wealth and domination. Madison doubted the truth and efficacy of both of these perspectives. For him, there were constraints that made the achievement of both Jefferson's and Hamilton's visions unachievable and dangerous. In Jefferson's case, the limits were psychological and built into the restrictions of human temperament. Regarding Hamilton's theory, Madison believed the boundaries were both environmental and political. The limits of Nature posed an insurmountable obstacle to endless economic expansion, and the policies Hamilton proposed to reach this unachievable goal were fraught with dangers to the lasting stability of the American political system. For Madison, there were limits to what we could expect from personal character and limits to what we could expect from the Earth.

But a political theory must be more than a series of objections. If Madison rejected Hamilton's unlimited government and Jefferson's civic humanism, what was it that he supported? Understanding the political thought of James Madison requires that we explore his perception of human nature and the severe limits that he saw to humans as rational, ethical citizens. Understanding the environmental legacy of James Madison requires an appreciation of how the writings of Thomas Robert Malthus blended with and influenced his philosophy. For Madison, the crunch between America's rapidly expanding population and the finite land mass of the continent carried with it dire consequences for society and politics. A world of limited resources and multiplying humans meant a struggle where gains for some could only be purchased by losses for others. The attempt to survive or prosper under these conditions could only result in strife among individuals or groups. The pursuit of self-interest in a world of limits would lead, therefore, to conflict and not community. It would reinforce all that was predictably dangerous in the human character. Creating a stable society within a world of limited resources, populated with passionate, self-interested individuals became the object of Madison's intellectual efforts. Jefferson's purchase of the vast Louisiana Territory might buy some time for the young republic, but it could not stave off the inevitable. The best humans could hope for, according to Madison, were "palliatives" that would create not a "more perfect" union as the Constitution promised but simply the "least imperfect" social order.[1]

For Madison, the goal of politics was not the creation of ethical communities, or the building of a powerful government, but rather the protection of individual rights and the pursuit of the common good in a world defined by the existence of limits. The purpose of this chapter is to isolate the assumptions and trace the logic that supports Madison's vision of politics. In the conclusion, we will consider the consequences that follow from this perspective and attempt to specify Madison's place in contemporary environmental thought. Before delving deeply into the political and environmental thought of James Madison, however, it is necessary to address the same issue that presented itself in the study of Alexander Hamilton: do the writings of Madison contain a consistent, coherent, single political philosophy that admits of analysis, or, as some commentators have suggested, did Madison change his philosophic stance so often that his only real legacy is a collection of ideological clichés displaying an opportunistic yet bewildering level of contradiction? Unlike Hamilton who is accused of one major shift in philosophy, Madison is charged with an almost continual waffling. Is James Madison America's great realist or America's great equivocator? Before attempting an analysis of James Madison's political philosophy, we must first ask: is there anything there to analyze?

Pragmatism, Flexibility, and Consistency in Madison's Thought

James Madison died at his home on June 28, 1836, after a prolonged period of steadily declining health. With him during his final moments were Paul Jennings, his slave-valet, and Madison's niece, Nelly Willis. As later reported by Jennings, the last surviving member of America's Constitutional Founders had some difficulty swallowing a bite of his breakfast and was asked by his niece, "What is the matter, Uncle James?" Madison quietly replied, "Nothing more than a change of mind, my dear." His head dropped, his breathing ceased, and Madison expired.[2]

For many scholars and commentators who have since written about Madison's political philosophy, these may be the most prophetic dying words in history, for "Nothing more than a change of mind, my dear" seems to epitomize the life of James Madison. Changing his mind and advocating contradictory policies were acts Madison performed with apparent frequency throughout his life. He was a slaveholder who despised slavery. He was an ardent nationalist whose Virginia Resolutions of 1798 argued for interposition and the right of states to declare acts of the national government unconstitutional (he would later deny that he ever advocated nullification). He joined Jefferson in a historic struggle against Alexander Hamilton over the creation of a national bank and then, as President in 1816, Madison signed a bill creating the Second National Bank. In <u>Federalist</u> 10, perhaps his most famous essay, Madison expounded the classic argument against "majority factions" as a threat to both individual rights and the public interest, yet toward the end of his life, he argued that "the constitutional majority must be acquiesced in by the constitutional minority …"[3] He initially opposed the addition of a bill of rights to the Constitution as a mere "parchment barrier" that "overbearing majorities" would ignore at their pleasure and then fought to secure the adoption of amendments he was largely responsible for writing.[4] In 1783, Madison joined with Hamilton and Oliver Ellsworth to draft an "Address to the States" suggesting remedies for the nation's economic crisis. The report contained virtually every policy proposal that Madison would later attack when initiated by Hamilton as Secretary of the Treasury, for example, assumption of state debts, a tariff, nondiscrimination between original and subsequent holders of national securities, and monetizing the debt through the sale of public lands.[5] At one point in time, he could be for judicial broad construction, public works, free trade, and national consolidation; and, at another moment, he could pen essays opposing all of these projects.

It is easy to see why Madison can be pictured as an ideological chameleon, changing his views to blend with those around him. When he was with Hamilton at the Philadelphia Convention and as a joint author of <u>The Federalist Papers</u>, Madison advocated one set of practices. When he was with Jefferson as cofounder of the Republican Party, creating opposition to the Federalist schemes of Hamilton and John Adams, he could champion policies in direct opposition.[6] Were there any deeply held core principles that Madison believed in, and was there a logically consistent philosophy to which Madison ascribed over the course of his life? Despite the superficial evidence to the contrary, a case can be made that James Madison did, in fact, have a coherent political philosophy.

Insight into that political philosophy may be found in <u>Federalist</u> Number 37, where, writing from behind the anonymity of the pseudonym "Publius," Madison engaged in some self-reflection regarding the interplay among policies, principles, and the Constitutional Convention of which he was a key member. Governmental policies and institutional arrangements, he suggested, were the surface manifestations of more deeply held philosophic positions. Madison speaks of the "superstructure" of the Articles of Confederation that must be transformed because the underpinning "foundation" of its "principles" is "fallacious."[7] His point seems to be that policies, and to some extent institutions, are merely reflections of the principles upon which they rest. Keeping in mind this distinction between facile "superstructures" and substantial "foundations" dissolves much of the ambivalence in his thought. The apparent contradictions in Madison's political life involve alternative policy prescriptions only. The arguments that support those policies undergo very little change throughout his life. Madison's writings over many years demonstrate his basic faith that policies and institutional arrangements may be subject to some modification as long as the foundational principles remain constant. In other words, means are defined by ends, for Madison, and while the means may be flexibly altered as the immediate circumstances require, the ends or goals of society must be constantly and consistently pursued. The Articles of Confederation must be abandoned, therefore, not only because of their failed policies, but, more importantly, because the principles upon which they rest were dangerously misguided. He was convinced that the new Constitution was built on a more sound philosophic foundation, and this, more than any specific action the new government might undertake, was the main justification for its adoption. During his subsequent experiences in political life, Madison would demonstrate this pragmatic pliability regarding policy while remaining steadfastly assured of the correctness of his philosophy. Madison's apparent inconsistency derives, therefore, from analyses that focus too narrowly on his historical policy positions. Moving one's intellectual attention below the surface to Madison's core principles reveals a remarkably steady set of political beliefs.[8]

Continuing the investigation of <u>Federalist</u> 37 provides a strong suggestion what the most important of those core philosophic beliefs might be. In the paragraph that immediately follows his discussion of superstructures and foundations, Madison begins an inquiry into "the difficulties encountered by the [Philadelphia] convention." The foremost of those he considers "very important" is "combining the requisite stability and energy in government, with the inviolable attention due to liberty, and to the republican forms."[9] This tension between stability and liberty, rather than any perfunctory debate over policy, is the foundation of Madison's constitutional thought. The individual values embedded in America's political culture, what was known in Madison's time as "the genius" of Americans, would not permit the establishment of any political system that did not guarantee a large scope of personal liberty, but as Madison demonstrates in this paragraph and, as we shall see, in numerous other places in his thought, it is the singular focus on stability that underpins all his thought. Social stability and governmental power (euphemistically referred to as "energy") form the *sine qua non* of his subsequent politics.

To use this paragraph from Federalist 37 as a succinct example, one of its eight sentences talks about "republican liberty," while five lengthy sentences are devoted to the "energy in government," which is the "the very definition of good government." Madison asserts that stability is "essential" to "that repose and confidence in the minds of the people, which are among the chief blessings of civil society." The absence of stability, on the other hand, produces "vicissitudes and uncertainties" and "irregular and mutable legislation" that are "odious to the people." Significantly, in this paragraph, Madison describes liberty as something that American culture "seems to demand." There is a note of indecision regarding liberty which is looked upon as a mere "form." When he talks about stability, on the other hand, Madison describes it as something that is "requisite," "essential," and a principle that good government "requires."[10] Given his prior discussion of superstructures and foundations, it is difficult to avoid the conclusion that Madison considers liberty to be the more superficial concept and stability the more basic principle.

James Madison did, therefore, have a coherent political philosophy that was built upon the need for social stability as its basic premise. In order to draw a complete picture of his thought, analysis must inquire into the reasons why stability was so important for him, and why alternative principles such as liberty, justice, or virtue were relegated to the position of superstructure, or at best, mid-range concepts located between core principles and superficial policies. Federalist 37 suggests the importance of stability in Madison's thought, but one essay does not a philosophy make. In order to comprehend the political and environmental legacy of Madison, it is necessary to focus on the role of stability in a world of human and ecological limits and demonstrate that this focus was a consistent, lifelong preoccupation of Madison's mind.

The Limits of Human Nature

In his acclaimed inquiry into Western political thought, J.G.A. Pocock writes about the "Machiavellian Moment;" which he defines as the time when a society comes to the realization that the slide into personal and political corruption has become irreversible. According to Pocock, the "Machiavellian Moment" occurs when concepts such as civic virtue, the public interest, justice, citizenship, and individual sacrifice for the common good are no longer seen as capable of guiding and shaping human behavior. Since the classical republican thought of Greece, Rome, the Italian Republics, and the English Commonwealth men all took it as axiomatic that virtuous citizens were essential for the existence of republican government, the loss of virtue essentially meant the demise of any hope for the republic's future.[11] When something akin to the Machiavellian moment might have occurred for James Madison, or if he ever believed in the assumptions of classical republicanism, is not known. It is clear, however, that by 1787, his writings and speeches reject the assumptions of classical thought and display a commitment to the premises and perspective of the emerging philosophy of liberalism.

In April of 1787, as part of his preparation for the Constitution Convention in Philadelphia, Madison penned a memorandum concerning his thoughts on human nature and politics that scholars now refer to as "Vices of the Political System of the United States." During the summer of 1787, he delivered several addresses to the Constitutional Convention, the most significant being his speech of June 6, where he again articulated his views, and, of course, throughout the remainder of 1787 and for most of 1788, he joined with Hamilton and John Jay in writing The Federalist Papers. These three sources, "Vices of the System," the Convention speeches, and his contributions to the Federalist, comprise the nucleus of Madison's thought on human nature and political order. In 1829, he revisited these themes when he addressed the convention to revise the Constitution of Virginia on December 2. This speech is significant since it both reaffirms the essential assumptions and logic of Madison's thought, and it confirms his lifelong commitment to these ideas. Together, these four documents can be used to analyze the political thought of James Madison and display its consistency over time. An examination of these works demonstrates how Madison restructured the central themes of classical republicanism in order to create his version of liberalism. Critical terms became redefined, the logical relationship among concepts became rearranged, and the fundamental perspective on human character became dramatically altered. Nowhere can this be seen more clearly than in the concepts of justice and the common good, for these were central notions within republican thought, and their reformulation by Madison both highlights his liberalism and manifests the role of social stability in his thought.

James Madison understood the philosophic importance of justice and the common good in America's political culture. In <u>The Federalist</u>, he alternately made each of these ideas the end goal of America's social order. In Number 51, Madison says, "Justice is the end of government. It is the end of civil society. It ever has been and ever will be pursued until it be obtained, or until liberty be lost in the pursuit." Note how this deviates from his position in <u>Federalist</u> 45, where he states that, "the public good, the real welfare of the great body of the people, is the supreme object to be pursued …"[12] What are the differences between justice and the common good, what is the philosophic basis for each, and how does an examination of this issue shed light on Madisonian thought? In Chap. 5, we remarked on the Lockean side of Madison's liberalism.[13] For both Locke and Madison, justice is seen as the protection of individual rights, and "rights" are described as entitlements whose existence and applicability rest upon rational deductions. The primary connections, therefore, are among justice, rights, and reason. Beyond the observation that this entails a shift from classical notions of justice as harmony within a community toward a much more individualistic concept, this reconceptualization also opens the way for Madison to adopt Locke's theory of property as the most basic of all human rights. For Madison, as for Locke, a "just" government is one that protects individual rights; the most important of those individual rights being the right to private property. In his "Property" essay, however, Madison took the notion of property beyond material objects and included all of those other entities that an individual might value. By blurring the distinction between rights and interests, Madison suggests that a just government must not only protect human rights, it must also defend individual interests. In <u>Federalist</u> 10, for example, he traces the source of property back to the "faculties of men," which may be seen as more extensive than merely the sweat and toil of labor.[14] In his "Property" essay, Madison includes within this category all the entities that a human might possibly value such as "a property in his opinions," "a property very dear to him in the safety and liberty of his person," and "an equal property in the free use of his faculties, and free choice of the objects on which to employ them."[15] Immediately following this broadened definition, Madison makes the connection between property, rights, and justice. "Government is instituted to protect property of every sort; as well that which lies in the various rights of individuals, as that which the term particularly expresses. This being the end of government, that alone is a *just* government which *impartially* secures to every man whatever is his *own*."¹⁶ The difficulty was that since "rights" were not empirical phenomena that could be seen, touched, or heard through the human senses, their existence could never be determined through experience. If rights existed at all, their recognition would have to depend on the power of rational thought, and, here, Madison was skeptical of the human capacity to exhibit enough reason to secure rights, property, and justice. This is the first point where Madison's philosophy confronts the notion of limits: justice appears to rest on the human capacity to reason, but humans are constrained in their rational capabilities. As Madison stated in a letter to Frederick Beasley dated 1825, "The finiteness of the human understanding betrays itself on all subjects ..."¹⁷

This is not to say that Madison did not ardently wish for the ascendancy of human rationality. In a different National Gazette essay titled "Universal Peace," he gave Enlightenment praise to "the rapid progress of reason and reformation, which the present day exhibits ..." Madison may have wished for the "empire of reason" to be spread around the globe, but his survey of the current state of the human condition revealed numerous "monuments of deficient wisdom."¹⁸ Madison provided two explanations for this gloomy appraisal. First, human reason was seldom able to withstand the corrupting influence of the passions and self-interest; and, second, even if it was possible for individual humans to act rationally, the effect of joining groups was to further weaken the power of reason and make it susceptible to the pressure of collective irrationality. Madison's clearest and most disturbing enunciation of his position comes in Federalist 55, where he states. "In all very numerous assemblies, of whatever characters composed, passion never fails to wrest the scepter from reason. Had every Athenian citizen been a Socrates, every Athenian assembly would still have been a mob."¹⁹

Perhaps no other single passage provides such insight into the mind of James Madison as this one. This exhibits Madison at his pessimistic worst, for here he is not only questioning the practical possibility of creating a community of rational, civic-minded, virtuous citizens, he is asserting that *even if* such a republic could be implemented, it would *still* fail. Even if rational human beings exist or could be created through education, Madison philosophically rejects the possibility that they could ever predominate in the formulation of collective policy. This is a restatement of the position he articulated in "Vices of the Political System" where he maintains, "The conduct of every popular assembly ... proves that individuals join without remorse in acts, against which their consciences would revolt if proposed to them under the like sanction, separately in their closets."[20] When brought together, particularly when brought together to debate and decide public policy, "The *passions* ..., not the *reason*, of the public would sit in judgment." Hence, "A nation of philosophers is as little to be expected as the philosophical race of kings wished for by Plato."[21]

Madison's "empire of reason" could never be built upon the wisdom of rational, active citizens. Several inferences presented themselves to Madison. Resting justice and property rights on human reason was a tenuous proposition. If reason was to be pursued, it would have to be more narrowly defined as mere instrumental reason, and its pursuit would have to be taken out of the hands of individuals as well as groups and placed in the cogs of institutions. Madison's lack of trust in human rationality, his desire to keep humans politically isolated, and his substitute faith in institutions over human reason suggest that although there is a large component of Lockean philosophy in his thought, a clearer picture can be gained by exploring the connections between his thinking and the Scottish Enlightenment of the eighteenth century. This can be brought to light by shifting our analysis away from justice, rights, and reason and focusing instead on Madison's perspective on the common good, the passions, and the interests.

There are, therefore, two versions of liberalism found in the writings of James Madison. One is the Lockean variant that seeks justice through individual rights and reason. The other strain can be traced to the eighteenth-century Scottish Enlightenment and, in particular, the writings of David Hume. Here, Madison's early training under the Scot, John Witherspoon, at Princeton's Nassau Hall reveals itself. The young Madison apparently studied closely with Witherspoon; so much so that after the Virginian's graduation, he remained in New Jersey for several months doing individual readings with Witherspoon.[22] In politics, the Scottish emphasis on passions, interest, habit, and institutions finds its way into all of Madison's work. Lockean liberalism is there in Madison's philosophy, but Humean liberalism predominates.[23] Epistemologically, the shift from Locke to Hume manifests itself in an altered focus that plays down the role of reason in human knowledge and behavioral motivation and increases the emphasis placed on experience and habit. Humans are seen less as rational creatures and more as sensuous, passionate animals. Learning is pictured less as a process of rational deduction and more as a mechanism whereby our senses record the world, induce patterns of cause and effect, and accustom individuals to expected perceptions of reality. The same empiricism that is described and advocated in Hume's A Treatise of Human Nature and "An Enquiry Concerning Human Understanding" is applied to the investigation of politics found in all his political essays. Madison followed Hume in consistently relying on the lessons taught by experience.

In his "Vices of the Political System of the United States," for example, Madison uses experience to challenge the common assumption of republican thought that a virtuous citizenry, operating according to majority rule, will respect and protect the rights of citizens. He says, "According to Republican Theory, Right and power being both vested in the majority, are held to be synonymous. According to fact and experience a minority may in an appeal to force, be an overmatch for the majority." In his June 6 speech to the Constitutional Convention, Madison asserts, "A prudent regard to the maxim that honesty is the best policy is found by experience to be as little regarded by bodies of men as by individuals." And in Federalist 62, Madison turns to history, the collected record of human experience, to make his case for the proposed Senate. "[H]istory informs us of no longlived republic, which had not a senate."[24] This last quotation is significant since it demonstrates Madison tying together knowledge (gained through experience) with institutions (here, the senate) in the pursuit of his goal of endurance (stability over time). To demonstrate Madison's consistency over time, we might turn to his address to the Virginia Convention of 1829, where he again used experience to undermine any classical republican thought of building a just community on civic virtue, moral conscience, or a respect for the common good. Experience taught James Madison not to trust his fellow humans. "Some gentlemen, consulting the purity and generosity of their own minds, without adverting to the lessons of experience, would find security against … danger, in our social feelings; in a respect for character; in the dictates of the monitor within; in the interests of individuals; in the aggregate interests of the community. But man is known to be a selfish, as well as a social being."[25] This may be a close as Madison ever came to openly and directly repudiating the theory of moral sentiments and the faith in republican government expounded by Thomas Jefferson. Jefferson's knowledge of the world and faith in the future lead him to believe that individuals could be educated to become good citizens who would respect each other's rights and work to achieve what was in the best interest of everyone. Madison looked only at past experience and concluded from that narrow understanding that human reason would not secure individual rights, and individual passions would never consciously seek the public interest.

For James Madison, therefore, the goal was to somehow produce a redefined form of republican government that excluded the need for concerned citizens. In the wonderfully concise expression of Sheldon Wolin, Madison's aim was to produce "democracy without citizens."[26]

James Madison followed Hume's belief that individual humans, driven by the "violent passions" would become attached to objects that both gratified and strengthened their private emotions. These objects of the passions could be material articles, ideas, other humans, political figures contending for power, or even oneself (this was true, for instance, with the emotion of pride). When a passion became attached to an object of the passion, an interest was created.[27] Violent passions were reflected in the private interests, and these private interests, Madison believed, were the real springs of human conduct. In his "Vices of the Political System" essay, for example, Madison puts forth his opinion why people seek public office. "Representative appointments are sought from 3 motives. 1. ambition. 2. personal interest. 3. public good. Unhappily the two first are proved by experience to be the most prevalent. Hence the candidates who feel them, particularly, the second, are most industrious, and most successful in pursuing their object ..."[28] This is an important quotation since it calls into question the argument sometimes raised that the object of Madison's politics is to produce an electoral "filtration" process that would guarantee the election of a publicly minded elite. Madison was willing to *use* representation, but he was far from certain elections could be trusted to produce "enlightened statesmen." It is wrong, therefore, to conclude that Madison sought to insure the reign of either publicly spirited elected officials or a wealthy ruling class.[29] James Madison was never optimistic enough to think that anyone could set aside their private interests and pursue the common good.[30]

When people become associated with others, their private passions become intensified, they convince each other of the righteousness of their cause, and they mistakenly confuse their particular private interest with the common good of the entire society. In other words, humans identify with those who share the same interests and thereby create factions. In his seminal essay, Douglass Adair compares David Hume's "Of Parties in General" with Madison's Federalist 10. Adair shows that Madison not only adopted Hume's theory of passions and interests, but when describing the various types of interest groups, Madison also directly copied Hume's inventory of factions in the exact order as Hume listed them. Adair concludes that the only reasonable explanation for this extraordinary concurrence is that Madison must have had a copy of Hume's <u>Essays</u> open next to him as he wrote <u>Federalist</u> 10. For both Hume and Madison, the essence of politics is the attempt to create a stable society in a world driven by private interest groups, all seeking to oppress each other by any means necessary, including violence. As Madison states, "The latent causes of faction are thus sown in the nature of man; and we see them everywhere brought into different degrees of activity, according to the different circumstances of civil society ... So strong is this propensity of mankind to fall into mutual animosities that where no substantial occasion presents itself the most frivolous and fanciful distinctions have been sufficient to kindle their unfriendly passions and excite their most violent conflicts."[31]

For Madison, human conduct has two essential constraints; we have a limited capacity to reason, and we have a limited ability to see or act beyond our immediate self-interest. Both of these psychological limitations are exacerbated when humans come together in groups. Collective behavior, according to Madison, becomes even more irrational, more passionate, and more narrowly focused on private interests. In <u>Federalist</u> 62, he restates both his distrust of collective behavior and his epistemological faith in historical experience (here supplemented with some a priori reasoning). Madison asserts, "the propensity of all single and numerous assemblies, to yield to the impulse of sudden and violent passions, and to be seduced by factious leaders into intemperate and pernicious resolutions. Examples on this subject might be cited without number, and from proceedings within the United States, as well as from the history of other nations. But a position that will not be contradicted need not be proved."[32]

The inability of humans to be guided by either reason or a concern for the common good is, for James Madison, not only verifiable, it is irrefutable! Here, we see the distinctively Madisonian outlook on politics. Thomas Jefferson distrusted elites but had a profound faith in the common people. Alexander Hamilton deeply distrusted the masses but had an ongoing confidence in the abilities of an educated, powerful, affluent elite. James Madison adopted Jefferson's suspicion of the elites and welded it to Hamilton's mistrust of the masses. Madison trusted no one. It was this Hobbesian belief in the "infirmities and depravities of the human character," and the unshakable fear that no group in the society could be relied on to pursue justice and the common good that lead Madison to seek stability in the workings of heartless, unthinking, unfeeling institutions. Institutions would do for humans what Madison thought humans were incapable of doing for themselves. Institutions would strive for justice, protect rights, pursue the common good, and take account of society's long-range, public interests. These institutions would be created by a group of disinterested, enlightened Founding Fathers who for one brief historical instant would be able to display "an exemption from the pestilential influences of party animosities ..." Madison believed that this world-historical moment had occurred in Philadelphia in the Summer of 1787, and he also believed that anyone studying the resulting Constitution would come to the conclusion that "It is impossible for a man of pious reflection, not to perceive in it a finger of that Almighty Hand, which has been so frequently and signally extended to our relief in the critical stages of the revolution."[33] Of course, the truth is that the hand that helped write the Constitution and thereby create America's political institutions was not God's, but Madison's. It was the perceived limits of human character that drove Madison to put his trust in institutions. What remains to be analyzed are the assumptions upon which those institutions are built, and whether other limits might cause one to question the institutions' ability to produce either stability or endurance.

Institutional "Palliatives" and Ecological Limits

Madison's political prescriptions are logically consistent and follow from his philosophical assumptions. His reliance on institutions rather than the civic virtue of citizens assumes not only "the infirmities and depravities of the human character" but also that these characteristics are so ingrained as to be unremovable.[34] The essential difference between Jefferson and Madison is not the former's naivete and the latter's pragmatism but Jefferson's unending search for ways to improve humans and Madison's resignation that the human character as he currently saw and interpreted it was the way humans had always acted and the way they always would act.[35] If we were to agree with Madison that humans are actually as corrupt as he makes them out to be; if it was granted that individual humans only rarely exhibit either reason or a concern for the common good; and if we were to accept that the formation of political groups only reinforces and worsens all that is bad in human nature, then the proper course would obviously be to socially alienate people from each other, pit citizen against citizen in order to forestall the creation of a sense of community, and minimize the possibility for popular political participation. And if the individuals who staff the institutions are assumed to be as corrupt as the masses, then the same fragmentation, competition, isolation, and frustration would have to be built into their institutional behavior as well. Of course, there are quite a few "ifs" in this argument, but recognizing the pattern of the logic explains both Madison's reliance on institutional behavior and the structure of the government he helped create. There are four basic principles upon which Madison builds his political scheme. They are fragmentation, conflict, alienation, and habit. These take as given the psychological limits of human nature we have just discussed. Their purpose is not to ameliorate the weaknesses of human character but to compensate for what Madison assumes cannot be improved. They are, therefore, most properly viewed not as solutions but merely as actions designed to alleviate predicaments that cannot be remedied. Any structures built on alternative assumptions were, he assumed, visionary utopias which failed to recognize "diseases … too deeply rooted in human society to admit of more than great palliatives."[36]

A reliance on political fragmentation, and the strategy of "divide and conquer," can be found everywhere in Madison's writings. It derives from his fear of humans acting in groups and is most prevalent when Madison addresses the issue of majority tyranny. The anxiety over group activity, however, was not limited to the masses. The belief in *divide et impera* is aimed against the general public in Federalist 10, but the same tactic is focused on the governmental elite in Federalist 51. Madison expressed his thoughts on this matter in 1787 to Thomas Jefferson. "*Divide et impera*, the reprobated axiom of tyranny, is, under certain conditions, the only policy by which a republic can be administered on just principles."[37] Madison placed a great amount of faith in the "reprobated axiom." In all of his major theoretical essays, Madison returns to the theme of dividing or "breaking" the society into as many groups as possible. In "Vices of the System" he advocates that the "society becomes broken into a greater variety of interests, of pursuits, of passions …" In his June 6 Convention speech, he reiterates that, "The only remedy is to enlarge the sphere, & thereby divide the community …" Beyond the classic defense for social pluralism in Federalist 10, Madison turns to fragmentation of the governmental structure in Federalist 51, where he acknowledges that the "constant aim" is "to divide." Where "the remedy" for concentrated Congressional power is "to divide the legislature into different parts," and where the "society will be broken into … many parts, interests, and classes of citizens …"[38] The extended republic, separation of powers, bicameralism, and federalism were all institutional manifestations of Madison's wish to divide and "break" individuals and groups.

It was not enough, however, simply to prevent unified actions by dividing the population. Conflict and competition among groups were also necessary in order for each segment to act as a check on the others. This belief in the redemptive qualities of competition permeates liberal thought. There is something Calvinist and feudal about the notion that those who win competitions do so because they are more worthy, but clearly there are many institutions in liberal society that rest upon a faith in the purifying value of competition. Our adversarial system of justice, the "marketplace of ideas" in liberal education, and the competitive economic marketplace all rely on this assumption. In political institutions, the idea is manifested in the system of checks and balances that sets interest group against interest group and one government branch against another. Madison's classic statement on this issue comes in Federalist 51, where he affirms the benefits of competitive checks, recognizes that this tactic is used throughout liberal society, and concedes that it is being used as a substitute for civic virtue ("the defect of better motives"). "This policy of supplying, by opposite and rival interests, the defect of better motives, might be traced through the whole system of human affairs, private as well as public. We see it particularly displayed in all the subordinate distributions of power, where the constant aim is to divide and arrange the several offices in such a manner as that each may be a check on the other ... These inventions of prudence cannot be less requisite in the distribution of the supreme powers of the State."[39] Unlike competition in the courts or in sport, however, political competition in the Madisonian system aims not at producing a definite winner but simply in forcing the adversaries to bargain, compromise, and settle for whatever short-term, incremental gains they may be able to establish. As Madison concisely put it, "Ambition must be made to counteract ambition." At the end of Federalist 51, he assures his readers, "In the extended republic of the United States, and among the great variety of interests, parties, and sects which it embraces, a coalition of a majority of the whole society could seldom take place on any other principles than those of justice and the general good ...."[40] This is a telling quotation that requires some analysis in order to reveal the social objectives of Madison's political scheme.

The key to understanding is found in the words "coalition" and "seldom take place." Madison's use of the word "majority" might suggest that he is advocating a democratic system, but such an interpretation would be deceptive. Keeping in mind that this is the man who spent a large portion of his efforts advocating the "dividing" and "breaking" of society, and whose writing are filled with references to "oppressive" and "overbearing" majorities, it is difficult to believe James Madison ever had a significant commitment to either popular participation or majority rule. If the term "democracy" is taken to mean citizen input into policy making or decision-making by 50 % plus one of the elected representatives, then Madison was always fearful and mistrustful of such a government. James Madison was neither a republican nor a democrat. He was a liberal to the marrow of his being.[41] What this means is that Madison made a clear distinction between the institutions of government and the values and habits that permeate society. Democratic government signifies majority rule, while liberal culture is characterized by an ethic of compromise and toleration. It would be fortunate to live in a country that exhibits both democracy and liberalism, but there are no indications Madison ever thought America was that kind of nation. Given the choice between democratic government and liberal society, Madison's clear preference was for practices that would promote, and require, the values of toleration and compromise. Democratic government without liberal values was, in Madison's view, a recipe for the tyranny of the majority.

In Madison's America, people would tolerate those with differing opinions not because they accepted and respected diversity but because negotiating with those you might personally despise was necessary in order to temporarily establish coalitions. Compromise would be encouraged not because politicians and the public were open-minded and cooperative but because they had no other option. Without toleration and compromise, the formulation of policy would "seldom take place." Given the institutional arrangements of Madison's Constitution, toleration and compromise are the values essential to avoid governmental paralysis. In the language of today, expressions with which Madison would surely concur, the only alternative to "bipartisanship" is policy "gridlock." In a government limited by enumerated powers and strict construction, everyone would be forced to compromise and exhibit some level of toleration. They would seldom achieve all of the policies they wanted, but they would get some incremental portion of their goals. The frustrating wait for the next round of compromises, the small and incomplete successes, and the forming of differing coalitions with advocates you did not respect but were forced to tolerate would all have the predictable consequence of calming the passions. Bargainers would repeatedly have to compromise their interests. As they worked inside Madison's bureaucratic process, self-interested, belligerent, emotional, and irrational Americans would start to behave as if they were reasonable.

There is so much in Madison's scheme that tears Americans apart that one begins to wonder what holds the society together. Madison addressed the Constitutional aspects of this issue in a long memorandum that, with later revisions added, is called "Notes on Nullification." In 1835–1836, as the crisis over slavery deepened and arguments in defense of secession spread, Madison, now a former President and an elder statesman in his eighties, pronounced, "Whether the centripetal of centrifugal tendency be greatest, is a problem which experience is to decide ...."[42] In these "Notes," he focuses on the Constitutional aspects of interposition and nullification (which he rejects) and goes to great pains to disassociate his late friend Thomas Jefferson from the idea of secession. These were questions that were threatening to, and eventually did, rip apart America's Constitutional system, but given the recognized primacy of liberal culture over Constitutional institutions in Madison's system, the deeper question must be what he thought would keep Americans tolerant and compromising. The centrifugal forces of self-interest, passion, and the irrational desire to "vex and oppress each other" needed centripetal forces to pull people toward the moderate center of politics and society. This tension between the fragmentation of society and the need for influences to hold America together had long been a concern for Madison.

Madison's liberal culture has three elements that seem to keep the society from flying apart; habit, individual liberty, and economic opportunity. In true Humean fashion, the first centripetal force that Madison depends upon is custom. Habit strengthens stability, for Madison, by leading people to respect their institutions. He warned that frequent changes or appeals to popular opinion "lessens their veneration for those fundamental principles, & makes them a more easy prey to ambition and self interest."[43] His strongest statement on this issue was delivered in <u>Federalist</u> 49, where he argued against the idea of frequent Constitutional Conventions to alter or correct the original document. Madison feared such a plan would disrupt the effects of time and habit. "[I]t may be considered as an objection inherent in the principle that as every appeal to the people would carry an implication of some defect in the government, frequent appeals would, in great measure, deprive the government of that veneration which time bestows on everything, and without which perhaps the wisest and freest governments would not possess the requisite stability."[44] On February 4, 1790, Madison wrote to Thomas Jefferson responding to Jefferson's suggestion that all laws and contracts cease their effectiveness every 19 years. This was part of Jefferson's scheme to assure perpetual participation through social renewal. Madison was not impressed. He wrote his friend asking, "Would not a Government so often revised become too mutable and novel to retain that share of prejudice in its favor which is a salutary aid to the most rational Government?"[45] Jefferson was seeking virtue though participation, but Madison was looking for deference through habit. A year earlier, Madison had addressed the House of Representatives and urged the adoption of a Bill of Rights. Speaking of the newly adopted Constitution, Madison conceded that, "the great mass of the people who opposed it, disliked it because it did not contain effectual provisions against the encroachments of particular rights, and those safeguards which they have been long accustomed to …"[46] If the new government could just keep operating long enough, the people of the United States would stay together because it was the system they had become use to. As a further guarantee of political stability, the direct involvement of citizens in the Madisonian system is usually limited to periodic and infrequent voting. In Federalist 49, Madison warns of, "The danger of disturbing the public tranquility by interesting too strongly the public passions …"[47] The *vita activa*, citizens' active participation in politics that Jefferson so strongly advocated, is dismissed by Madison as a "danger" to the "public tranquility" that would result from "interesting too strongly the public passions." Compliance with the system would depend on adherence to the rule of law, and the rule of law would be reinforced by the force of custom or habit.

But no matter how much Madison's liberal society divided groups, promoted competition, discouraged public participation, and habituated individuals to accept the rules of these institutions, the overall system could guarantee neither stability nor endurance, neither toleration nor compromise, unless it also guaranteed both individual liberty and economic opportunity. Rather than expressing their "public passions," Americans would be encouraged to focus on their private concerns. Rather than the republican concept of liberty as life inside a harmonious, just community, freedom in Madison's liberal society would be closely tied to the self-perception of individualism. Ironically, one of the strongest ideals that holds Americans together is the ubiquitous cultural belief that each person is unique in their opinions, their interests, and their dreams. The foundation for toleration in America is the conviction that individuals should be permitted to lead the life they choose as long as they do not interfere with the freedom of others. Social diffusion would reinforce individualism, individualism would reinforce toleration, and toleration would contribute to holding people together in a liberal culture.

Perhaps the most important glue that held the society together would be the promise of economic opportunity. Individuals in the American political system would have to become convinced that the rewards of working within the system were greater than the potential benefits from leaving the system. Only a growing, expanding system could hope to maintain the loyalty of its citizens while seldom calling on them to sacrifice their passionate commitment to their personal interests. Otherwise, without civic virtue and a commitment to a recognized common good, individuals might see little reason to either tolerate or compromise. Madison was trapped within his own logic. Since he rejected the idea that individuals could be motivated by anything besides self-interest, his system had to gain cohesion by a mechanism that presumed self-interest. Economic expansion would provide the individual benefits to all who participated in the system, at the same time that it removed the need for either a common interest or personal sacrifice.[48] Without the rewards of economic growth, it is difficult to see what holds the Madisonian system together.

During his State of the Union Message on March 3, 1817, Madison lauded the representatives for supporting public works and economic growth. "No objects within the circle of political economy so richly repay the expense bestowed on them …," he says, and then goes on to describe the opportunities open to Americans, "Nor is there any country which presents a field where nature invites more the art of man to complete her own work for his accommodation and benefit." For Madison, economic and population growth carried with them multiple benefits. The individual passions could be harmlessly exercised, private property would be protected, the government could tax the excess revenue, and individuals would be habituated to a life of private pleasure and public stability. "Under other aspects of our country the strongest features of its flourishing condition are seen in a population rapidly increasing on a territory as productive as it is extensive; in a general industry and fertile ingenuity which find their ample rewards, and in an affluent revenue which admits a reduction of the public burdens without withdrawing the means of sustaining the public credit, of gradually discharging the public debt, of providing for the necessary defensive and precautionary establishments, and of patronizing in every authorized mode undertakings conducive to the aggregate wealth and individual comfort of our citizens."[49]

Self-interested individualism, shaped by the competitive institutions of market capitalism and interest-group liberalism, would, it was hoped, produce consequences similar to those promised by classical republicanism (justice and the common good), but without the destabilizing and misplaced trust in civic-minded citizens. This has been termed the "theory of spontaneous order," but that expression is misleading.[50] The order produced by the American Constitution and liberal culture is not spontaneous, it was planned by Madison, but it is *unintended* by those operating within the system. This mechanism of unintended consequences seeks to produce stability and endurance through institutions which facilitate and demand economic growth. The argument thus far developed is as follows: because James Madison was convinced of the rational and public-spirited limitations of human character, he advocated a political system that would substitute rule-guide institutional behavior for active citizen participation. These institutions, in turn, depend upon the values of toleration and compromise dominating the society, and these principles presuppose continued territorial and economic growth in order to fulfill their function of producing a stable and enduring social system. Within Madison's system, growth was needed both to separate and to unite the nation. This was Madison's defense of an extended republic, but as the nation's population grew, the land area of the nation would also have to grow. As he famously said in Federalist 10, "Extend the sphere and you take in a greater variety of parties and interests; you make it less probable that a majority of the whole will have a common motive to invade the rights of other citizens; or if such a common motive exists, it will be more difficult for all who feel it to discover their own strength and to act in unison with each other."[51] Madison's "extended republic" would have to be an *extending* republic. The critical questions remaining are these: how long can territorial and economic growth be assumed in a world of finite resources and expanding population? What would happen to Madison's system if it was forced to confront ecological limits that precluded further growth? How would individuals in Madison's system, habituated to self-interested behavior, react if they were called upon to make sacrifices in the name of some ethical goal larger than their immediate emotional concerns?

Addressing these possibilities drove the perennially pessimistic Madison to turn to Malthus' Essay on the Principle of Population. Jefferson also had "perused" Malthus' work but discounted its significance for America.[52] Madison, on the other hand, read and accepted what Malthus had to say and gave considerable attention to its implications for America. Even before Thomas Robert Malthus published his Essay in 1798, Madison had recognized the importance of an expanding population and the pressures such expansion put on the social system. On June 19, 1786, Madison wrote to

Jefferson responding to Jefferson's observations on the condition of the unemployed poor in Europe. Madison acknowledged that population density was as important as political institutions in securing a prosperous and stable society. "I have no doubt that the misery of the lower classes will be found to abate wherever the Government assumes a freer aspect, & the laws favor of subdivision of property. Yet I suspect that the difference will not fully account for the comparative comfort of the Mass of people in the United States. Our limited population has probably as large a share in producing this effect as the political advantages which distinguishes us. A certain degree of misery seems inseparable from a high degree of populousness."[53]

Malthus' essay simply confirmed what James Madison already believed regarding population and limits. In May of 1818, Madison delivered an extended address to the Agricultural Society of Albemarle, Virginia. In it, he addressed the issue of increasing human population from a biological perspective. Even if all plants and animals were converted to human food, the number of humans the world could support was still limited. There was, Madison argued, an "economy of nature" which would require the halt in population growth far below the number that could be fed. Limits to growth were not only nutritional, for Madison, they were ecological in the sense that a "balance of nature" required that a large portion of the earth must remain in a wild state, not subject to conversion to food. In his report, Madison continued and argued that there were physical limits to growth in addition to food production. In a fascinating section on air quality, he explores the relationship between human respiration and plant photosynthesis, notes that humans cannot rebreathe their own expired air, and concludes that if the earth became too populated, the air would become unfit to breathe. "An increase not consistent with the general plan of nature arrests itself. According to the degree in which the number thrown together exceeds the due proportion of Space and of atmosphere, disease and mortality ensue. It was the vitiated air alone which put out human life in the crowded Hole at Calcutta."[54] Although Madison observes that human intelligence and ingenuity are capable of increasing the conditions that might support a larger population, beneath this appeal for ongoing agricultural science is the expressed recognition that no amount of science, technology, or political power is capable of overcoming inherent biological and physical limits to growth.

"Nature has been equally provident in guarding against an excessive multiplication of any one species, which might too far encroach on others, by subjecting each when unduly multiplying itself to be arrested in its progress, by the effect of the multiplication itself... All animals as well as plants sicken & die when too much crowded."[55] In the first decades of the nineteenth century, James Madison was articulating a concept that would later become known as the ecological principle of "carrying capacity."

In addition to these biological limits to growth, Madison also recognized that his political system face institutional restrictions on population size. In the latter part of his life, he found himself reflecting with some gloom on the future of his country. One object of his attention was the right of suffrage, since Madison knew that important relationships existed among suffrage, property, and the hope of maintaining a stable social system. Specifically, the American most closely associated with the idea of an extended republic wondered how long the republic could keep extending itself, and what political consequences would follow the end of that extension. Returning to a theme he had enunciated as Publius 42 years earlier, Madison offered in 1829 his "more full and matured view on the subject."[56] If those citizens who were without property ever became the majority of voters in America, it was the "lot of humanity" that they would use their power to promote "agrarian laws and other levelling schemes." Such programs of "injustice" seemed to suggest that voting rights ought to be restricted to those who hold property, but such a plan violated the republican principle that those who are bound by the law must have a voice in its making. Fortunately for Madison, the problem of suffrage and property was not of immediate political concern. He believed that the United States had a "precious advantage" in the immensity of its territory. There was so much land in America that virtually everyone was either a property holder or could hope to become one. "There may be at present a majority of the nation who are even freeholders, or the heirs and aspirants to freeholds; and the day may not be very near when such will cease to make up a majority of the community."[57] As long as the growing population could find land in the extending republic, respect for the property rights of others could be built into a system of compromising interests. But when Madison looked forward in time, he saw trends that caused apprehension.

In 1829, as part of the notes he made during the convention to amend the Constitution of Virginia, Madison applied Malthus to America and concluded that the abundance of agricultural land meant a temporary surplus of food and, as a result, a rapidly expanding population. Madison had been Secretary of State when Meriwether Lewis, William Clark, and the Corps of Discovery returned from their expedition up the Missouri River to the Pacific and reported on the immense territory America had acquired from the French.[58] This "precious advantage" would buy time for America's Constitutional system and cultural values by providing new land that an ever-increasing source of agricultural labor could seek out to put under the plow. Eventually and inevitably, however, the entire continent would be filled to the population density of Europe. At that point, the ratio of voters without property to those with property would begin to increase until they became a majority and began to execute their "schemes of oppression." Calculating from the estimated amount of arable land in the United States in 1829, and judging the growth of population unrestricted by the food supply, Madison produced a very Malthusian-looking chart which led him to conclude that America would reach its limits to growth somewhere around the year 1929.[59] It would be presumptuous to suggest that James Madison predicted the Stock Market crash and the Great Depression, but what is not disputable is his appreciation that his theoretical liberalism depended upon a large and growing extended republic and his recognition that this growth must come to an end sometime in the future.

The Constitution, which he had played such an important role in writing, was predicated on the possibility of infinite growth and inexhaustible resources, but the aging statesman now realized that physical reality would someday intervene. The liberal in Madison understood the cultural base of his system relied on people exhibiting toleration and compromise, but those values would come under threat as limited space and resources made individual liberty and economic opportunity increasingly tenuous. As the fact of limits began to impinge on Madison's America, conflict among interest groups would become more bitter, rights would become less secure, individuals would be less willing to compromise, tolerance would decrease, and the decision-making institutions of the government would exhibit gridlocked paralysis more frequently. A perpetually extending liberal republic was a hopeless utopia. Commenting on Robert Owen's "New Harmony" community, Madison made observations that could equally apply to his own social scheme. "[Y]our plan is to cover, and that rapidly, the whole earth with flourishing communities. What is then to become of the increasing population?" Nature had given America historical breathing space for its experiment with limited government, individualism, and economic growth, but the days remaining for that experiment were numbered.

Somewhere in the future, Madison understood America would have to confront limits to ever-expanding property ownership, and the plan which he had helped adopt in Philadelphia would have to be fundamentally altered. Madison warned his countrymen, "To the effect of these changes, intellectual, moral and social, the institutions and laws of the country must be adapted; and it will require for the task all the wisdom of the wisest patriots."[60] Environmental pressures would test America's capacity for change.

Conclusion: The Environmental Legacy of James Madison

James Madison's political legacy as the "father of the Constitution" has long been established. His classic defense of limited government with enumerated powers, a Federal system built upon separation of powers, bicameralism, checks and balances, and an extended republic than enshrines the pluralism of interest-group politics is well known to even the most casual student of the American political system. The social legacy of the culture he sought to create is more easily felt than explicitly acknowledged. The pursuit of the values of toleration, compromise, and rights upon which our society ultimately seeks stability all harken back to Madison's liberalism. Perhaps the greatest bequest Madison has left us, however, is the omnipresent and virtually omnipotent power of American individualism to shape and constrain our perceptions, habits, and decision-making. America is the most individualistic nation in the world. Any analysis or understanding of the United States must take as its most foundational starting point this incredibly entrenched individualism. Americans truly believe that every individual is entitled to pursue their own dreams in their own way; to seek their private, idiosyncratic and relativistic interests; and to express their individual viewpoints on reality (regardless of how crackpot) which they rigorously insist are all "a matter of personal opinion." Any discussion of America's capacity to react to challenges or muster the will for collective action must take into account this extreme individualism. If our political system has difficulty generating the collective will to respond, one reason for this is that we are, in many ways, a nation of Madisonian individualists who seldom feel the need or inclination for *any* collective, political action. Madison's legacy lives on in a nation that both distrusts and relies upon the institutions of government; that gives voice to the ideals of unity and community but has misgivings regarding the amount of coercive conformity such communities sometimes demand. Madisonian anxiety is found expressed in a mistrust of Alexander Hamilton's powerful central government coupled with a skepticism regarding the potential of Thomas Jefferson's moral sense. Liberalism wants and requires what it cannot have—democracy without citizens, stability without community, and institutions without ethics. What it needs most of all is unlimited growth in a world of limits.

Still, 200 years since the Republican Party of Jefferson and Madison took control of the Federal government, and 80 years after the date Madison predicted our nation would surpass its optimum population, the vision of James Madison continues to shape, guide, and mislead America.

His environmental legacy can be seen everywhere. It is there in the writings of environmental economists who continue their unquestioned rationalizations of the competitive market and who steadfastly refuse to question the perspective that humans are self-interest individuals instinctively incapable of becoming good citizens. It is there in the outlook of reform environmentalism, which believes that interest-group liberalism can formulate policy, and bureaucratic regulation can implement rules that will protect the environment without seriously looking at the deep philosophic assumptions upon which liberalism rests.[61] There is one environmental perspective, however, where the legacy of James Madison can be seen most clearly. Just as Madison filled his political writings with the language of "stability" and "endurance," in today's literature, one persistently encounters talk of "sustainability." Sustainability, and the viewpoint of sustainable development, is the contemporary world's version of a system that is stable and that endures. On the one hand, sustainable development recognizes that profound environmental damage can be caused by human overpopulation, resource depletion, and pollution, but at the same time, mainstream advocates of sustainable development regularly refuse to take on the tough issues of ecological limits to growth and the critical need to halt population increase.[62] Like Madison, they might believe the day of political and ecological reckoning is still somewhere down our historical road. But unlike the supporters of mainstream sustainability, James Madison knew over 200 years ago, that the end of the road of unending growth must surely come one day.

The legacies of Thomas Jefferson, Alexander Hamilton, and James Madison have evolved over the years. Elements of each have blended with other, once contradictory, perspectives. At the end of the nineteenth century, as development through space confronted the end of the American frontier; as the corporatist, neo-merchantilism of Hamilton gained in popularity; as managed capitalism replaced the benign faith in the workings of Madison's institutionalized "heartless empire of reason," and as desperate farmers in the Midwest and South sought some condolence in the strength of their Jeffersonian communities, American political thought became rearranged and redrawn. The political, social, and environmental legacies articulated during America's Founding remained, but they mutated and evolved into a collection of quite distinct creatures.

Notes

1. Madison's comments on the "least imperfect" government are found in "Majority Governments" [1833] in Marvin Meyers, The Mind of the Founder: Sources of the Political Thought of James Madison (Hanover and London: University Press of New England, 1981), 412.
2. See Ralph Ketcham, James Madison: A Biography (Charlottesville, University Press of Virginia, 1990), 668–670; and Irving Brant, The Fourth President: A Life of James Madison (Indianapolis and New York: The Bobbs-Merrill Company, 1970), 641–642.
3. "Majoritarian Government," in Meyers, Mind of the Founder, 416.
4. Madison to Jefferson, October 17, 1788, in Meyers, Mind of the Founder, 157.
5. Madison, "Address to the States, by the United States in Congress Assembled," in Meyers, Mind of the Founder, 19–25.
6. For the "Hamiltonian" Madison see Garry Wills, Explaining America: The Federalist (Garden City, New York: Doubleday and Co., 1981). The "Jeffersonian" Madison is described and analyzed in Drew McCoy, The Jeffersonian Persuasion: Political Economy in Jeffersonian American (New York: W.W. Norton and Company, 1980).
7. James Madison, Alexander Hamilton, and John Jay, The Federalist Papers, edited with an introduction by Isaac Kramnick (New York: Penguin Books, 1987). Unless otherwise specified, all further references to the Federalist are to the Kramnick edition and will be cited by author, number, and page. Madison, 37, 243.
8. A further defense of the existence of a consistent philosophy in Madison's thought is found in Richard K. Matthews, If Men Were Angels: James Madison and the Heartless Empire of Reason (Lawrence: University Press of Kansas, 1995), Chapter 1. Readers who are familiar with Matthews' work will recognize my debt to him. See my review of If Men Were Angels in Perspectives on Political Science, Winter 1996.
9. Madison, Federalist 37, p. 243.
10. Madison, Federalist 37, 243.
11. J.G.A. Pocock, The Machiavellian Moment: Florentine Political Thought and the Atlantic Republican Tradition (Princeton: Princeton University Press, 1975). See also Isaac Kramnick, Bolingbroke and His Circle: The Politics of Nostalgia in the Age of Walpole (Ithaca: Cornell University Press, 1968).
12. Madison, Federalist 51, 322; and Federalist 45, 293.
13. These themes are particularly explored by David F. Epstein, The Political Theory of The Federalist (Chicago and London: The University of Chicago Press, 1984).

14. See Epstein, <u>The Political Theory of The Federalist</u>, 73; and Madison <u>Federalist</u> 10, 124.
15. Madison, "Property" in Meyers, <u>Mind of the Founder</u> 186. This essay is reviewed by Matthews in Chapter 5 of <u>If Men Were Angels</u>.
16. Madison, "Property," in Meyers, <u>Mind of the Founder</u> 186–187.
17. Madison to Frederick Beasley, November 25, 1825, cited in Richard K. Matthews, <u>If Men Were Angels</u>, 34.
18. "Universal Peace" in Meyers, <u>Mind of the Founder</u>, 193. "Empire of Reason" quotation is from "Consolidation" in <u>ibid.</u>, 183. "Monuments of deficient reason" in Madison, <u>Federalist</u> 62, 367.
19. Madison, <u>Federalist</u> 55, 336.
20. Madison, "Vices of the Political System" in Meyers, <u>Mind of the Founder</u>, 63.
21. Madison, <u>Federalist</u> 49, 314–315
22. See Ralph Ketcham, <u>James Madison: A Biography</u>, Chapter III.
23. Although I tend to agree with much of the analysis found in Richard Matthews, <u>If Men Were Angels</u>, we disagree on the relative importance of interests versus rights in Madison's thought. See Matthews, <u>If Men Were Angels</u>, 81.
24. "Vices of the Political System" in Meyers, <u>Mind of the Founder</u>, 59, Madison, "Speech before the Constitutional Convention," June 6, 1787, in James Madison, <u>Notes of Debates in the Federal Convention of 1787</u> (New York: W.W. Norton and Company), 76. <u>Federalist</u> 62, 148.
25. Madison, "Speech in the Virginia Constitutional Convention," December 2, 1829, in Meyers, <u>Mind of the Founder</u>, 402.
26. Sheldon S. Wolin, <u>The Presence of the Past: Essays on the State and the Constitution</u> (Baltimore and London: The Johns Hopkins University Press, 1989), Chapter 10, "Democracy Without Citizens."
27. For a fuller exposition of the Scottish theory of the passions, see Franklin A. Kalinowski, "David Hume on the Philosophic Underpinnings of Interest Group Politics," <u>Polity</u>, Vol. XXV, No. 3, Spring 1993, 355–374.
28. Madison, "Vices of the Political System of the United States," in Meyers, <u>Mind of the Founder</u>, 62.
29. Those who argue for the "filtration" thesis include David Epstein, <u>The Political Theory of the Federalist</u>, op. cit.; and Garry Wills, <u>Explaining America" The Federalist</u> (Garden City:Doubleday & Company, Inc. 1981). Authors who support to belief that Madison sought the rule of a wealthy ruling class are Charles Beard, <u>An Economic Interpretation of the Constitution of the United States</u> (New York: Macmillan, 1913 and 1935), and Vernon Louis Parrington, <u>Main Currents in American Thought: An Interpretation of American Literature from the Beginning to 1920</u>, 3 vols. (New York: Harcourt, Brace, & Company, 1927-1930. In rebuttal, see

Forrest McDonald, <u>Novus Ordo Seclorum: The Intellectual Origins of the Constitution</u> (Lawrence: University Press of Kansas, 1985).
30. Franklin A. Kalinowski, "David Hume and James Madison on Defining 'The Public Interest,'" in Richard K. Matthews (ed.) <u>Virtue, Corruption, and Self-Interest</u> (Bethlehem: Lehigh University Press, 1994), 172–201.
31. Douglass Adair, "'That Politics May Be Reduced to a Science': David Hume, James Madison, and the Tenth Federalist," in Trevor Colbourn (ed.) <u>Fame and the Founding Fathers</u> (New York: W.W. Norton and Company, Inc., 1974). Madison, <u>Federalist</u> 10, 124.
32. Madison, <u>Federalist</u> 62, 366.
33. The quotations in this paragraph are all from Madison, <u>Federalist</u> 37, 247.
34. Madison, <u>Federalist</u> 37, 247.
35. For a contrast of Jefferson and Madison, see Richard K. Matthews, <u>If Men Were Angels</u>, Chapter 7.
36. Madison to Nicholas P. Trist, January 29, 1828, in Meyers, <u>Mind of the Founder</u>, 355. Madison's use of the word "palliatives" may be an indication of his familiarity with Malthus' writing.
37. Madison to Jefferson, October 24, 1787, cited in Kramnick's "Introduction" to <u>The Federalist</u>, 54.
38. Madison, "Vices of the Political System of the United States" in Meyers, <u>Mind of the Founder</u>, 64. June 6 Convention speech in Madison, <u>Notes on the Federal Convention</u>, 77. <u>Federalist</u> 51, 320–321.
39. Madison, <u>Federalist</u> 51, 320.
40. <u>Ibid.</u>, 325.
41. All three of these positions were debated in <u>The Review of Politics</u> (Winter 2005), with Alan Gibson, "Veneration and Vigilance: James Madison and Public Opinion, 1785–1800," arguing for an essentially democratic reading of Madison; Colleen A. Sheehan, "Public Opinion and the Formation of Civic Character in Madison's Republican Theory," contending there are strong civic humanist elements in Madison's thought; and Richard K. Matthews "James Madison's Political Theory: Hostage to Democratic Fortune" defending a liberal reading of Madison. The strongest and most convincing argument seems to belong to Matthews.
42. Madison, "Notes on Nullification," in Gaillard Hunt, ed., <u>The Writings of James Madison</u> (New York: G.P. Putnam's Sons, 1900–1910), Vol. IX, 573–607; reprinted in Meyers, <u>Mind of the Founder</u>, 417–442, the quotation is at 440.
43. Madison to Caleb Wallace, August 23, 1785, in Meyers, <u>Mind of the Founder</u>, 32.
44. Madison, <u>Federalist</u> 49, 313–314.
45. .Madison to Jefferson, February 4, 1790, in Meyers <u>Mind of the Founder</u>, 176.

46. Madison, Speech in the House of Representatives, June 8, 1789, in Meyers, Mind of the Founder, 164.
47. Madison, Federalist 10. 314.
48. See Edward Ehler, "The Problem of the Public Good in The Federalist," Polity 13 (1981).
49. Madison, "Seventh Annual Message to Congress," December 5, 1815, in James D. Richardson, ed., A Compilation of the Messages and Papers of the Presidents, 1789–1897 (Washington: Government Printing Office, 1896) I, 562–569.
50. See Ronald Hamowy, The Theory of Spontaneous Order. The argument for unintended consequences is in Franklin A. Kalinowski, "David Hume on the Philosophic Underpinnings of Interest Group Politics."
51. Madison, Federalist 10, 127.
52. Thomas Jefferson to Jean Baptiste Say, February 1, 1804, in Merrill D. Peterson, ed., Jefferson: Writings (New York: The Library of America, 1984), 1143–1144.
53. Madison to Jefferson, June 19, 1786, in Robert A. Rutland, ed., The Papers of James Madison (Chicago: University of Chicago Press, 1975), Vol. 9, 76.
54. Madison, "Address delivered before the Albemarle, Va., Agricultural Society," May 12, 1818, in Saul K. Padover, The Complete Madison: His Basic Writings (New York: Harper and Brothers, 1953), 276–290.
55. Ibid., 286.
56. James Madison, "Notes on Suffrage," in [William C. Rives and Phillip R. Fendall, eds.], The James Madison Letters (New York: Townsend MacCoun, 1884), Vol. 4, 21–30.
57. Ibid., 23.
58. Stephen E. Ambrose, Undaunted Courage: Meriwether Lewis, Thomas Jefferson, and the Opening of the American West (New York: Simon and Schuster, 1996).
59. Madison, "Notes on Suffrage," 29. Madison says,
 "The ration of increase in the U.S. shows that the present
 12 Millions will in 25 years be 24 Mils.
 24   "   "   " 50   "   " 48 "
 48 " " " 75 " " 96 "
 96 " " " 100 " " 192 " "
 For Madison's favorable comments on the work of Malthus and his application of Malthusian theory to other issues in America, see The James Madison Letters, Vol. 3, 348–350 and 614–616. See also Marvin Meyers, The Mind of the Founder, 354–357. Compare Madison's chart with those found in Chapter VII of Malthus, An Essay on the Principle of Population (New York: W.W. Norton and Company, 1976), 48–50. According to the

census of 1930, the population of the USA was 122,775,046. In 2010, the population was 308.7 million.
60. Madison in The James Madison Letters, 30.
61. For a review of these perspectives, see Franklin A. Kalinowski, "A Practical Solution to the Environmental Crisis: The Range of Options," Public Works Management and Policy Vol 2, No. 1 (July 1997), 66–78.
62. See, for example, World Commission on Environment and Development (Brundtland Commission), Our Common Future (New York: Oxford University Press, 1987); The President's Council on Sustainable Development, Sustainable America: A New Consensus (Washington, D.C.: U.S. Government Printing Office, 1996); Al Gore, Earth in the Balance (Boston and New York: Houghton Mifflin Company, 1992); J. Stockdale, "Pro-growth, Limits to Growth, and the Sustainable Development Synthesis," Society and Natural Resources, 2(3), 163–176.

PART III

The Heritage: Environmentalism and the Evolving Constitution

CHAPTER 9

The Constitution After 100 Years: Environmental Theory in the Gilded Age

As America prepared to commemorate the first 100 years of its Constitution's existence, philosophic forces were underway that would direct the nation's theoretical debate and political agenda for the next 100 years. American political thought became more focused and divided. Perspectives and values that were initially reflected in the debates of the Founding period—positions that contained profound consequences for humanity's relationship with the nonhuman environment—became more explicitly aimed at controversies surrounding resource conservation, the limits of state regulation, and the future of the then clearly emerging industrial mass society. At the same time that the environment became seen as an overt political issue, classical liberal thought began to take the form of contemporary conservatism as issues such as individualism, free market laissez-faire capitalism, and the sanctity of private property rights resisted efforts to define a larger "public interest." During the same period, Progressive liberalism emerged as the voice of a nation-wide civic humanism, now coupled to what Samuel Hays has called "the gospel of efficiency" as a group of intellectuals, scientists, and reformers called for regulation of capitalist markets, conservation of natural resources, and a unified national spirit that would beckon the country away from the pursuit of individual self-interest and toward sacrifice for the common good. Finally, agrarian Populism, with its deep distrust of both large-scale industrial corporations and a distant Eastern elite claiming to speak for all Americans, put forth its vision of a rural, small-scale economic and political system built around family farms, direct democracy, and a constrained financial sector. In many remarkable ways, contemporary environmental politics as characterized by the intellectual and policy struggles among conservative anti-environmentalists or free marketeers, liberal reform environmentalists or supporters of a managed "sustainable development," and decentralist bioregionalists or ecological economists can all be seen coming forth during the period that Mark Twain dubbed "the Gilded Age".

American Conservatism: Nature Against Nature

When Clinton Rossiter looked at American political thought as it appeared in the closing decades of the nineteenth century, he was impressed by the similarities in language being used across the political spectrum. "While the Left fought for social reform in state and nation with words like 'democracy,' 'liberty,' 'equality,' 'progress,' 'opportunity,' and 'individualism,' the Right struck back from its privileged position with the very same words ... [F]ew moved outside the Liberal tradition in their search for a persuasive rhetoric."[1] The language and theoretical logic being used by conservatives at the end of the nineteenth century were remarkably similar to the language and logic used by liberals at the close of the eighteenth century.

In 1883, Yale professor William Graham Sumner wrote the defining treatise of Gilded Age conservatism, What Social Classes Owe to Each Other.[2] If the title is restated as a question ("what *do* social classes owe each other?") Sumner's response can be cogently summarized in two words—absolutely nothing. There were no class-based or communal duties, obligations, or interests because communities and classes were merely collections of individuals, and the only commitments they had to each other were those that they contractually and voluntarily agreed to as individuals. Freedom and liberty, the twin spires of the liberal edifice, were both built upon the faith that human nature was essentially personal and idiosyncratic. Although a slim book, What Social Classes Owe to Each Other is filled with examples of the classical liberal mind set. "A society based on contract is a society of free and independent men, who form ties without favor or obligation, and co-operate without cringing or intrigue. A society based on contract, therefore, gives the utmost room and chance for individual development, and for all the self-reliance and dignity of a free man." "A free man in a free democracy has no duty whatever toward men of the same rank and standing, except respect, courtesy, and good-will ... He is, in a certain sense, an isolated man." For Sumner, equality, freedom, opportunity, rights, and democracy are all mutually defined and interplay with each other in a circle that has at its center the autonomous individual. "Certainly, liberty, and universal suffrage, and democracy are not pledges of care and protection, but they carry with them the exaction of individual responsibility. The State gives equal rights and equal chances just because it does not mean to give anything else. It sets each man on his feet, and gives him leave to run, just because it does not mean to carry him."[3]

As indebted as Gilded Age conservatism was to the classical liberal tradition, they did not hesitate to borrow concepts and logic from the civic humanist past. An important feature of that republican tradition was the ongoing struggle between power and liberty. When eighteenth-century republicans spoke of corruption, however, individual self-interest was invariably seen as the source of that evil, and the communal or civic interest was pictured as the potential victim of the assault. With their focus on the individual, American conservatives in the last decade of the nineteenth century managed a reversal in this civic logic. The terms "power," "liberty," and "corruption" remained, but now, it was the community that was seen as threatening corruption by using its power to invade the liberty and rights of the individual. For Sumner, it was not the community that was *being* threatened (in his logic, "the community" had very little meaning). It was a small, governmental elite, speaking *in the name of* a national community that was doing the threatening. The objects of this corrupt trespass of power were the rights and interests of the individual, and the instrument for perpetuating this violation became state regulation. In Sumner's reasoning, late nineteenth-century social policy and governmental regulations bore an ominous likeness to the eighteenth-century royal usurpations decried in documents such as the Declaration of Independence. "The history of the human race is one long story of attempts by certain persons and classes to obtain control of the power of the State, so as to win earthly gratifications at the expense of others ... They developed high-spun theories of nationality, patriotism, and loyalty. They took all the rank, glory, power, and prestige of the great civil organization, and they took all the rights." He continued, "[W]hat we need to do is to recognize the fact that we are face to face with the same old foes—the vices and passions of human nature ... The new foes must be met, as the old ones were met—by institutions and guarantees."[4]

To his dismay, however, Sumner perceived the trend of his time was not toward supporting the institutions of free competition (read capitalist economic markets) or to providing guarantees (most notably by securing individual property rights). Instead, he looked around and saw what he believed was a dangerous tendency toward government paternalism, regulation, and interference (an obvious reference to the growing Progressive movement). He was convinced, however, that in the not-too-distant future such ill-designed efforts were doomed. "The fashion of the time is to run to Government boards, commissions, and inspectors to set right everything which is wrong. No experience seems to damp the faith of our public in these instrumentalities ... The system of interference is a complete failure to the ends it aims at, and sooner or later it will fall of its own expense and be swept away. The two notions—one to regulate things by a committee of control, and the other to let things regulate themselves by the conflict of interests between free men—are diametrically opposed; and the former is corrupting to free institutions ..."[5]

Upon these theoretical foundations, Sumner was able to build his political program. Working from the premises of individualism, rights, and self-interests, he deduced the essential need for a political and economic system centered on *laissez-faire* government, the protection of private property rights, and an unfettered capitalist market. All of these concepts had been part of the political discussion during the Founding period, but it is perhaps surprising that the first two received relatively little endorsement. Jefferson's advocacy of limited government was aimed primarily at the national level of administration. At the local level—his "ward republics"—he was willing to concede rather broad powers to regulate property and influence personal behavior. Hamilton was certainly no supporter of *laissez-faire*. In his "Report on Manufactures," he had specifically rejected that physiocratic doctrine in favor of a corporatist model of explicit governmental support for America's infant industries. Only Madison's notion of the Federal government as an "umpire" can be seen as an early promotion of limited government.[6]

In the interim between the Founding and the Gilded Age, the issue of private property rights underwent considerable change. Andrew Jackson took the Jeffersonian notion of self-sufficient, yeoman farmers, stripped it of its connection to communal "ward republics," and reshaped private property into a democratic, individualized right to be protected from the interference of government and bankers while at the same time being encouraged through public policies such as westward expansion, Indian removal, and the cheap sale of public lands to homesteaders. Jefferson's concept of western "space" as the arena for new communities that would foster active citizens by allowing property holders to display their virtuous stewardship became the Jacksonian promotion of individualized, self-interested property owners vigorously resisting any communal oversight of their activities and demanding the privatization of whatever land or resources might still be held in collective hands.[7] Jefferson's yeoman farmer slowly mutated into the "forgotten man" of Sumner's Gilded Age conservatism on his way to becoming a member of Ronald Reagan's Sagebrush rebellion and Wise Use movements. The genesis may have been in classical republican thought, but the creature that eventually evolved bore a much closer resemblance to a liberal progeny.

Sumner's role in all of this was to emphasize the essentially individualistic and competitive aspects of property by making a distinction between the right to *have* and the right to *possess* property. He was shrewd enough to recognize that if all citizens were granted the right to have private property as part of some inalienable entitlement or as a precondition for their full development as human beings, then the argument could be made that it was the duty of society or the state to provide them with some requisite amount of property to fulfill those ends. This was obviously a route that Sumner did not want to travel and one, which he believed, had lead Progressives and socialists astray. In Sumner's mind, confusion and injustice were avoided by strictly delineating between the opportunity to struggle in order to acquire property and the protection of whatever property was acquired by those few who may have been successful in the contest on the one hand and an absolute entitlement to own property that could be extended to all members of the nation. Society should guarantee the opportunity to struggle, it should protect ownership of the booty thus acquired, but it could not, and should not, provide anyone any assurance that they would be successful in their quest for private property. "It cannot be said that each one has a right to have some property, because if one man had such a right some other man or men would be under a corresponding obligation to provide him with some property," argued Sumner. "Each has a right to acquire and possess property if he can. It is plain what fallacies are developed when we overlook this distinction. Those fallacies run through *all* socialistic schemes and theories."[8]

Pulling together the various pieces of this logic, William Graham Sumner pushed it through to its conclusion. If each individual had the right to pursue and to possess property, and if no one had any right to have property, then some individuals would emerge from the competition with considerable quantities of property, some with moderate amounts, and some with no property at all. Furthermore, if justice lay in the protection of the results of this competition, and if the workings of the capitalistic market produced wildly unequal results in terms of who possessed what, then the pursuit of justice lay in the defense of unequal social conditions. For Sumner, the more extreme the social and economic inequality found in any nation, the more impartial and just was its political and economic system. Inequality was the surest indicator of justice: the more unequal the outcome, the more just the process. "We each owe it to the other to guarantee rights. Rights do not pertain to *results*, but only to *chances*. They pertain to the *conditions* of the struggle for existence, not to any of the results of it," Sumner declared. "This, however, will not produce equal results, but it is right just because it will produce unequal results ... Therefore, the greater the chances the more unequal will be the fortune of these two sets of men. So it ought to be, in all justice and right reason. The yearning after equality is the offspring of envy and covetousness ..."[9]

Since this reasoning purported to apply to everyone in the country, Gilded Age conservatives took particular care in explaining its benefits to members of the middle and lower classes (its benefits to members of the privileged class were abundantly obvious). Economic competition, private property rights, and a highly stratified social order required workers, farmers, and indeed all citizens, to play their appropriate roles in the system. That meant competition between worker and worker, between workers and management, and between managers of one enterprise and another. The opportunity to "get ahead" and move up the scale of material possessions, increased consumption, and higher social standing required the unrestricted pursuit of this competition. With regard to those whose only means of acquiring property was through the sale of their labor, it was argued that their participation in the contest required the opportunity to compete, bargain, and agree to the best conditions of labor they could achieve. "Freedom of contract" became the centerpiece of the conservatives' program for industrial relations. It meant that each worker should approach his or her manager, negotiate the best arrangement possible regarding salary and working conditions, and either accept the agreement or move on to another employer and another set of negotiations. The stunningly naïve assumption, particularly given the always contending arguments rationalizing extreme levels of inequality, was that worker and employer were somehow negotiating from positions of relative equality. When the issue was the distribution of wealth, conservatives argued that inequality was good, and the pursuit of equality was mere "envy and covetousness." When wages and the working conditions of the masses were the topic, conservatives argued that equality substantially existed and was to be tacitly acknowledged as beneficial.

Providing some historical, practical, or moral justification for this ideology became Sumner's next task. Two defenses were offered. The first was an historical rationalization dating back to the early days of classical liberal thought. Sumner reused Adam Smith's argument of "the invisible hand" as a mechanistic, institutionally bound, pragmatic assertion that free market competition and private property rights were the surest, most efficient means to convert the pursuit of private interest into the attainment of the public good. Sumner followed this same logic, declaring "The modern industrial system is a great social co-operation. It is automatic and instinctive in its operation ... The parties are held together by impersonal force—supply and demand. They may never see each other; they may be separated by half the circumference of the globe. Their co-operation in the social effort is combined and distributed by financial machinery ... All this goes on so smoothly and naturally that we forget to notice it."[10] Neither Smith nor Sumner made explicit the exact character of this "public interest" or "common good" that is being unintentionally achieved, but significant hints surround the comments noted earlier. Immediately preceding his "invisible hand" comment, Smith opines "every individual necessarily labours to render the annual revenue of the society as great as he can ..." and Sumner follows his version of the invisible hand by commenting that "by the great social organization the whole civilized body (and soon we shall say the whole human race) keeps up a combined assault on Nature for the means of subsistence. Civilized society may be said to be maintained in an unnatural position, at an elevation above the earth, or above the natural state of human society." For Smith, the public interest consists in maximizing economic wealth through the unending growth of industrial production. Sumner's only minor variation on this theme sees the goal of social organization as the pursuit of "civilization," defined by the unremitting conquest, subjugation, and domination of Nature.

Sumner's second ethical defense takes advantage of changes in the intellectual climate of the nineteenth century, for between the writings of classical liberals such as Adam Smith and the vindication of *laissez-faire* capitalism being put forth by apologists for the robber barons had come the writings and thought of Charles Darwin and the Social Darwinist interpretations put forth by right-wing spokesmen such as Herbert Spencer and T.H. Huxley. As we saw in Chapter 3, Social Darwinism could take either right-wing forms (the positions of Spencer and Huxley) or left-wing communitarian forms (the collectivist libertarian anarchism of Peter Kropotkin). At the end of the nineteenth century, Sumner clearly used Social Darwinism to right-wing ends. Although the Yale professor resorted to Adam Smith's invisible hand defense upon occasions, the main underpinning of his theory rested upon his version of Social Darwinism. In 1870, he read a series of essays by Herbert Spencer that was later to become <u>The Study of Sociology</u>, and, from then on, the doctrine provided the unsentimental, organic, and ostensibly scientific basis that Sumner obviously found extremely attractive, since he employed it so often. "Nature's forces know no pity. Just so in sociology," argues Sumner. "The forces know no pity … Nature's remedies against vice are terrible. She removes the victims without pity. A drunkard in the gutter is just where he ought to be, according to the fitness and tendency of things. Nature has set up on him the process of decline and dissolution by which she removes things that have survived their usefulness."[11] In adopting Social Darwinism to vindicate the harshness of human life in Gilded Age capitalist society, Sumner chose a particular turn of words that are significant because they point deep toward the foundations of his conservative thought. In the opening chapter of <u>What Social Classes Owe to Each Other</u>, Sumner makes the following assertion, "[G]od and Nature have ordained the chances and conditions of life on earth once and for all. The case cannot be reopened. We cannot get a revision of the laws of human life … Certain ills belong to the hardships of human life. They are natural. They are part of the struggle with Nature for existence."[12]

As Sumner saw it, the "natural" thing for humans to do was to dominate, control, and subdue "Nature." Beneath all the talk about individualism, liberty, rights, private property, *laissez-faire*, and competitive markets, lay a peculiarly paradoxical perception of humanity's place in the ecosystem. Our internally fixed destiny was to manipulate, alter, and (ultimately) destroy the external biosphere we found around us. Human nature was locked in a war against pristine Nature. It was nature against Nature. In an essay bearing the alternative titles "Socialism" or "The Challenge of Facts," Sumner gave repetitive declaration that the underpinnings of his Social Darwinism rested upon a dedication to the conquest of Nature. "Man is born under the necessity of sustaining the existence he has received by an onerous struggle against nature, both to win what is essential to his life and to ward off what is prejudicial to it. He is born under a burden and a necessity. Nature holds what is essential to him, but she offers nothing gratuitously." For Sumner, this world view of domination required the development of technology which, for him, was seen as armaments in the war against Nature. "The next great fact we have to notice in regard to the struggle of human life is that labor which is spent in a direct struggle with nature is severe in the extreme and is but slightly productive. To subjugate nature, man needs weapons and tools." Finally, he pulls the various components of his ideology together. In his view, private property, inequality, competition, and the domination of nature combine to form the essential elements of existence. "Private property, also, which we have seen to be a feature of society organized in accordance with the natural conditions of the struggle for existence produces inequalities between men. The struggle for existence is aimed against nature. It is from her niggardly hand that we have to wrest the satisfactions for our needs, but our fellow-men are our competitors for the meager supply. Competition, therefore, is a law of nature."[13] The "natural conditions" of conservative capitalism are necessary in order to pursue the "struggle ... against nature."

Sumner is saying that it is "natural" for humans to suppress and destroy "nature." Although he was apparently unconscious of the fact, Sumner utilized the terms "natural" and "nature" in two senses with historically different roots. In one instance, "natural" can be employed with its Aristotelian meaning to signify that which is in accordance with the full development of human potential. Hence, human reason (and with an important distortion in logic), science, technology, and industrial society can be seen as being "natural." In the other instance, "nature" can be used in its Rousseauian sense to refer to an original, pristine condition that is self-willed and has not been corrupted. "Nature" (both human and nonhuman) is all that is wild, unmanipulated, and operating exclusively on signals from its instinctive, savage, inner self. "Nature against nature" really suggests a dispute between these Aristotelian and Rousseauian connotations that persists to this day as scholars, students, and commentators become increasingly uncomfortable when discussing what is "natural" and what is not. The once critical distinction between the natural and the artificial has been confused and lost, and in that sad recognition is expressed the depth of alienation that human nature has undergone from Nature.[14]

The ideological origins of contemporary conservatism can be pictured as coming in three layers or geological strata of argument. The middle layer consists of the key concepts that form the basis of their political faith. Individualism, liberty, rights, the pursuit of self-interest, and the advocacy of limited government all act as evidence of conservatism's derivation from the classical liberal tradition. At the upper or surface layer is found the policy implications of this perspective. Private property rights, unregulated capitalist markets, and "freedom of contract" between employer and employee demonstrate the linkages between Sumner's philosophy and the political program that we now refer to as American conservatism. Beneath these assumptions and policy prescriptions, however, lies the deep ethical justification without which classical liberalism and contemporary conservatism lose all coherence. It is here where the logic of perpetual economic growth, right-wing Social Darwinism, and the conquest of Nature reveal the fundamental anti-environmentalism of this trend in American social thought. Gilded Age conservatism is a bridge from the classical liberalism of the Founding period to the anti-environmentalism of the twentieth century, and it is a mistake to view this cornucopian/Promethean agenda as merely a tactical policy sidelight to the conservative cause. Without an unwavering commitment to anti-environmentalism, contemporary conservatism loses its rationalization and makes no logical or ethical sense. The challenge is to seek out and isolate alternative perspectives that may lead to a more environmentally healthy and politically just social order. Gilded Age conservatives believed that the Progressive movement represented the most fundamental challenge to their ideology. A deeper look at the environmental theory of those Progressives may reveal that the fears of the conservatives were, unfortunately, unwarranted.

Progressive Resource Conservation: Managed Domination

In a simple schema of political thought during the Founding period, Thomas Jefferson could be characterized by his distrust of elites and his enduring faith in the American masses. In contrast, Alexander Hamilton exhibited a faith in the elites and an intense distrust for the common American citizen. James Madison represents a third perspective, which can be portrayed as distrusting the granting power to either the elites or the masses. Madison placed his confidence in the impersonal workings of law and political institutions. Pictured in this light, the American Progressive movement can be seen as beginning with the political thought of Abraham Lincoln, and his philosophy can be briefly summarized by setting it apart from each of the towering geniuses of the Founding. Abraham Lincoln distrusted the masses (they lynched people); he distrusted the economic elites (they were corrupt and only interested in money); and, perhaps strangely for a lawyer and politician, Lincoln distrusted the Constitution. It was, after all, the law and the Constitution that established and protected chattel slavery. According to Lincoln, the central flaw of the blind workings of Madison's institutions, checks and balances, separation of powers, and competitive adversarial markets was their lack of a distinct moral vision. This nation, he asserted, "must have a new birth of freedom," and that meant the establishment of a spiritual, ethical vision of national unity, sacrifice, and moral purpose. It also meant the creation of an elite, based not on economics and the pursuit of self-interest but a small group of intellectual and disinterested leaders, who would articulate this national purpose, preserve the indestructible Union of the states, call forth the American public to sacrifice for the common good, and reform the tainted Constitution so it could truly become the instrument of national greatness it was meant to be. Jefferson's moral ward republics would become a continental community, and virtuous statesmen would direct Hamilton's centralized state. Madison's extended empire would get a heart. The organizational and philosophical model was the Union army—huge, disciplined, hierarchical, under the command of a leader who understood its moral purpose, and willing to give "the last full measure of devotion."[15] It was the veterans of Lincoln's Union army, organized into the politically powerful Grand Army of the Republic, or G.A.R. (which some claimed stood for "generally always Republican"), that formed the backbone of the Progressive movement during the Gilded Age.[16] The overall pattern established in Abraham Lincoln's political thought was extended and sharpened by Progressive thinkers, the most significant of which was Herbert Croly.

Herbert Croly's political philosophy is found in his first, and most important, book, The Promise of American Life (1909), his second book, Progressive Democracy (1914), and the numerous articles he wrote while editor of The New Republic magazine from 1914 to 1928.[17] Within all of these writings, a consistent pattern of argument, which may be divided into three parts or stages, is evident. First, he attacks. Croly historically relates the problems of modern society to what he believes are their source: a failed or defected ideological foundation. In The Promise, the premises of classical liberalism take the brunt of his assault. Here, he repeatedly asserts his view that individualism and the rampant, incessant acquisitiveness characteristic of liberal society are the origins of America's slide toward political and moral degeneracy. Croly's second stage is the formulation of an alternative social structure (he uses the term "reconstruction"). He creates a verbal image of his ideal society and provides a blueprint of a moral community based on "the brotherhood of humanity." His solution to the problems of American society, put simply, is civic humanism. In the final stage of his social analysis, Croly tries to design some means for moving from the fallen contemporary condition to renewed civic morality and national community. The tone of this section marks a distinct change from the normal self-confidence of Croly's argument. Its vagueness reflects a fundamental incommensurability between the republican and liberal idioms.

There are, however, two significant differences between Herbert Croly's interpretation of politics and the civic humanist/liberalism debate of the eighteenth century. During the Founding period, many who spoke in the voice of civic humanism argued for decentralized government. By contrast, it was the liberals who promoted an extended republic built on a diversity of interests. By the time of the Progressive movement, Croly reordered these positions, believing a true continental community was possible. The executive would be the spokesman for "the public interest" as America became one politically homogeneous "nationalized democracy." For Croly, it was in local government and in legislative branches everywhere that one could find the conflictual clash of private interests. The second change from the earlier debate involved a new philosophical element that had been added to political discourse during the nineteenth century. The writings of Henri Saint-Simon, Auguste Comte, and others had instilled in many people a faith in the potential for science, planning, and organization. Through his father, David Goodman Croly, Herbert Croly had been thoroughly indoctrinated in this faith. "From my earliest years," Croly would later write, "it was [my father's] endeavor to teach me to understand and believe in the religion of Auguste Comte ..."[18] The result was a civic humanism projected onto the entire nation and blended with a trust in bureaucratic planning. The emotional, spiritual, and moral would be combined with the specialized, disinterested, and rational. Corruption, for example, would be attacked as being both ethically wrong and "inefficient." Aristotle and Harrington would meet Saint-Simon and Comte.

For Croly, America was on the road to destruction, and the seriousness of this predicament suggested the existence of some radical disease. He sought to prove this premise by analyzing the history of American thought, which he characterized as being dominated by two different and antagonistic groups of ideas: those of Alexander Hamilton and those of Thomas Jefferson which he turned into archetypes (some might say caricatures) and ascribed to each of their philosophies a single, driving force. For Hamilton, it was the force of nationalism, a drive that Croly came more and more to equate with community. Jefferson, on the other hand, became the representative of individualism, and, for Croly, individualism epitomized corruption and communal dissolution. Croly believed that Hamilton's policy created a definite theory of the function of government. The central government was to be used, not merely to maintain the Constitution but to promote the "national interest" and to consolidate the "national organization." This implied an interference with the course of American economic and political business. Most importantly, it implied a "conscious and indefatigable attempt on the part of the national leaders to promote the national welfare."[19]

The only flaw in Hamilton's national plan, according to Croly, was its fear of and lack of faith in the general public. Hamilton, Croly explained, "did not seek a sufficiently broad, popular basis for the realization of [his] ideas." Instead, Hamilton bestowed upon the central government the support of a strong special interest: the rich. "He succeeded in imbuing both men of property and the mass of the 'plain people' with the idea that the well-to-do were the peculiar beneficiaries of the American Federal organization, the result being that the rising democracy came more than ever to distrust the national government."[20] Croly went on to explain that as soon as Hamilton's opponents discovered that his ideas and plans were in some respects inimical to popular democracy, they confronted him with "one of the most implacable and unscrupulous oppositions which ever abused a faithful and useful public servant."[21] This opposition was led by Jefferson, who Croly attacked mercilessly. Jefferson's triumph, for Croly, meant that the democratic political system was made tantamount to "extreme" individualism, or "rampant" individualism, or sometimes "excessive," "automatic," or "blind" individualism. Croly never used the noun without an accompanying derogatory adjective, and as Byron Dexter has observed, "in the adjective lay the meaning of the word for him."[22] In this way, Croly united Hamilton, nationalism, active government, and community and singled them out for praise, while targeting Jefferson, localism, individualism, and a limited Federal government for censure. Other elements of the classical liberal theory were also attacked.

Croly had no faith in an "invisible hand" or "checks and balances." In The Promise of American Life, he says, "The plain fact is that the individual in freely and energetically pursuing his own private purposes has not been the inevitable public benefactor assumed by the traditional American interpretation of democracy."²³ To believe that social progress could be achieved through the inevitable workings of a self-correcting system was to advocate what Croly contemptuously called "the politics of drift." If progress were to be achieved, it would be the result of struggle directed by a conscious will.²⁴ The alternative to classical liberalism's pluralist system would be an organic, corporatist, concentrated state which would actively lead toward a moral, idealistic vision of a common good. At the beginning of <u>The Promise of American Life</u>, Croly outlined his alternative: "In becoming responsible for the subordination of the individual to the demand of a dominant and constructive national purpose, the American state will in effect be making itself responsible for a morally and socially desirable distribution of wealth."²⁵ Croly's image of a reformed America rested upon the restraint of big business, the promotion of organized labor, and the creation of a strong centralized government. Within his solution rested the promise of American life. At no time did Croly argue for the decentralization of huge social organizations; he merely wanted them managed for national purposes. But if Croly's image of the problem and the solution was clear, he vacillated wildly when considering the means to move from America's fallen state to its promised future.

Over the course of his writing career, Croly considered and rejected a number of possibilities for transition. In <u>The Promise of American Life</u>, Croly placed his faith in the charismatic hero who would change America through the force of his political leadership and the emotional appeal of his personality. Theodore Roosevelt was his model.²⁶ When he wrote <u>Progressive Democracy</u>, however, Croly shifted his faith and reliance to the average citizen as a vehicle for transformation (Roosevelt was out and Taft was President). With the outset of World War I, Croly argued that the Progressive agenda could best be realized by bonding it to the nationalism of the war fever (Taft was out, and Wilson was President). This uncertainty over the means for effecting change within Croly's vision specifically and in Progressive ideology more generally reveals a tension between anxiety toward the future (the republican fear of entropy and chaos) and hope for a better society to come (the liberal faith in progress). Insight into this quandary may be had by looking at the application of Progressive thought to the issue of environmental protection.

As the nineteenth century drew to a close, signs were abundant that the relationship between humans and Nature was not good. Earlier instances of environmental destruction and "waste" seemed to be getting worse. In 1832, artist, traveler, and naturalist George Catlin had written an account of his travels in the early American West. Catlin decried the massive slaughter of buffalo he had witnessed and bemoaned the "profligate waste" as natives took only the tongues of the animals, which they traded for the quickly consumed whiskey of the white man. In 1877, Secretary of Interior Carl Schurz (German immigrant and Division commander in Lincoln's Army of the Potomac) submitted his annual report deploring the "rapidity with which this country is being stripped of its forests" and warning that if the cutting were to continue at its present rate "the supply of timber in the United States, will, in less than twenty years, fall considerably short of our home necessities." By 1908, in the Chicago Zoo, the last passenger pigeon in the world—that species that once darkened the skies across America—died. And by way of summarizing all this ecological destruction, Frederick Jackson Turner first wrote of the importance of the frontier in shaping American culture, then the 1890 census reported that the frontier was closed, and finally Turner, in his 1910 presidential address to the American Historical Society, commended as "a wonderful chapter, this final rush of American energy upon the remaining wilderness," and then noted that it was "peculiarly the era when competitive individualism in the midst of vast unappropriated opportunities changed into the monopoly of the fundamental industrial processes by huge aggregations of capital as the free lands disappeared."[27] Although the connection to Founding ideology was never explicitly enunciated, the message was apparent: the boundless room for Jeffersonian yeoman communities and the extended republic of fragmented Madisonian factions were gone. That left only the political vision of Hamilton's efficient state and Lincoln's moral Union.

Anxiety, apprehension, and fear were powerful tools in the Progressives' philosophical kit. If the fabled existence of a boundless, wild Eden was essential to the classical liberal's faith in freedom, opportunity, and individualism (and had it not been Locke himself who declared that, "in the beginning all the world was America"?), and if that cornucopian myth was now gone, then the liberal dream was exposed as either an outdated delusion or a sentimental pining for social and ecological conditions that will never exist again.[28] The century-long experiment with Madison's extended republic would have to come to an end. The *status quo* faith in markets, checks and balances, and institutions would have to be discarded. Anxiety was the result of taking present trends in demographics, resource extraction, or water consumption and projecting them linearly into the future. This is what Progressives meant by the "politics of drift," and this is why their writings are so full of appeals to "future generations," "those yet unborn," "our children and our children's children," or simply "posterity."[29]

Even if one accepted the argument that the ever-expanding, market-driven, individualistic, self-interested liberal society could no longer continue, two courses of action remained open: either there could be a radical reconsideration and reconstruction of the basic assumptions of the liberal faith, or new means would have to be found to achieve the same ends. Rather than engaging in a radical critique of existing social patterns, the Progressives adopted the latter option and constructed a program to achieve the same fundamental ends of liberalism but with new economic and political methods. The approach was that of reform and not radical reconsideration. The program became known as Resource Conservation, and the intellectual force behind it was Roosevelt's confidant and head of the Division of Forestry, Gifford Pinchot. To Pinchot fell the task of redesigning the means of reaching the liberal ends of economic growth and the domination of Nature while rejecting the discredited methods of individual self-interest and free market competition. "If we, as a Nation, are to continue the wonderful growth we have had," wrote Pinchot, "it is forethought and foresight which must give us the capacity to *go on as we have been going.*"[30]

Pinchot took the basic principles of Progressivism and applied them to the specific issue of environmental policy. An important component of the new philosophy and policy of Progressive Resource Conservation (in Pinchot's mind Progressivism and Conservation were both philosophy and policy and were inextricably linked) was the Forestry Director's recognition that the prevailing methods of policy fragmentation and decentralization would not work. Pinchot believed that any policy, but particularly environmental policy, needed to be coordinated on a national scale in order to be effective. Removing management of resources from state or local political jurisdiction and thereby effectively nationalizing the regulatory process (or, as was eventually made the case with the Tennessee Valley Authority, regionalizing the level of control), and moving toward policy integration that cut across particular ecological media such as forests, water, wildlife, and soil became the foundation points for Pinchot's Conservation philosophy. The now-famous story of Pinchot's epiphany in Rock Creek Park shows not only how Conservation began but also where it might be heading. During his celebrated horseback ride, "in the gathering gloom of an expiring day," Gifford Pinchot was bringing his work home and brooding over the seemingly insurmountable difficulties he faced. "The forest and its relation to streams and inland navigation, to water power and flood control: to the soil and its erosion; to coal and oil and other minerals; to fish and game; and many another possible uses or waste of natural resources—these questions would not let him be." Then the revelation struck. "Suddenly the idea flashed through my head that there was unity in this complication ... Here were no longer a lot of different, independent, and often antagonistic questions, each on its own separate little island ... here was one single question with many parts." For Pinchot, the goal toward which all these separate issues pointed suddenly became obvious; it was "the use of the earth for the good of man."[31] Eventually, the approach would be called "sustained-yield multiple use" and the insight would morph into the trite expression "a holistic approach," but Pinchot had no doubt that he had stumble upon something profound in that city park.

Viewed from alternating perspectives, the recognition that ecological issues were interconnected and mutually reinforcing could be seen as either simplistically self-evident, or religiously apocalyptic, or politically revolutionary. If Pinchot and subsequent supporters of mainstream Conservation ever saw the complete political ramifications of what they were saying, they never voiced their conclusions, for at the deepest level the logic carried with it disturbing consequences. The facts are that Nature is a totally integrated system, that "tinkering" with one aspect at a time is very likely to create a problem for every one solved, and that only a political system with totalitarian control could possibly hope to direct all aspects of such a complex organism. Since establishing a political system with the potential for totalitarian control was precisely the type of decision-making and administrative apparatus that the men in Philadelphia in 1787 sought to preclude, one of the possible consequences of Pinchot's revelation was the awareness that the Constitution of the United States cannot be used as a framework for addressing environmental crises. A genuinely comprehensive approach to ecological problems would have to operate against, or at least around, the fragmented processes and restrictions of the Constitution.

Pinchot did acknowledge, however, that on the surface, there was something prosaic in his observation. "But, you may say, hadn't plenty of people before that day seen the value of Forestry, or irrigation, of developing our streams, and much besides? Hadn't plenty pointed out that forests, for example, affect floods, and many other cases in which one natural resource reacts upon another?" And yet, it was the amalgamation of facts, goals, and rationalizations that made the apparently banal so momentous. What made Resource Conservation so different as a public policy was this simple fact of ecological interconnection, coupled with an intense ethical bifurcation built upon a division between humans and Nature, and joined with a divine endorsement for anthropocentric hegemonic manipulation. Pinchot's Conservation combined a holistic perspective, a humanistic ethic, and the morality of subjugation into a single course of action. Tennessee Valley Authority Director David Lilienthal would bring these aspects together when he proclaimed, "What God had made one, man was to develop as one," and Gifford Pinchot was equally biblical in his unification of ecological principles, administrative cohesion, and sanctimonious domination. "[H]ere was one question instead of many, one gigantic single problem that must be solved if generations, as they came and went, were to live civilized, happy, useful lives in the lands which the Lord their God had given them."[32]

Packaging this policy to make it palpable to the American public (and the elites in the American government) drew Pinchot and his fellow Conservationists in the direction of seeking a scientific rationalization for what was, in effect, a shift in political theory. Just as Sumner used Social Darwinism to cover his politics with a veil of scientific acceptability, Pinchot pictured Conservation as being in accord with the best practices in science, and therefore, it was argued, obviously preferable to any alternative that Progressives would depict as unreasonable, subjective, lacking in empirical justification, based on false premises, or, more simply, sentimental. In the best book-length study of Progressive Resource Conservation, <u>Conservation and the Gospel of Efficiency</u>, Samuel P. Hays emphasizes the alliance of politics and science in the Theodore Roosevelt–Pinchot policy. "The deepest significance of the conservation movement, however, lay in its political implications: how should resource decisions be made and by whom? Each resource problem involved conflicts. Should they be resolved through partisan politics, through compromise among competing groups, or through judicial decisions?" Of course, to say "partisan politics," "compromise," and "judicial decisions" is to describe classical liberal decision making. Hays, recognizes, however, "To conservationists, such methods would defeat the inner spirit of the gospel of efficiency. Instead, experts, using technical and scientific methods, should decide all matters of development and utilization of resources, all problems of allocation of funds … The crux of the gospel of efficiency lay in a rational and scientific method of making basic technological decisions through a single, central authority."[33]

Hays is astute to call the pursuit of efficiency a critical element in the Progressive's "gospel," for the term became something of a ritualized response to a social catechism. It seemed to point toward a national value consensus; for who would defend the pursuit of its opposite—"waste"? It appeared to bridge the gap between science, technology, and economics and make them all parts of the same enterprise. More than the mere advocacy of technocracy (although that Comptean idea was certainly part of it), the worship of efficiency evolved into an apparently scientific justification for the entire Progressive agenda. To the engineers and foresters in the Roosevelt administration, efficiency entailed rational, coordinated, centralized planning, and since the economic marketplace seemed incapable of doing this, the economic market place became anathema to the best scientific and economic allocation of resources. To be efficient, an economic system needed to be corporatist. Technocratic and economic efficiency also suggested a political system that was elitist and antidemocratic. Citizenship became reduced to mere membership in a cheering section as the national leadership pursued the most efficient means of achieving "the greatest good for the greatest number for the longest period of time." As Hays puts it, "Roosevelt drew closer to a conception of the political organization of society wherein representative government would be minimized, and a strong leader, ruling through vigorous purpose, efficiency, and technology, would derive his support from a direct, personal relationship with his people."[34] Theodore Roosevelt's "stewardship theory" began the transition to the contemporary plebiscitary presidency.[35]

There was, therefore, as much politics as there was science and economics wrapped up in the gospel of efficiency, and to these ends, Progressive Conservationists put forth a united front of support. Pinchot closed Breaking New Ground with a summation of his philosophy; "The first duty of the human race on the material side is to control the use of the earth and all that therein is. Conservation means the wise use of the earth and its resources for the lasting good of men ... Nationally, the outgrowth and result of Conservation is efficiency."[36] Theodore Roosevelt drew together the themes of conservation, republican civic virtue, and efficiency in his address to the 1908 Conference of Governors. "[L]et us remember that the conservation of our natural resources, though the gravest problem of today, is yet but part of another and greater problem to which this Nation is not yet awake, but to which it will awake in time, and with which it must hereafter grapple if it is to live—the problem of national efficiency, the patriotic duty of insuring the safety and continuance of the Nation."[37]

Perhaps the man who best embodied these various themes was WJ McGee—self-made scientist, member of the Geological Survey, organizer of the 1908 Governor's Conference and the Inland Waterways Commission, and speechwriter for Pinchot and Roosevelt. Gifford Pinchot called McGee "the scientific brains of the new movement," and Samuel Hays argues that he was "the chief theorist of the conservation movement" who in his speeches and writings "formalized the spirit of efficient planning." Hays contends that for McGee, "The ultimate goal was absolute efficiency."[38] John Rodman, however, points out that McGee's obsession with efficiency was only part of his greater objective, which was the absolute subjugation of Nature. Coordinated, comprehensive state planning and regulation were the means of achieving efficiency, and efficiency was the way to achieve the liberal goals of growth and consumption. McGee declared that in the long run (and wasn't a concern for distant posterity a guiding principle of Progressivism?), the aim was "a conscious and purposeful entering into control over nature, through natural resources, for the direct benefit of mankind." Unable or unwilling to manage *human* population and *human* consumption, Progressive Conservationists decided to manage everything else instead.

Pinchot and McGee represented the technocratic side of Conservation and Croly/Roosevelt the theoretical and political side of Progressivism. Yet another feature of Progressivism is illustrated in the writings of an early sociologist and advocate of political reform, Lester Frank Ward. Once again, the biographical profile of Progressives repeats itself: Ward was a veteran of Lincoln's Union army (seriously wounded at Chancellorsville and discharged), a professional bureaucrat in the Federal government (Treasury Department), and an Ivy League college professor (Brown University). In his most widely read writings, Ward articulated a version of evolutionary sociology that is highly critical of Spencer's Social Darwinism. While giving wholehearted acceptance to Darwinian evolution *as applied to nonhumans*, Ward argued that the human mind was capable of moving beyond "the economy of nature" to an "economy of mind." Ward rejected the use of terms such as "survival of the fittest" or "natural selection" to rationalize the unrestricted competition of market capitalism, but, interestingly, his renunciation of Social Darwinism is based on the belief that it makes humans *too attached to Nature*! Right-wing social Darwinism, for all its intellectual shortcomings that emphasized only the dominating side of behavior, still believed that humans were governed by natural processes. Ward argued that civilized humans were capable of severing even this frail link with Nature, and through science, technology, and comprehensive planning could artificially manufacture an Earth that was far superior to anything mere evolution could provide.[39] This version of Progressive thought sheds light on the fact that at its foundation, Progressivism is as dedicated to economic growth and the domination of Nature as any of William Graham Sumner's writings. Ward's thought takes Progressivism's antinaturalism and humanistic ethics to their final extreme in arguing that the complete subjugation of Nature through scientific planning stands as proof of the moral superiority of humans. "Any good human engineer, in other words, could do a better job of designing the environment than nature has," asserts Donald Worster in summary of Ward's belief. "Only in a world totally under its own control could any one species pursue its private goals in the rigidly efficient, straight-line fashion Ward admired."[40] For Lester Frank Ward, what was formerly a bestial struggle for existence could become an anthropocentric moral crusade.

Progressives were as committed to the ethical gulf between humans and Nature as any conservative ever was. In fact, if there is one sentence that most succinctly summarizes the environmental theory of Gifford Pinchot, it is his proclamation in <u>Breaking New Ground</u>, "There are just two things in this material earth—people and natural resources."[41] Progressives would extend some degree of protection to Nature, but the justification for those safeguards was always *human* development, *human* welfare, and *human* economic growth. If demands for ecological health ever came into conflict with the increase in human population or the increase in human material consumption, Progressives unequivocally gave preference to their own species. Again, it was Pinchot who stated the case with blunt honesty. "The object of our forest policy is not to preserve the forests because they are beautiful ... or because they are refuges for the wild creatures of the wilderness ... but ... the making of prosperous homes ... Every other consideration comes as secondary."[42]

Although it was thought that a focus on efficiency could act as a bridge between science and economics, in practice that bridge was always a one-way street. Scientific management adopted concepts from mainstream economics, but economics never adopted such common ecological principles as "carrying capacity." What made "efficiency" so attractive was the idea that we as a society could keep doing what we had always been doing as long as we got better at doing it. The serious questioning of our goals was avoided, and only incremental adjustments in our means were considered. In this way, Progressives could continue to trick themselves into believing that no irreconcilable contradictions exist between genuinely long-term environmental health and exponential human economic growth. But if there is more to the environmental crisis than efficiency, if the scale of human activity and the environmental values of citizens are important, then Progressive Conservation in both its Gilded Age and its contemporary "sustainable growth" phases may prove equally incapable of resolving those fundamental concerns. At that point, America will have to search its cultural roots for a perspective that is radically different from both Sumner's conservatism and Croly's Progressivism. In the nineteenth century—and possibly also today—that third alternative was provided by agrarian Populism.

Populism: The Unfulfilled Promise of American Agrarianism

When the term "populism" is used today, two facts usually become apparent: first, people using the term often do so without a clear understanding of the theoretical components and historical origins of the Populist movement; and second, "populism" today is associated with both a conservative anti-government, anti-environment faction; and a radical, local, pro-environment form of bioregionalism that is sometimes called "civic environmentalism." Gilded Age Populism is an important bridge with one anchor in eighteenth-century agrarian republicanism and the other end of the bridge branching in two distinct directions. In order to appreciate the historical significance of the Populist movement and to understand Populism's important, albeit ambivalent, role in modern day environmentalism, its nineteenth-century roots must be explored and its evolution traced.

With the creation of the People's Party in St. Louis, Missouri, in February of 1892, the Populist movement was given an official umbrella organization that could nominate candidates for national office and seek to gain control of the government. These men and women can be grouped into two overarching organizations, the National Farmers' Alliance and Industrial Union (often called the Southern Alliance) and the National Farmer's Alliance (commonly called the Northern Alliance or the Northwestern Alliance). These alliances were composed of a multitude of organizations which focused either on specific issues, such as disenchanted former members of the Grange, the Greenback Party, the single taxers, the Silverites, and the Prohibitionists or those that were geographically structured at the state level around a growing movement for agricultural cooperatives.[43] The programs and platforms of these organizations are a valuable source for Populist ideology. So, too, are the speeches and writings of major Populist activists. Gilded Age Populism seemed to produce more than its share of colorful and eccentric spokespersons. These include the long-bearded US Senator from Kansas, William Peffer; the powerful orator and self-trained lawyer, Mary Elizabeth Lease; and the "sage of Nininger" (Minnesota), the outspoken, always radical, master of "unsparing denunciation and encomium," Ignatius Donnelly.[44] A list of the major Populist leaders would also include Kansas Congressmen Jerry Simpson (called "Sockless Jerry" for a variety of obscure and contradictory reasons); Georgia Congressman and Senator, Tom Watson; Texas Alliance man Charles Macune; and the 1892 People's Party Presidential Candidate, General James B. Weaver. None of these people, it must be noted, can be considered a political theorist in a formal or professional sense of that term. They were, above all else, activists who embodied a zealous commitment to reform and who used ideas as a means of inspiring rather than simply enlightening. There is no *magnum opus* of Populist thought; not even a single exegesis of their philosophy.[45] In order to discover what Populists believed in, multiple sources and differing levels of argument must be picked through. Speeches, platforms, pamphlets, newspaper articles and editorials, letters, the Congressional Record, and even a novel or two contain pieces of the Populist vision.

If Gilded Age Populism could be distilled down to a few essential ideological elements, the most basic of those would be a commitment to a class struggle between the haves and have-nots—the oppressors and oppressed—the forces of evil and forces of good. The Populist message is one that pictures America divided between the elites and the masses, those who wield and gain the benefits of power, and those who are subjected to the will of others. In this context, Populism is the revolt of the underdog. The Populist goal, repeated in various formats and diverse prescriptions, is a call for social justice, which they saw as achievable only in a society of economic and social equality structured upon an active political system rooted in direct democracy, citizen participation, and the virtues of communal responsibility. If these objectives sound to the modern ear like a reformulation of Jeffersonian democracy, nineteenth-century Populists were also quite aware of the linkages. In his Fourth of July speech in 1893, Tom Watson made the connection explicit. "I believe in the Jeffersonian creed with all my heart, and think that all the aims of good government can be covered by that one sentence. EQUAL AND EXACT JUSTICE TO ALL MEN! To the rich and to the poor, to the farmer and to the merchant; to the Banker and the miner; to the scholar and the ditcher."[46]

It is said that Texas Populist James H. ("Cyclone") Davis would carry volumes of Jefferson's collected works to the podium when he delivered his fiery speeches, and if it is true that Davis' interpretation of Jefferson's legacy was often imaginative and self-serving, it is also true that no Populist ever gave Hamilton the type of glowing support found in Croly. Populists, especially southern Populists, were proud to consider themselves the heirs of Jeffersonian politics.[47]

In many ways, the Populists were both more superficial and more complex than the Progressives. On the one hand, the passion and religious fervor that Donnelly, Watson, Lease, and the thousands of Alliance lecturers brought to their speeches and rallies cannot be accounted for simply by the desire to relieve debt and bring money to economically depressed farmers. There was more at work than that. Beneath the specific grievances and the various schemes to ameliorate the plight of the agrarian class—beneath all the talk of subtreasuries, free silver, railroad passes—there was a frustration and an anger that cannot be explained in simple economic terms. For many Populists, the real problem was the betrayal and loss of what they saw as the American dream. The West was supposed to be a place where individuals could find independence, where hard work would bring reward, and where communities could be built that would be strong and self-sustaining. Instead, Southern, Midwestern and Western farmers found themselves completely at the mercy of institutions over which they had no control. The furnishing merchants traded goods and credit for forced monocropping, eventual foreclosure, and life as a landless tenant. The banks controlled their mortgages and their loans. The railroads controlled the shipment and prices of their goods. The government clearly cared more about the security and profits of the furnishing merchants, banks, and railroads than the farming families. Finally, the Democratic and Republican Parties seemed only to care for their votes without ever addressing the gnawing inside that kept saying to the farmers, "it wasn't supposed to be this way."

This is why explanations that claimed the farmers' plight was the result of "overproduction" generated such rage. To argue that the agrarian economic crisis was the result of depressed farm commodity prices and that low prices were created by overproduction on the part of farmers was equivalent to saying that farmers were the source of their own problems by being too good at what they were doing. It was, from the farmer's point of view, tantamount to asserting that the entire American agrarian dream—some land to call your own, hard work, studying and working with the forces of Nature, frugality, and self-sacrifice—amounted to nothing more than a sham and a cruel hoax. If this was the conclusion toward which objective economic analysis pointed, then, for the farmers, objective analysis be damned. Alternative explanations were called for, and it was here that the agrarian movement reached back into America's long republican heritage and came to see these outward grievances as simply signs of a deeper, more nefarious, conspiracy against the rights of the innocent. Just as revolutionary colonists came to view the Stamp Act, the appointment of an American bishop, or the quartering of troops as an insidious plot by George III to reduce his colonies to absolute despotism, so too, Populists pictured the crop-lien system, agreements between railroads and grain elevator operators, "the crime of 1873," the repeal of the Sherman Silver Purchase Act, and the growth of monopolies and trusts as signs— indeed, as *proof*—of a secret, planned, coordinated, and unspeakably evil conspiracy. Populism became considerably more than simply a series of reforms to the transportation and monetary systems; it was a moral crusade to save all that was decent and worthy in American society. "[I]f there is no remedy in a new party," thundered Ignatius Donnelly, "then is this nation lost—ruined—damned beyond redemption, dead of dry-rot and universal corruption." Far from being an alien force in American culture, or a temporary primitivist reaction, Populism was a Gilded Age expression of political values dating back at least to Jefferson's Notes on the State of Virginia, Paine's "Agrarian Justice," or Dickinson's "Letters From a Farmer in Pennsylvania."

Mary Elizabeth Lease lashed out at what she saw as the combined forces of evil represented by the financially privileged, the urban elites, and corrupt politicians. "Wall Street owns the country. It is no longer a government of the people, by the people, and for the people, but a government of Wall Street, by Wall Street, and for Wall Street. The great common people of this country are slaves, and monopoly is the master. The West and South are bound and prostrate before the manufacturing East. Money rules, and our Vice President is a London banker. Our laws are the output of a system which clothes rascals in robes and honesty in rags. The parties lie to us and the political speakers mislead us … We want money, land and transportation … We want the accursed foreclosure system wiped out … We will stand by our homes and stay by our fireplaces by force if necessary, and we will not pay our debts to the loan-shark companies until the Government pays its debts to us. The people are at bay, let the bloodhounds of money who have dogged us thus far beware."[48]

The anonymous newspaper reporter, writing in 1893, said more than he probably intended when he call Mary Elizabeth Lease "the Patrick Henry in petticoats."[49]

But there was more to Populism than passion and rage. Along with the Jeffersonian sensibilities went a vision of an alternative society that was also remarkably Jeffersonian. The complexity of Populism is found in the fact that they produced policy proposals and substitute institutions that challenged the mainstream programs and thinking of both conservatives and Progressives. In their most basic form, many grievances of the Populists eventually became centered on what they saw as a flawed foundation of America's economy. Economic wealth and political power were concentrated in railroads, banks, and Wall Street investment houses because capital was assumed to be the core economic asset. Whether that capital was represented by gold or silver was almost beyond the point. Capitalists dominated the American economy because Lockean assumptions regarding wealth, labor, money, and Nature set the parameters for economic thought. In the definitive history of American Populism, Democratic Promise, Lawrence Goodwyn makes the case that the Alliance system and the subtreasury plan represent the essence of Gilded Age Populism.[50] The financial crises of 1873 and 1893 hit hardest those who were in debt and since these tended to be farmers much more than industrial laborers, Populists' stronghold was always rural agriculturalists. Furthermore, although Populists were sometimes smeared by their enemies with the accusation of being "socialists" and "anarchists," they never extended their attacks to a full-scale assault on the institution of private property. It was not property, *per se*, that Populists denounced, but monopolistic power. This was not a socialist attack on private ownership, but rather it was a defense of the small farmer against the abuses of predatory capitalism.

The Populist Alliance system consisted of two components designed with no less radical a task than reconfiguring American economics. In order to counter the economic control of the furnishing merchants and the crop-lien system, Populists began organizing a series of cooperatives across the South and the Midwest. These Alliance-owned cooperatives would use their power to collectively bargain for best prices from manufacturing businesses and pass these savings on to farmers who would pay a fee to join the cooperative. As important as they were, however, these "farmers exchanges" would serve more than economic ends. Attached to each farmers alliance or suballiance, and each cooperative farmers exchange would be several "lecturers" who would travel around the region, organizing meetings of local citizens who would vent their grievances, analyze their causes, and formulate plans for reform. The cooperative Alliance system was more than a method of purchasing goods at reduced and fair rates; it was a program to bring direct democracy back to America's heartland.[51]

The ability to purchase goods at cheaper rates meant little if farmers still lacked credit. It was on this point that Populists produced a truly radical alternative to America's financial system that, if adopted, would revolutionize our received cultural assumptions regarding wealth, finance, power, and credit. The plan was called the "subtreasury," and it was the brainchild of Texas Populist Charles Macune. Under Macune's scheme, farmers would be permitted to store their nonperishable products in government-owned warehouses. Rather than selling their products immediately after harvest, this system of warehouses would allow farmers to hold out for better market conditions and sell their goods intermittently throughout the year. Most importantly, the stored produce could be used as collateral in order to provide low-interest loans from the Federal government to the farmers. The loans would be come in the form of subtreasury "certificates" that would be "full legal tender for all debts, public and private" and since farmers could draw certificates in value up to 80 % of the value of their crops, as agricultural production increased, so would the money supply of the nation. Both gold and silver would be demonetized but rather than a return to strict fiat currency, agricultural production would be the commodity underpinning the value of currency. The critically significant aspect of Macune's subtreasury plan was the fact that it would completely undermine the crop-lien system, strip banks and mortgage companies from their control of America's credit, and go a long way toward making agricultural produce the true indication of wealth.[52] Put simply, the subtreasury can be seen as replacing John Locke's capitalism with Françoise Quesnay's physiocracy. Together, the Alliance cooperatives and the subtreasury credit system were the economic complements of Jefferson's ward republics.

As economically sound as the plan might have been, and as compatible as it might have been with the values of the Jeffersonian republican tradition, it almost goes without saying that Populism's alternative to corporate America's financial culture did not meet with enthusiasm by banks, financiers, or their agents in the Republican and Democratic Parties. In order to implement the Populist agenda, two criteria would have to be met; the Federal government would have to enact the appropriate legislation to put the subtreasury plan into practice, and vast numbers of Americans would have to become economically informed and politically mobilized in order to pressure the Federal government into action. As the nineteenth century drew to a close, neither of these conditions materialized. One clear lesson that emerges from the study of Gilded Age Populism is the extreme difficulty in creating cultural change in the face of entrenched liberal values, obstinate capitalist institutions, and the Madisonian procedural republic. Those few Populists who were elected to the Federal Congress were usually forced into the role of vocal champions for American working-class values and antagonists to the mainstream parties. The opportunity to sponsor successful legislation rarely presented itself.[53] Paralleling these events at the local level, the Alliance lecturers made notable achievements in explaining finance capitalism, monetary policy, and the subtreasury alternative to citizens around the nation, but as important as an informed citizenry was, some form of political organization was essential to spreading the message. Here, the crucial role of the People's Party as an educational and political focus became apparent, and here, perhaps, the decision of the People's Party to combine or "fuse" with the Democratic Party in the election of 1896 spelled doom for any radical cultural option. With William Jennings Bryan heading the "fusion ticket," the innovative, far-reaching subtreasury and Alliance system of the 1892 Omaha Platform was downplayed and eventually ignored in favor of a focus on "free silver" and reform of the monetary system within an unchanged structure of banking, finance capital, and corporate-dominated credit.[54] The attempt to broaden their appeal by joining a coalition with mainstream Democrats who were not interested in basic economic, political, and cultural reform—indeed who were openly hostile to any kind of radical change—proved disastrous for the Populist movement. There may possibly be important lessons here for contemporary radical environmentalists. In any event, it is now possible to garner a clear understanding of Populist ideology; describe the profound differences between Populism, conservatism, and Progressivism.

In their platform, proposals, and pronouncements, Gilded Age Populism displayed an unwavering hostility to concentrated economic wealth and finance-based capitalism. Present-day analysts who attempt to draw connections between Populism and American *laissez-faire*, free market conservatism miss the ideological roots of American political philosophy and the intense theoretical conflicts between the classical liberal, pro-corporate, competitive, monetary-based political economy and a civic humanist, local agrarian, land-based system of republican communities.

William Graham Sumner and Tom Watson had no more in common than did Alexander Hamilton and Thomas Jefferson, nor for that matter, than did John Locke and James Harrington. In "an open letter to millionaires," Henry Demarest Lloyd, whose 1894 work, <u>Wealth Against Commonwealth</u> became an instant best seller, issued an ominous warning to the giants of American industry in terms that echoed Jefferson's philosophy regarding stewardship, the limits of private property rights, and the ability of communities to hold ownership accountable to a larger common good. "Political economy gives you private property only that the interest of all may be served by your self-interest," Lloyd states. "[T]he law gives you your franchises and estates only for the general welfare and the public safety; religion holds you to be only stewards of your riches. If you usurp for your private profit all these trusts and grants, if you withdraw yourself from serving and protecting the public and take to oppressing and plundering them from your points of advantage, you will but repeat the folly of your mediaeval exemplars whose castles now decorate a better civilization with their prophetic ruins."[55] For Populists, private property was an important element in developing communities which needed to be widely dispersed and publicly managed for the common good. Their goal was decentralized corporatism. It was a direct nineteenth-century restatement of Jefferson's philosophy in "the earth belongs to the living."

In a manner and language remarkably similar both to Founding Era republicanism and contemporary ecological economics, Gilded Age Populism also attacked the financial system that underpinned corporate capitalism. Whether it was Henry George's assault on rent or Charles Macune's proposal for a subtreasury, radical economic thinkers during the last quarter of the nineteenth century believed that America's economic system was structurally flawed. The economy was sending the wrong signals. Those who held large wealth gained it in ways that did little to help the society at large, while those who worked the hardest and deserved returns the most were economically penalized. Individualism, greed, and corporate concentration were reinforced. Frugality, community, simple living, and honest labor were discouraged. In an ironic reversal of Jefferson's reflections in <u>Notes on the State of Virginia</u>, Gilded Age society treated corporate capitalists as if they were "the chosen people of God" at the same time the hard working farmers in America's rural areas were dealt with as if they were "mobs" and "fit instruments for tyranny."[56] Tom Watson noted this role reversal. In his Labor Day speech in 1891, he publicly wondered, "Labor asks of capital, 'Why is it you have so much and do so little work, while I have so little and do so much?' What is capital and what is labor? Originally they were the same, to the extent that cause and effect are the same. There was a time when there was no capital. There never was a time when there was no labor ... Yet there is this queer thing. Everybody wants labor protected when it becomes capital, while most people laugh you to scorn if you propose to protect it while it is still labor."[57]

Clearly, Populism had little in common with the conservative thought of men such as Sumner, but Populists also contested many of the tenets of the Progressive agenda in both the Gilded Age and today. Like many present-day neo-Malthusians and unlike most Progressive liberals, Populists resisted attempts to open America borders to floods of immigrants. Populists favored direct democracy and a delegate theory of representation rather than the Progressive tendency toward representative democracy built on a trustee theory of representation. The Omaha Platform endorsed immigration restrictions to hold down the size of the population as well as the referendum and initiative in order to give citizens a direct say in legislation. Unlike the Progressivism of Croly and Roosevelt, Populists distrusted any shift in power toward the Executive branch, much preferring the more democratic decision making of the Legislature. Even then, however, Populists understood the Democratic and Republican parties controlled Congress, and they knew corporate Wall Street wealth controlled the Democratic and Republican parties. Radical Arkansas Populist W. Scott Morgan charged both political parties with reactionary hostility to revolutionary change, antidemocratically ignoring the wishes of the people and collusion with the financial forces of oppression. If the twentieth-century language of C. Wright Mills were available to him, Morgan would have accused the established parties with a shared consensus built upon crackpot realism. "The men who control the policies of the old parties are opposed to either a change of policies or the formation of a new party. Thus the contest widens and deepens. There are only two sides to the question, and both the great political parties, by their policies and their acts, occupy one side. Look at their policies and their records ... On every vital issue of the day they occupy a position with the capitalist."[58]

The most striking difference between Populists and Progressives was over the appropriate size of the national government. It is primarily this disparity that continues to be the clearest sign of theoretical incommensurability between the two ideologies. Progressives always viewed the Federal government as the appropriate locus of problem solving.

Their argument was that big problems require big solutions from big government. Alexander Hamilton, Gifford Pinchot, and Al Gore are alike in seeing large scale as an answer to, rather than a cause of, crises. In contrast, Founding era republicans, Gilded Age Populists, and contemporary radical environmentalists including ecological economists, social ecologists, bioregionalists, and deep ecologists all unite around the belief in diminishing returns, diseconomies of scale, "the impossibility of thinking globally," and the conviction that ever-increasing scale generates at least as many problems as it solves. From Thomas Jefferson, to Jerry Simpson, to Wendell Berry, the call has been for greater decentralization. During his campaign for governor of Texas in 1892, Populist Thomas L Nugent declared his belief, "A great thinker has said that 'as institutions grow larger, men grow small.' It is so." Henry Demarest Lloyd in <u>Wealth Against Commonwealth</u> stated the issue in words that would draw nods of approval from any present-day neo-Malthusian or bioregionalist. "Liberty produces wealth, and wealth destroys liberty ... Our businesses, cities, factories, monopolies, fortunes, which are our empires, are the obesities of an age gluttonous beyond its power of digestion. Mankind are crowding upon each other in the centers, and struggling to keep each other out of the feast set by the new sciences and the new fellowships. Our size has got beyond both our science and our conscience."[59] Herman Daly would surely extend appreciative approval to such sentiments.

At its center, the dilemma for Populists was a clash of cultures. On the one side stood monopoly, finance capital, gigantic industrial corporations, centralized government, and the consumer culture of Gilded Age urbanism. Populism's alternative culture advocated rural simplicity, direct democracy, independence and equality, citizen empowerment, strong decentralized communities, and the values of the agrarian lifestyle. It was the struggle of republican virtue against liberalism's pursuit of power. In this battle, Populists knew as well as anyone in the Gilded Age that the American Constitution had stacked the theoretical deck in favor of liberalism. Speaking on the floor of the House of Representatives, Kansas Populist Jerry Simpson aimed the pointing finger of accusation directly from the political and economic crisis of 1893 to the institutional arrangements of 1787. "To my mind, Mr. Speaker, the causes of the condition of our people today are numerous; they did not begin yesterday or the day before, or last year or the year before. This [depression] … had its rise in the bad institutions of government with which we started out. We began wrong. We have failed to secure to human society and to individuals the rights that belong to them. This great nation in the course of its progress has created enormous powers, and instead of fortifying the rights of the people, has granted vast powers to the privileged class."[60]

American political thought during the Gilded Age both perpetuated and altered the political discourses of the American Founding. Civic humanist thought and the ideology of classical liberalism were carried forward but recast into what we today recognize as conservatism, Progressive liberalism, and Populism. Conservatism stresses individualism, competition, private property rights, and the domination of Nature. Progressive liberalism emphasizes community, nationalism, public regulation of private property, and the domination of Nature. Populism is committed to cultural change in the pursuit of agrarian democracy. The extent to which American Populism also partakes in the domination of Nature depends on whether the libertarian "Tea Party" wing reconnects with its roots, and how the bioregionalist wing addresses the tension between rural lifestyles, restricting the growth of the human population, and the preservation of external and internal wildness. Populists were by no means Nature lovers in the sense of caring about wilderness or wild animals. What they loved was the land, but in that strong emotional attachment to place resides the possibility for an extension of the notion of community. Gilded Age Populism is the obvious link between Jeffersonian republicanism and modern environmental bioregionalism, the question is the location of deep ecology in the agrarian, democratic mindset.

Notes

1. Clinton Rossiter, Conservatism in America: The Thankless Persuasion, 2nd Edition (New York: Alfred A. Knopf, 1962), 128–129.
2. William Graham Sumner, What Social Classes Owe Each Other (Caldwell, ID: The Caxton Printers, Ltd., 1989), first published in 1883.
3. Ibid., 24, 34, and 36.
4. Ibid., 88, 94, and 95.
5. Ibid., 85–86.
6. Rossiter, Conservatism in America" The Thankless Persuasion.
7. Lawrence Kohl, Politics of Individualism: Parties and the American Character in the Jacksonian Era (New York: Oxford University Press, 1989). Robert Vincent Remini, The Legacy of Andrew Jackson: Essays on Democracy, Indian Removal, and Slavery (Baton Rouge: Louisiana State University Press, 1988).
8. Sumner, What Social Classes Owe to Each Other, 141.

9. Ibid., 141 and 145.
10. Sumner, What Social Classes Owe Each Other, 98 and 58.
11. Sumner, What Social Classes Owe Each Other, 133 and 114.
12. Ibid., 14 and 17. It is interesting to note Sumner's strange rules of capitalization ("strange" given his political philosophy). He never capitalizes "earth" but consistently begins "Nature" with a capital letter, for reasons that are not apparent (the name of the deity?).
13. William Graham Sumner, "Socialism" (alternatively titled "The Challenge of Facts"), in Social Darwinism: Selected Essays of William Graham Sumner (Englewood Cliffs, NJ: Prentice-Hall Inc., 1963), 70, 72, and 76.
14. For an excellent discussion of this topic, see Bill McKibben, The End of Nature (New York and London: Anchor Books/Doubleday, 1989).
15. Lincoln's fear of mob rule and lynching is found in "Address Before the Young Men's Lyceum of Springfield, Illinois," January 27, 1838, in Roy P. Basler, ed., Abraham Lincoln: His Speeches and Writing (Cleveland, Ohio: De Capo Press, 2011), 76–85. Regarding corporate elites, Lincoln stated in a letter to Col. William F. Elkins (November 21, 1864), "As a result of the war, corporations have been enthroned and an era of corruption in high places will follow, and the money power of the country will endeavor to prolong its reign by working upon the prejudices of the people until all wealth is aggregated in a few hands and the Republic is destroyed." See Archer H Shaw, ed., The Lincoln Encyclopedia (New York: The Macmillan Company, 1950), 40. See also Gary Wills, Lincoln at Gettysburg: The Words That Remade America (New York: Simon and Schuster, 1992); and John P. Diggins, The Lost Soul of American Politics (Chicago: University of Chicago Press, 1984), especially Chapter 9.
16. For an analysis of the role of veterans' organizations in shaping the Progressive political program, see Stephen Skowronek, Building a New American State: The Expansion of National Administrative Capacities, 1877–1920 (Cambridge and New York: Cambridge University Press, 1982).
17. Herbert Croly, The Promise of American Life (Boston: Northeastern University Press, 1989): and Progressive Democracy (New York: The Macmillan Company, 1914).
18. Croly, as quoted in Charles Forcey, The Crossroads of Liberalism—Croly, Weyl, Lippman and the Progressive Era 1900–1925 (New York: Oxford University Press, 1961), 16. In his recent biography of Croly, Edward A. Stettner, Shaping Modern Liberalism: Herbert Croly and Progressive Thought (Lawrence, KS: University Press of Kansas, 1993) downplays the importance of "narrow Comtean focus" in Croly's life (see 17).
19. Ibid., 40.
20. For Croly's review of Hamilton's thought, see "Federalism and Republicanism as Opponents," The Promise of American Life, Chapter II, 38–42.

21. Jefferson's influence on American political thought is discussed in Section II, Chapter II, 42–46, The Promise of American Life.
22. Byron Dexter, "Herbert Croly and the Promise of American Life," Political Science Quarterly, Vol. 70 (1955), 202.
23. Croly, The Promise of American Life, 17 and 106; cf. 22
24. For examples of the use of the terms "struggle" and "will," in ibid., see 21 and 93.
25. Ibid. 23. On this topic, see also James Weinstein, The Corporate Ideal in the Liberal State: 1900–1918 (Boston: Beacon Press, 1968), and Stephen Skowronek, Building a New American State, op. cit. For a strong defense of the link between Progressivism and civic humanism, see Patrick Diggins, "Republicanism and Progressivism," American Quarterly, 37 (Fall 1985).
26. Croly, The Promise of American Life, Chapter VI, "Reform and Reformers."
27. The Catlin reference is from "An Artist Proposes a National Park," and the Schurz quotation is from the Annual Report of the Secretary of the Interior on the Operations for the Fiscal Year Ended June 30, 1877, both found in Roderick F. Nash (ed) American Environmentalism: Readings in Conservation History, 3rd Edition (New York: McGraw-Hill Publishing Company, 1990). The Turner quotation is from Edward Stettner, Shaping Modern Liberalism, pp. 80–81.
28. The revealing Locke quotation is in the Second Treatise of Government, Chapter V, section 49. The argument here contrasts with that of Roderick Nash, "The American Cult of the Primitive" American Quarterly XVIII (1966) who contends that some conservationists (perhaps he should have said preservationists) did believe that "the primitive was worth getting excited about."
29. For an excellent analysis of the role of "posterity" in the Conservation movement, see John R. Rodman, "Resource Conservation: Economics and After," a paper presented at the American Political Science Association meeting, Chicago, IL, August 1976.
30. Gifford Pinchot, The Fight for Conservation (New York: Doubleday, Page and Co., 1910), 72, emphasis added. "Conservation," as Rodman has pointed out, "does not function to question the paradigm but to perpetuate it." John Rodman, "Resource Conservation: Economics and After," 33.
31. Gifford Pinchot, Breaking New Ground (New York: Harcourt, Brace, and Company, 1947), 322–323.
32. The Lilienthal quotation is from TVA: Democracy on the March as citied in Roderick Nash, The American Environment: Readings in the History of Conservation, 133. For the Pinchot quotation, see Breaking New Ground, 323.

33. Samuel P. Hays, Conservation and the Gospel of Efficiency: The Progressive Conservation Movement 1890–1920 (New York: Atheneum, 1969), 271.
34. Ibid., 269–270.
35. Craig Rimmerman, The Rise of the Plebiscitary Presidency (Boulder, CO: Westview Press, 1993).
36. Pinchot, Breaking New Ground, 505.
37. Theodore Roosevelt, "Opening Address by the President," Proceedings of a Conference of Governors in the White House, cited in Roderick Nash, The American Environment: Readings in the History of Conservation, 52.
38. Pinchot, Breaking New Ground, p. 325; Hays, Conservation and the Gospel of Efficiency, 124.
39. Lester Frank Ward, "Plutocracy and Paternalism" as included in Alpheus Thomas Mason and Gordon E. Baker (eds.) Free Government in the Making (New York: Oxford University Press, 1985), 517–522; and Lester Frank Ward, The Psychic Factors of Civilization (1893) as cited and discussed in Donald Worster, Nature's Economy: A History of Ecological Ideas, 174–176.
40. Donald Worster, The Economy of Nature 175.
41. Pinchot, Breaking New Ground, 325.
42. Pinchot as quoted by Hays, Conservation and the Gospel of Efficiency, 41–42.
43. John D, Hicks, The Populist Revolt (Minneapolis: The University of Minnesota Press, 1931.
44. For biographies of these, and other Populists, see Ibid. and O. Gene Clanton, Kansas Populism: Ideas and Men (Lawrence, KS: The University Press of Kansas, 1969).
45. The Omaha Platform can be found as Appendix F in Hicks, The Populist Revolt, 439–444.
46. Tom Watson speech in Douglassville, Georgia, July 4, 1893. First reprinted in People's Party Paper (Atlanta), July 7, 1893, and found in Norman Pollack (ed.), The Populist Mind (Indianapolis and New York: The Bobbs-Merrill Company, Inc., 1967), 398 (emphasis in original reprint).
47. An excerpt of James H. Davis, A Political Revelation is found in Pollack (ed.), The Populist Mind, 203–227.
48. Mary Elizabeth Lease, quoted in John D. Hicks, The Populist Revolt, 160.
49. Ibid., the Donnelly quotation is on page 290, and the reference to Lease is on page 286.
50. Lawrence Goodwyn, Democratic Promise: The Populist Movement in America (New York: Oxford University Press, 1976), 2 Volumes. Goodwyn's thesis is restated in a single volume in Lawrence Goodwin, The Populist Moment: A Short History of the Agrarian Revolt in America (New York: Oxford University Press, 1978).

51. A discussion of the Alliance cooperative system can be found in Hicks, The Populist Revolt, Chapter IV, and Goodwyn, The Populist Moment, Chapter 3.
52. Hicks outlines the subtreasury plan in The Populist Revolt, Chapter VII, but the best (and most sympathetic) discussion is Goodwyn, The Populist Moment, Chapter 3 and passim.
53. For a review of the part played by Populism in the national government, see Gene Clanton, Congressional Populism and the Crisis of the 1890s (Lawrence, KS; University Press of Kansas, 1998).
54. The best analysis of the 1896 election is Lawrence Goodwyn, The Populist Moment. For a critique of "fusion" by one of the Populist politicians, see William A. Peffer, Populism: Its Rise and Fall, edited and with an Introduction by Peter H. Argersinger (Lawrence, KS: University Press of Kansas, 1992).
55. Henry Demarest Lloyd, A Strike of Millionaires Against Miners, in Norman Pollack, The Populist Mind, 423.
56. See Robert L Heilbroner, The Worldly Philosophers: The Lives, Times and Ideas of the Great Economic Thinkers (New York: Simon and Schuster, Inc., 1986. Jefferson's original citation is in Notes on the State of Virginia, Query XIX, "Manufactures."
57. Tom Watson, "A Labor Day Message" in People's Party Paper. Reprinted in Norman Pollack, The Populist Mind, 424.
58. W. Scott Morgan, History of the Wheel and the Alliance, and The Impending Revolution (1889), excerpted in Norman Pollack, The Populist Mind, 282.
59. Thomas L. Nugent, campaign speech of 1892, in Norman Pollack, The Populist Mind, p. 286. Henry Demarest Lloyd, Wealth Against Commonwealth, in Alpheus Thomas Mason and Gordon E. Baker, Free Government in the Making, 4th edition (New York: Oxford University Press, 1985), 545.
60. Jerry Simpson, Congressional Record, 53rd Cong., 1st sess., August 18, 1893, cited in Gene Clanton, Congressional Populism and the Crisis of the 1890s, 45.

CHAPTER 10

Living with the Legacies: Our Culture Confronts Our Environment

In 1944—roughly half way between the Gilded Age and the present—Karl Polanyi published his extraordinary treatise, The Great Transformation.[1] The book is many things. In its broadest, most explicit meaning, The Great Transformation is a study of the rise of modern market economies, an analysis of how they operate, and a critique of the social failings of market-driven economic orders. In the wide-ranging genre of economic history, Polanyi is correctly viewed as an intellectual counterweight to the individual libertarian thought of men such as Ludwig Von Mises and Friedrich Hayek. With regard to ecological issues and an attempt to recover the environmental legacies that have led us to our present condition, Karl Polanyi's scholarship is frequently cited as the starting point for looking into a future that extends beyond market-driven, exponentially growing capitalist society. Robyn Eckersley, Kirkpatrick Sale, Chet Bowers, and Gus Speth all acknowledge Polanyi as a source for their environmental thinking.[2] In his definitive study on environmental education, Ecological Literacy, David Orr recommends The Great Transformation as one of two critical sources for understanding "the religion of economics."[3] The other source mentioned by Orr under this heading is For The Common Good by Herman Daly and John Cobb, and here Orr does well to connect Polanyi with Daly and Cobb, for the two books represent the intellectual bookends of thought regarding market economics. Polanyi describes how the market became the central organizing principle for Western societies, while Daly and Cobb explain how a future society organized around ecological principles might operate.[4] The Great Transformation is, therefore, a good place to pivot from a historical analysis of the origins of our predicament to an attempt at foreshadowing where these various legacies may be leading us.

As carefully detailed by Polanyi, the "great transformation" to a capitalist market economy occurred when "the market" as a tangible place was altered into the Market as a conceptualization of humans and Nature. For Polanyi, the market (small "m") began as an actual, objective, physical location where humans gathered to exhibit what Adam Smith called the "propensity to barter, truck and exchange one thing for another." It was, in a literal sense, a market *place*. As capitalism took over the Western mindset, this changed and changed profoundly. The Market (spelled by Polanyi with a capital "M") became a perceptual proclivity to value everything by the price it could command in a hypothetical set of exchanges where theoretically postulated "willing sellers" met assumed "willing buyers." The world of exchange where one thing was traded for another thing was converted into a world devoid of any value that could not be expressed in monetary terms. Prior to the advent of capitalism, the market had been a relatively small part of human life that was largely disconnected from the more intimate, ongoing, and important aspects of human society. "Market day" was exactly that—a single, brief portion of a person's time that was otherwise typified by a host of other activities. The great transformation reversed these priorities. With the conquest of capitalism, social life became the small, disconnected, and relatively trivial component of human existence while economic concerns—the training for and acquisition of a job, the selling of one's labor for a salary, and the frenetic search for ways to convert Nature into consumable products—became the primary methods by which nations and individuals determined their relative level of "development." Instead of the market being embedded within human social life, society became a subset of, and embedded in, economics and the Market. The greater part of Polanyi's Great Transformation is an account of when and how this took place and an analysis of the implications that follow from these historical events. Moving forward into a more ecologically healthy and socially sustainable society may very well involve understanding how we got to where we are and then going back and undoing the policy choices that drove us into our present dysfunctional predicament.

The movement to a capitalist market economy began in England where a series of Parliamentary acts both signaled and caused critical shifts in human economic relationships. For Karl Polanyi, the subtle, but decisive, first step was the conversion of Nature and human beings from entities with intrinsic worth inside of communal social orders to mere commodities for sale to the highest bidder. In order to appreciate Polanyi's argument, it is necessary to understand an important distinction he makes between *real commodities* and *fictitious commodities*. As he characterizes them, "Commodities are here empirically defined as objects produced for sale on the market; markets, again, are empirically defined as actual contacts between buyers and sellers."[5] Real commodities are goods produced for sale and manufactured by intentional human effort. It follows, then, that treating something as a commodity that was not created by humans for market transactions is to create a fictitious commodity. This, according to Polanyi, is what capitalist market economies do with Nature and with human labor since it is self-evident that neither Nature nor human beings are consciously produced for the sole purpose of being sold. In Britain, this momentous transformation was achieved by laws such as the Prescriptions Act, the Inheritance Act, the Real Property Act, and, most importantly, by the Enclosure Act of 1801 and its successors. Land, which as Polanyi observes, is "only another name for nature" and "which is not produced by man" was denied its traditional meaning as being either the sovereign's holding or the commons of the people and was, instead, privatized as a possession of individual property owners to be used and dispensed with as they saw fit.[6] At the same time the Enclosure Acts converted land (Nature) into a commodity by destroying the idea of a "commons" and making it saleable at a price determined by the Market, human beings were also transformed into commodities. Driven from the land, pushed into the rapidly expanding cities, and cut off from any direct, legal interaction with Nature, the only remaining resource available to the masses was their labor, and that, too, was only valued at the price it could command on the Market. Nature and humans were stripped of any intrinsic, noneconomic value they may have possessed and became fictitious commodities—entities with only utilitarian, instrumental value.

For Polanyi, the next step in the creation of modern market economies occurred when economics and politics became seen as separate and distinct activities. The Speenhamland Act of 1795 had provided social guarantees for worker's incomes and had thereby established a "right to live" by affirming society's responsibility for setting a minimum material base for existence. A series of Parliamentary acts culminating in the Reform Bill of 1832 and the Poor Law Reform of 1834 repealed these protections. Henceforth, workers were on their own to find employment at whatever terms the owners of capital were willing to offer. The political domain of social life became severed from, and impotent against, the economic realm where commodities were exchanged for the cheapest price possible.[7] This is what we now mean when we speak of the "free market:" the economic sphere is independent of—and resistant to—pressure, control, or regulation by the social sphere. The immediate consequences of these perceptual and policy shifts were far-reaching and ominous. Society became defenseless to protect itself from the periodic distortions, dislocations, and depressions of the economic system and was driven to seek some means for overcoming both the commodification of humans and the decoupling of politics and economics. This last point is particularly relevant for the discussion of political economy beyond capitalism.

As argued by Polanyi, market capitalism presents society with a totally unrealizable "utopia," for it expects what no rational system should anticipate, namely, that in the face of cyclical or permanent economic hardship, society will not take the steps necessary to protect itself. According to Polanyi, market capitalism is an unworkable system because turning humans into commodities undermines life in ways that cause humans to react, and those reactions will take the form of rejecting the autonomy of the economic realm by reestablishing the primacy of politics over economics. The utopian aspect of market capitalism is built on the unrealistic assumption that humans will passively submit to wrenching dislocations and horrible fluctuations in the quality of their lives as markets go through periodic booms and busts. It assumes humans will not defend themselves against these economic onslaughts when they have the actual, or potential, power of self-defense. At this point, <u>The Great Transformation</u> turns from being a historical analysis to a projection of future trends for the political economy. Polanyi argues there are three broad regimes that can be used to reject the autonomy of the Market, reassert the primacy of politics over economics, and establish economic activity as being embedded in the social life of the community. As he saw the situation in 1944, the available alternatives were socialism, fascism, and some variety of national corporatism on the model of Franklin Roosevelt's New Deal. Each of these systems is built upon the rejection of liberal free market capitalism and each seeks—albeit in dramatically different forms—to make the direction of the economy subordinate to the social demands of the community, nation, or interests of the dominant social class.

If Polanyi's otherwise brilliant analysis has a shortcoming, it is his failure to explore fully the relationship between the commodification of Nature and the commodification of human beings, for the history of Western societies over recent decades has demonstrated liberal capitalism's ability to use the domination of Nature to produce the economic growth necessary to ameliorate the commodification of human labor. Put more simply, citizens in Western cultures have appeared willing to see themselves as merely workers and consumers if market capitalism generates enough growth to provide material conditions pleasant enough to compensate for their lost social community. The market-based economic order will remain "free"—that is capitalism will remain resistant to control by the political order—as long as the production of goods and services can continue to grow exponentially, and that false perception will apparently prevail as long as the commodification of Nature seems to be capable of infinite continuation. The autonomy of the capitalist market system relies on a faith in the limited liberal state and that, in turn, depends on a perception of Nature as limitless, devoid of nonutilitarian value, and capable of boundless substitution and exploitation.

Karl Polanyi's The Great Transformation is prescient in pointing out that the failure of market-based capitalism is dependent upon the recognition of humans and Nature as fictitious commodities. Here, he foreshadowed many of the environmental discourses that began to appear in the early 1960s with publication of Murray Bookchin's Our Synthetic Environment and Rachel Carson's Silent Spring.[8] At the end of World War II, Polanyi understood that markets fail for reasons that are simultaneously social and ecological. What he did not, and probably could not, flesh out was the extent to which limitations of the environment as a resource fund, a waste sink, and an energy supply would be critical in exposing market capitalism's failure to exponentially and infinitely grow. As a European with intellectual roots in British economic history, Polanyi never articulated how American political thought would constrain and shape the alternatives available to liberal, free market capitalism once the failures of that system were fully understood.

If America and the world are now beginning the next "great transformation," an analysis of social alternatives open in the future must be built upon an acknowledgment of this interacting and mutually supportive relationship among economics, politics, and ecology. The time has come to join the environmental discourses described in Part I of this book with the perspectives on American politics analyzed in Chapters 5, through 8 and use these united patterns to construct a picture where we might be headed. The twenty-first century will be a time when economic theory, political ideology, and environmental discourses become grouped together into more or less consistent sets or patterns of perception. In America, the role of informed citizens will be to understand those patterns, acknowledge the consequences that flow from each, and choose the future we want for our society and our nation. Regrettably, one of the alternatives available to us is to deny the existence of these interconnected realities and to delude ourselves into believing we can continue to separate economic markets and social control from ecological truths. When, and if, Americans come to recognize these interconnections, the range of alternative futures will begin to narrow.

The Challenges We Face

Given everything that has been said thus far, we can now return to the challenges discussed in Chapter 1 and see these environmental crises as surface indicators of deeper contradictions that have their origins in the way Americans perceive problems and conceive potential solutions. Responses to the ecological destruction we face need to be viewed through the political and social legacies that are part of our cultural heritage and national character. A multitude of environment threats from climate change, to air pollution, to water availability and pollution, to topsoil loss, to hazardous, toxic, and nuclear waste, to the increase in the human population, and the decrease in the populations of many other species might be cited. The environmental legacies of America's past are contributing factors in these threats. Hamilton's expressed wish to "ransack" America's environment is seen as many applaud the march of national power, and Jefferson's naïve belief that this country possessed "vacant lands" that would last for thousands and thousands of generations is reflected in the passive apathy with which many Americans ignore these threats.

Perhaps the strongest connection between our current challenges and the legacies of America's Founders lies with James Madison's pursuit of a stable and enduring political system and what is presently stated as the goal of "sustainability." If sustainable means stable and enduing over time, then the questions need to be asked: sustainable for whom ..., and exactly what is it that we are trying to sustain? It is certainly evident that for a great many species who are either extinct or headed in that direction, the present level of human population and technological affluence is not sustainable. Or are we trying to sustain an optimal size of the human population? That figure for America is probably between 150 and 200 million, and for Earth, somewhere around two billion (the population levels before fossil fuels allowed us to temporarily overshoot the planet's carrying capacity). Are we trying to sustain ecosystem flexibility, resilience, and long-term stability? Species extinction, atmospheric carbon overload, the loss of prime agricultural land, and wilderness destruction are all indicators that we have exceeded that measure of sustainability. Or are we trying to sustain biological diversity and the wonder of living in a complex but interconnected world where our lives can find excitement and meaning from acknowledging the presence of something bigger and grander than the mundane pursuits of our everyday lives? It is heartbreakingly clear that our contemporary world of dwindling diversity is neither stable nor enduring according to this definition. Or are we trying to sustain our present political and economic system? If that is our goal, then it can only be achieved by further destroying the biological and ecological systems upon which that system depends. Of all the various notions of sustainability, this is the one that is unambiguously not possible. A system cannot be sustained if it destroys the foundations upon which it is built. At a certain level, James Madison understood this, as his 100-year time limit on his Constitution and as his recognition that his Constitution would have to be significantly revised make clear. Beyond the twin goals of stability and endurance, Madison's legacy contains another aspiration, for we need to remember his observation that "Justice is the end of government. It is the end of civil society. It ever has been and ever will be pursued until it be obtained, or until liberty be lost in the pursuit."[9] Even if it were physically possible to continue our current population and affluence, doubts exist as to whether our contemporary political and economic systems are moving us toward a more just society. Perhaps the most daunting challenge we face is building a sustainable society that is just, not only for humans but also for all members of the ecosystem.

Beneath the immediate issues of public policy lies this fundamental challenge which needs to be faced: most, if not all, of America's environmental dilemmas have as their root cause the attempt to achieve unending, exponential economic growth. This is more than simply trying to achieve a more efficient economy, for no matter how efficient we may become at allocating resources (and the Laws of Thermodynamics limit that ability), the fact is that the scale of human economic activity has grown so large that it is impinging on the Earth's ability to provide resources, sinks, and energy. Ecological sustainability requires that we halt the growth of economic activity and redesign our economy around institutions and policies structured to perpetuate a no-growth, stationary state system. Recognition of this fact, however, brings into stark relief the central conundrum facing Western civilization, for a growing, expanding economy is now considered essential to our definition of what an economically just society looks like and provides. How to accomplish both of these goals—building an environmentally sustainable society that is at the same time economically just—is the critical challenge we face in the twenty-first century. We need to take a fresh look at what "justice" means.

As we have seen, Lockean liberalism defines justice as the protection of individual rights, while Humean, interest-group liberalism, sees justice as the product of a rule-guided adversarial system. "Justice" for this branch of liberalism means establishing a process to channel conflict that insures the participants have the incentives, the rule of law, and the opportunity to compete successfully for their individual development, while using expressions such as "level playing fields" and "fair starts" to characterize a society pictured more as a competitive game than an organic community. Justice, according to interest-group liberalism, means objective and neutral rules that are equally administered. In other words, justice means procedural fairness. It is safe to argue that justice as protection of rights and/or justice as fairness encompass what most contemporary Americans mean when they use the term.

But justice as the protection of individual rights has always been fraught with confusion and the potential for discord. Which rights are to be protected, and for whom are those rights to be recognized? Individual rights are based on reasoned arguments, but since reasonable people can disagree, conflicts over rights are seldom rationally resolved. Does a fetus have a right to life, or does a woman have a right to control her own uterus? Does the public have a right to safety or do individual citizens have a right to own handguns? Assault weapons? Is there a right to clean air and safe drinking water? Do animals have rights? Do rocks have rights?[10] Since there is no objective method for resolving these disputes, bitter wrangling and disharmony generally ensue until one side or the other manages to garner enough votes in either the legislature or the courts to bring a (temporary) resolution. These quandaries, however, lead to the uncomfortable conclusion that in too many cases might does make right. Alexander Hamilton's misgivings about a Bill of Rights have often been demonstrated to have historical substance.

If justice as the protection of rights has caused trouble for liberal thinkers, justice as fairness and equality within a social process also exposes limitations of this ideology. Lawyers are likely to assume that if the legal system is equally administered and the law is assiduously applied, the outcome of the proceedings will be declared to be just. The connection between legal justice and economic justice is more difficult to establish since conservatives from William Graham Sumner to Dick Cheney consistently defend gross inequalities in economic outcomes and philosophical references to Adam Smith's invisible hand, or Herbert Spencer's Social Darwinism are likely to fall on deaf ears to those middle-class and poor Americans struggling to make ends meet. The quandary for conservative politicians is that given a nation where the poor and middle classes are enfranchised, how can right-leaning politicians hope to garner votes with their logic? The answer, deeply engrained in American perspectives, political values, and environmental thinking, is that continuous, exponential, unending economic growth will "make the pie larger" and allow everyone the chance to compete for the opportunity to "get ahead" and move up the scale of material possessions, increased consumption, and higher social standing. Capitalist economics and dramatic income inequality can be considered just only in a nation that has committed itself to unending exponential growth in production and consumption. Economic growth has traditionally underpinned justice in capitalist systems by holding out the promise of a distribution arrangement where "everyone can have more," where income inequality is converted from a zero-sum game to a "win-win" competition, and where the potential conflict between economic groups is transformed into an all-out human assault on Nature. The benefits of adopting this strategy should not be ignored. Economic growth removes from the political agenda the contentious debate over income distribution and produces a kind of passive tranquility, if not actual harmony, within the culture. If America has never had a really significant, national discussion about alternative meanings of social and economic justice, it is because the promise of unending, exponential growth seems to make that discussion unnecessary. The political debate we do have—free market growth versus government managed growth—revolves around the much shallower issue of how to achieve an end that is acknowledged by both parties. Capitalist growth mixes well with the idea of private property rights, and as long as some modicum of opportunity appears to be within the reach of most people, a perpetually expanding economic base gives the impression of our best, and certainly most favored, option. It seems clear, therefore, that as long as our contemporary national debates are reduced to which party can best promote growth through lax labor standards, weakened environmental regulations, and increased incentives to rich "job creators," environmentalists and advocates for the nonprivileged will be fighting a losing battle where definitions and logic stack the odds against them. The option of "everyone gets more" has, however, one major drawback. On a planet with finite resources, finite sinks, and finite energy, exponential growth is ecologically unsustainable and doomed to eventual failure. That failure, becoming more evident every day, makes it imperative that nations clarify what sustainability means, and it requires a reconsideration of the social, environmental, and ecological dimensions of justice.[11]

Prior to our modern notions of justice as protection of rights or equal treatment before the law, there existed a premodern, classical conceptualization dating back to the Greeks and Romans and continued in civic humanist or classical republican thought. Within these traditions, justice meant unity, collaboration, and harmony within a self-conscious community. It meant citizens sacrificing for the common good and performing those functions for which their inborn specialty, calling, or *arête* selected them. Since the creation of man-made positive law was taken to be a sign of conflict and disharmony, it was axiomatic among classical republicans that the more laws a society had, the less just that community could be considered. Thomas Jefferson pays homage to this premodern definition of justice when he speaks of the need for unity and harmony, and when he expresses discomfort at the multitude of laws being passed by his contemporaries. Today, premodern beliefs in justice as communal harmony are seen when civil rights protesters carry signs proclaiming "no justice—no peace." Note the priorities here. For republicans, justice must precede stability and peace, while for modern liberals, stability, rules, and law are the essential preconditions for the process needed to move toward their concept of justice. The differences between the two versions of what constitutes a just society could not be more glaring. Each carries ramifications for human and ecological order, but if modern, liberal notions of justice have limitations, the classical view also had restrictions.

Justice, as the Greeks taught Western civilization, is bound up with the belief in an organic community. The problem was that since these early societies had a narrow view of what constituted their community, they also had a deficient view of what justice might mean. In classical thought, justice only applied to humans and, even then, only to certain members of the human community. Human conduct toward animals, plants, streams, mountains, and (by extension) the atmosphere and the ecosystem might be considered prudent or imprudent, productive or unproductive, sustainable or unsustainable, but, according to classical thought, it cannot be considered just or unjust. And even within the human species, the premodern conception of justice did not apply to everyone. Within the first page of The Republic, for example, Plato acknowledges that human slavery existed in the Athens of his time. He saw no contradiction between this reality and the pursuit of a just community.[12] Similarly, Jefferson could believe he was a just man while owning slaves, and Gilded Age America could think it was pursuing justice while simultaneously denying civil liberties to African Americans or voting rights to women. The classical definition of justice has been limited by a constricted vision of who was in and who was out of the just community.

Historically, the promise and potential for classical justice has been realized by subsequent generations expanding the boundaries of their community, including previously excluded but functionally connected entities within the community, and subsequently recognizing that unity, harmony, and stability necessitated the just treatment of these parties. This is how ethical progress has been made in the past, and it is probably how it will be made in the future. Environmentally, the philosophical breakthrough came when Aldo Leopold coupled the scientifically verifiable fact that humans have an interdependent relationship with their surrounding ecosystem with the teaching of the moral sense theoreticians that ethics were based on emotional ties of "approbation," compassion, pity, and love. Since humans are codependent with Nature, and since (at least some) humans can love Nature, feel compassion for animals and places, and suffer emotional loss at the destruction of their environment, it follows that community, and hence justice, can be extended beyond simply humans. The classical republican definition of justice as communal harmony began with a narrow scope of membership, but the objective, verifiable facts of environmental interdependence and the possible extension of our emotional sympathy give this principle the best chance for creating a "land ethic" that could heal our equaling empirically unhealthy planet. As Leopold succinctly and superbly put it, "A thing is right when it tends to preserve the integrity, stability, and beauty of the biotic community. It is wrong when it tends otherwise."[13]

The fundamental challenge we face is to move toward some form of no-growth, stationary state economy that is maintained within the renewable carrying capacity between humans and their nonhuman environment and that recognizes humans and the ecosystem as interconnected, mutually supporting members of the same community. This, or something very much like this, is what "ecological justice" must mean in the future. Given these definitions and distinctions, we might inquire what options are available for a society that seeks the protection of rights and fair equality for all while promoting harmony both among humans and between humans and the rest of Nature. We might also, following the pattern of Karl Polanyi, speculate on the possible futures available to America and the world as the transition to a new political, economic, and social system unveils itself.

The Next Transformation

Great Britain's, and subsequently the rest of Europe's, transition from mercantilism to capitalism, from shared commons to private holdings, and from classical republicanism to modern liberalism holds within it lessons that may provide insights into the next social transformation currently underway. Although the United States, as Louis Hartz observed, avoided this movement and began its existence as a commercial, liberal society, the historical process detailed by Polanyi had elements that may very well be repeated as the world moves out of its high-consumption, free market capitalist, limited liberal nation states to whatever may be coming next.[14] The first thing to note is that the process took time. From the Glorious Revolution (1688), to the Speenhamland Act (1795), to the Enclosure Act (1801), to the Reform Bill (1832), to the Poor Law Reform (1834) decades past as British society transitioned. The second point is that the movement was neither spontaneous nor inevitable. A powerful and well-positioned minority fought to have certain acts of Parliament repealed and others adopted. These men had a vision of where they wanted Britain to go, they had a plan to put that vision into practice, and they executed that plan with diligence and perseverance. These events expose a fourth message; the fact that the transformation was strongly resisted during its entire period. On the left, workers, agrarians, and eventually socialists struggled to prevent, or at least soften, the harsher effects of the new system, while on the conservative side, the landed, titled aristocracy viewed with open disdain the rise of the upstart capitalists and brazen newcomers in the House of Commons. And it must be acknowledged that the great transformation carried with it a large amount social disruption and violence. The cultural shift denounced by Romantic authors, the deplorable working conditions described by writers such as Charles Dickens, the Luddite industrial sabotages between 1811 and 1813, as well as the numerous riots, political upheavals and acts of civil disobedience that swept Britain all through this period are testaments to the wrenching dislocations undergone.[15] Yet with all these laws, protests, and resistance, it was probably possible at some time in British history to see the outlines of the future and understand where all of these cultural, political, and economic changes were headed. The full picture might not have been clear, but by—say—1832 or 1834, enough of the details were in place to give a fairly accurate description where the course of British society was going. Given these historical lessons, we might turn to present-day America and apply them. With an appreciation of the challenges we now face and the American cultural legacies bequeathed to us from our past, a noticeable, albeit fuzzy, outline of the next great transformation may be attempted.

The one future that is unquestionably beyond the realm of possibility is to continue on our present path of limited liberal government, an exponentially growing economy, a human population continuing to expand, and the overshooting of the planet's carrying capacity. Although this may be the tacitly preferred option for many Americans, the fact that it denies inescapable ecological limits makes it predestined for failure. As atmospheric carbon buildup, species extinction, dwindling clean water supplies, and a spreading list of toxic pollutants demonstrate, our ecosystem is already beginning the process of collapse. Since—protestations to the contrary notwithstanding—our economy is a connected subset of our environment, and our political stability is linked to our economic productivity, social collapse will follow ecological breakdown.

UCLA Professor of Geography, Jared Diamond, made something of a scholarly sensation in 2005 with the publication of his book, <u>Collapse: How Societies Choose to Fail or Succeed</u>.[16] Diamond traces the downfall of several ancient civilizations from Easter Island to Norse Greenland and the environmental crises currently confronting Rwanda, Hispaniola, China, and Australia and concludes, "the best predictors of modern 'state failure'—i.e., revolutions, violent regime change, collapse of authority, and genocide—prove to be measures of environmental and population pressure ...."[17] How a society chooses to respond, or not to respond, to these threats depends on its decision-making institutions and its cultural values. With respect to current conditions, he repeats Al Gore's nostrum regarding collective will being the essential missing ingredient in an effective response, but Diamond exhibits a bit more reticence than the former Vice President. "We don't need new technologies to solve our problems," states the geographer, "while new technologies can make some contribution, for the most part we 'just' need the political will to apply solutions already available. Of course, that's a big 'just'."[18] According to Diamond, success in preventing collapse is a product of effective planning and institutions that recognize and react to dangers before they become overwhelming. Institutional failure, he argues, is the product of several factors, including "the failure to anticipate a problem, failure to perceive it once it has arisen, failure to attempt to solve it after it has been perceived, and failure to succeed in attempts to solve it."[19] He continues, "Perhaps a crux of success or failure as a society is to know which core values to hold on to, and which ones to discard and replace with new values, when times change." Since reinventing cultural values is such a nearly impossible undertaking, the more feasible undertaking for a nation would appear to be deciding, "which of their core values were worth fighting for, and which no longer made sense."[20] His inquiry does not penetrate into the environmental legacies that might point toward collapse or sustainability, and he skirts the interesting question of how far "fighting" might extend, but Diamond is resolute in his conviction that archeological and historical research demonstrate that many societies have collapsed, that America carries no immunity to drastic disruption, and that our current trends in population and ecological destruction cannot long continue. Dramatic changes in our numbers and methods of doing things are underway and will certainly persist. What is uncertain is the direction those responses will take. Jared Diamond is correct when he asserts, "the world's environmental problems *will* become resolved, in one way or another, within the lifetimes of the children and young adults alive today. The only question is whether they will become resolved in pleasant ways of their own choice, or in the unpleasant ways not of our choice ...."[21]

One option for the future in a context of decreasing energy, deteriorating climate, and increasing stress on ecological services is to ratchet up social control in the pursuit of the well-established values of economic growth and political stability. This is the path to what we might call "authoritarian capitalism." The elements of authoritarian capitalism are emerging and discernable. There would be private ownership of the means of production coupled with an extreme (and expanding) disparity in wealth both within and between nations. Social stability would be maintained, and unrest would be quelled by an omnipresent and purportedly omnipotent police and military "security system." Consistent with Diamond's factors for failed decision-making, environmental crises would be ignored or denied, responded to with superficial policies that sought only to palliate the surface manifestations of the issue without addressing the underlying causes, or worsened by intensifying the conditions producing the devastation. Climate change being brought on by burning fossil fuels would be denied, declared a nonthreatening event in need of further research, or subject to incremental policies designed to mitigate the effects of a collapsing atmospheric system. All the while, expanded fossil-fuel extraction and a rush to build more nuclear power plants would become the hallmarks of energy policy.

Most significantly, the power and reach of the government would become less restricted and more obvious. Maintaining the stability of the economic and social systems would be the state's primary concern. Dissent and the call for change would be suppressed and surveillance of possible agitators would be seen as a necessary precaution. Under a system of authoritarian capitalism, citizens advocating significant, effective responses to ecological threats could be labeled extremists, radicals, or terrorists and placed on governmental "watch lists." An authoritarian capitalist regime would probably not preclude every attempt at environmental management, but the efforts that were officially sanctioned would be limited to superficial rules to make energy more efficient, combat the results rather than the causes of climate change, place a few endangered species in zoos, or perhaps treat and store the vast quantities of hazardous, toxic, and radioactive pollutants being produced. The essential characteristic of such a system, however, is the fact that it would only address the efficient allocation of resources and not the overall scale of economic activity. Such a political economy would actively promote economic growth and wealth disparity. It would exist not to regulate corporations or redistribute the wealth of the super-rich, but to protect the power and privilege of those groups.

The political and economic systems of present-day China and Russia can probably best be described as authoritarian capitalism. Both regimes have private corporations, stock exchanges, an important class of extremely wealthy individuals, and an apparatus of state authority committed to suppressing opposition and perpetuating the status quo. It would be as inaccurate to call these systems free market liberal states as it would any longer be true to call them communists. In their recent transitions, China and Russia could have adopted the best of Western values—liberty, toleration, and compromise—with the best of socialist/communist values—at least a verbal commitment to the working class, social mobility, and support for a widespread social safety net. Instead, the elites in Beijing and Moscow decided to combine the worst aspects of communism—intolerance, repression, autocracy—with the worst characteristics of capitalism—a culture of greed, gross income inequality, and a neglect of the unorganized poor. The 1989 brutal suppression of pro-democracy demonstrators in Tiananmen Square represented a defining moment in history as Chinese leaders revealed to the world that although their economic system might resemble capitalism, they would not tolerate any form of democracy. In America, the faith of liberal internationalist, from Progressive idealists such as Woodrow Wilson to neoconservative ideologues such as Francis Fukuyama, that capitalism and liberal democracy were self-supporting institutions proves to be unfounded. The argument had been that a capitalist economy required property rights and competition, which required diversity and the rule of law, which required a pluralist distribution of power, which inevitably resulted in liberal democratic political institutions. Events in Vladimir Putin's Russia, post-Tiananmen China, and America's tacit acquiescence to these actions have demonstrated that faith to be misplaced. Liberal, democratic capitalism has not proven to be "the end of history."[22] A new form of social organization is emerging, and one shape that organization may take is a Chinese/Russian model of authoritarian capitalism.

Since the values embedded in a nation's cultural past shape its political will and constrain its collective choices, the question is whether America has beliefs that might make us susceptible to this system of political economy. Authoritarian capitalism, the combination of capitalist economics and a national security state based on political corruption and "metadata" surveillance programs is the "unpleasant way" America might respond to environmental challenges. Unfortunately, there are selected elements in our cultural legacies that give us the capacity for this outcome. Madison's fixation on stability and his fear of mass democratic movements, Hamilton's cynical animosity toward the American public, his reliance on a powerful military establishment, his creation of a rich and well-placed financial elite, and his broad interpretation of Federal Constitutional authority to respond to what that authority declares are "national emergencies" are embedded in our past and make us prone to accept virtually any remedy that promises to protect the "homeland." America has a tradition of valuing security, stability, and deference to the police, military, and government. Conservatism's worship of technological capitalism can join with Progressivism's faith in governmental management to produce a system where the state promotes a moderate rate of economic growth, protects the wealth and political access of wealthy corporations and individuals, and monitors the activities of those who choose to challenge the legitimacy or goals of this arrangement.

How might the material wealth of the nation be distributed under a system of authoritarian capitalism? As "everyone gets more" becomes less and less of a feasible or convincing scheme, a limited number of alternative options appear, at least in theory, to be possible. The wealth distribution option with the least support in America is for everyone to get the same equal share. Given the long experience of Americans tolerating, and desiring, some level of income differential, it is hard to imagine this option achieving any popular foothold in our cultural debates. Much more likely is the possibility that any environmental or social group advocating a socialist or communist equal distribution of wealth will be culturally ostracized to the point of political irrelevance. Environmentalists need to acknowledge the fact that socialist regimes have been some of the most pro-growth and anti-environmental states in recent times, and we must, therefore, be very explicit in renouncing any connection between socialist theories of economic justice and justice for the nonhuman ecosystem.

One unpleasant, but possible, alternative may be for everyone to receive the material wealth they inherit. Those who are born rich get to remain rich, and those who are born poor stay in that condition. Historically, this income distribution pattern has been referred to as "feudalism." It was sustained for centuries probably because aristocrats managed to instill the idea that a frozen class structure was ordained as just and fair by divine right, and because they were able and willing to exert significant violence against anyone who challenged their static distribution scheme. There is evidence that contemporary America is moving closer toward something like a feudal distribution of wealth, as economic mobility tends to stagnate, and the upper class commits a significant portion of its wealth to the corrupt manipulation of the political system in order to guarantee its continued predominance. Once the growth economy is no longer a choice, much more convincing or much more violence will be needed to keep American feudalism going.

Another speculative possibility is for everyone to get the material wealth they can seize or steal from others. The have-nots might grab from the haves, or, more likely, the have-mores might grab from the have-lesses and become the new, or richer, propertied class. This distribution scheme is the model of economic class warfare, or as Derrick Jensen so nicely labeled it in a recent Orion piece, it is a "kleptocracy," with the rule of thieves who steal "from the poor, from the land, and from the future."[23]

There is, once again, evidence that present-day America is trending in this direction toward a kind of legalized, institutionalized version of "Mad Max." As politics in Washington grows more polarized and malicious, proposals for addressing contentious economic issues, such as the national debt, have brought forth plans that bear all the markings of class warfare. Right-wing advocates have recommended cutting programs for seniors, students, the impoverished, and those in poor health, while supporting economic growth policies that extend oil drilling in ecologically sensitive areas, promote and protect shale fracking, authorize the Keystone oil pipeline and further support Canada's disastrous tar sands project, deny the scientific evidence demonstrating climate change, species extinction, and the peaking of fossil fuels, while eviscerating the EPA, and many of the statutory protections passed over the past 40 years. Further policies regarding women's health, abortion, contraception, and sex education will, if adopted, have appalling consequences on any attempt to achieve a population plan aimed at stabilizing the number of humans. These conservative, pro-growth, anti-environment policies represent an indisputable example of economic class warfare waged by the plutocrats against middle- and low-income Americans, and they are an unabashed attack on Nature and anything that might look like a harmonious and ecologically just relationship between humans and this Earth we inhabit. It is cynically ironic that the same individuals and corporations who complain about "redistributing wealth" toward the poor have no compunction against—and, indeed, fight mightily for—income redistribution that takes from the lower and middle classes and pumps it into upper class portfolios. This *chutzpah* is only exceeded by the hypocritical shouts of "class warfare" when Progressives make the meek suggestion that the wealthy should pay a tax rate *equal to* that of middle-income Americans. Contradicting the political smokescreen being put out by conservatives in this country, there is no evidence that American politics is sliding toward socialism and an equal division of property. Quite the opposite, the division of wealth is becoming more extreme, more static, and more the product of successful class warfare by the very rich. Economic injustice under a system of authoritarian capitalism might be the model for Diamond's "unpleasant" way to address environmental challenges.

A better, more "pleasant" way of responding to environmental, economic, political, and social challenges would be to construct a vision of what a sustainable, healthy, and just society might look like and work to transform that vision into reality. The question would then become whether America has the institutional capability and the values necessary for the will to actualize the transformation. The chances for transforming to a better ideology would be strengthened if the vision specified which values were to be emphasized and which were to be downplayed or rejected. Placing these ideals into a consistent, coordinated image of what a sustainable, just system would look like could move further toward actualization if the selected values were those long established within the cultural legacies of America's past. As the debates of the Gilded Age demonstrate, American politics is capable of taking theoretical elements from our Founding and rearranging them in different patterns that appear unique but are, in fact, reformulations of the basic models first enunciated by men such as Jefferson, Hamilton, and Madison.

The perceptions and values of the cornucopian and free market environmental discourses must first be singled out for rejection. It is unlikely America can react with anything like an adequate response to our challenges as long as we deny the existence of those threats, refuse to acknowledge the limits of technological innovation, and pursue a course of human population and economic growth. Myopic ignorance is not the pathway out of our current dilemma. Rather, it is the road to authoritarian capitalism and collapse. And if our goal is to resolve and not merely respond to environmental, economic, and social challenges, then the fatal limitations of reform environmentalism, environmental economics, and mainstream sustainable development must also be recognized. John Dryzek's "democratic pragmatism," the belief that government structures as currently established can pass regulations and laws that adequately address our multitude of risks fails to take account of the fragmentation, incrementalism, and short-term bias of the American Constitutional system. His "administrative rationalism," the trusting reliance on experts within the Executive Branch is premised on the questionable assumption that these experts have not only the specialized knowledge but also the broad theoretical understanding to implement a fundamental reordering of our goals and values. Scientists can provide many much needed services to the public. They can tell us what we are doing to the planet, they can warn us of forthcoming disasters if present trends continue, and they can describe what a healthy, sustainable ecosystem might look like. But scientists, with their deeply instilled professional standards of value-free objective neutrality are ill-prepared to analyze the conflicting values within our culture or create a vision of a reconstructed pattern of American principles. Environmental economics, with its appeal to make our present system more efficient by "total cost accounting" and "internalizing externalities," is a superficial reaction by a profession engrained with the ideas of the economic system as freestanding from the environment. Rather than the time-honored tradition in scholarship of critical analysis, mainstream economics in this country has too often seen its role as apologist and advocate for the liberal capitalist system. Finally, mainstream sustainable development, the Brundtland Commission smart growth approach to environmental politics, goes about as far as conventionally possible without addressing the fundamental causes of ecological and social devastation. Their misplaced reliance on the demographic transition to tackle the population explosion, their faith in managed growth rather than a no-growth stationary state, and their anthropocentric ethics which talks about equity for the human species while ignoring ecological justice for the vast remainder to the Earth's inhabitants mark this discourse as too shallow to serve as a genuine guide for the future. There is good reason why three decades after Brundtland, Rio, Al Gore, and Thomas Friedman so few nations have move toward genuine sustainability. Mainstream environmentalism has done little more than slow down the process of destruction because it fails to significantly describe either the causes of the destruction or paint a picture of humanity's place in a healthy, harmonious ecological community.

Various elements of the radical environmental discourses, therefore, need to be selected out and brought together into a coherent, comprehensive vision that can be the basis for individual, political, and social action. Ecological economics must be the basis for any truly practical, workable response to the materialistic components of our environmental crisis. The United States, and the entire world, owe these scholars a debt of gratitude for their painstaking work in describing the failures of growth-based capitalism and their meticulous description of the specific policies necessary to move to a genuinely sustainable, no growth, fair trade, economically just community. Such an economy might looks like this: individuals could own and have exclusive use of property they acquired, and there would be gradations in wealth status. People would own differing amounts of material wealth based on their education, skills, diligence, or good fortune. This is not socialism. The overall scale of the economy, however, would be fixed and steady in order to keep human economic activity within the sustainable, ecologically replenishable carrying capacity of the community, and the term "community" would be extended beyond humans to include other species of animals and plants, soil, water, air, geology, and climate that encompass the interconnected ecosystem. Humans would be citizens of the community, but only citizens. They would not be despots or the sole possessors of intrinsic worth. The scale or overall size of the economic system would be established by the ecological requirements for a sustainable system and managed by the public, democratically determined limits set by the human political system. There would be a lower income limit (e.g., $12,000 a year) and an upper income limit (say, $500,000), but within that range, an individual's annual earnings could rise or fall. In other words, there would be income and social mobility. This is not feudalism. The actual implementation of the policy might work this way; each year, a person would file their income tax statement, declaring their annual income. If that income was below $12,000, they would receive a check for the difference between what they earned and the socially determined lower limit (such a system might have the added benefit of reducing administration costs by eliminating the need for social welfare programs such as food stamps). This represents an extension of our present Earned Income Credit, or the negative income tax plan advocated by conservatives such as Milton Friedman and Richard Nixon, with this notable amendment: at the upper end of the wealth spectrum, individual earnings would be progressively taxed until they reached the established upper limit (in our example, $500,000) when they would be taxed at a rate of 100 %. Over time, citizens would probably refrain from the rat race to earn and consume more and more, while lower income earners might take advantage of their income security to learn new and useful skills. Class warfare would be intercepted by everyone's knowledge that the spread in wealth distribution was a collective decision made by the community at large. The lower income limit promotes fairness, and the upper income limit institutionalizes a sustainable, no growth economy. There would be private property rights, procedural fairness, a good chance for harmony among humans, and, perhaps most importantly, harmony between humans and the rest of the planet. This is not to say, however, that ecological economics represents a complete picture of where we need to head. Missing from the work of ecological economists is a genuine appreciation of the role of ethics in human affairs, the necessity to move beyond anthropocentric humanism, and the recognition that a real harmonious community must include the nonhuman and nonsentient in our understanding of an interconnected and interdependent Earth.

This same inability or unwillingness to extend ethics beyond humans also characterizes social ecology and represents the basic weakness of that discourse. Still, social ecology has elements that should be part of our sustainable, just future. As critical as individual ethics may be, socially structured behavior is likewise essential. The importance of institutions in the shaping of human conduct is a key contribution made by social ecology, along with the recognition that these institutions can be judged to be alienating or liberating, in accordance with human instinctive first nature or dysfunctionally repressive. The analysis of left-wing Social Darwinists—from Peter Kropotkin to Murray Bookchin—arguing that mutual aid, cooperation, and small-scale communities are a part of humanity's inherent needs and are just as "natural" as conflict or competition points toward a genuinely satisfying system of human interaction. Humans are not genetically flawed, but the history of our species shows that we have made, or had forced upon us, institutional choices that reinforced and perpetuated the worst, most dysfunctional aspects of our potential. That same analysis also gives the heartening news that something better lies beyond our current condition. The technological and economic destruction of the planet is not our only option, and it certainly is not the end of history, but to reach that end, institutions must change and change dramatically.

The vital juncture where our cultural values confront our interconnected, limited ecosystem manifests itself with the issue of controlling the size of the human population. The failure of technological solutions, the need for America to check illegal immigration, the inadequacy of the "demographic transition," and the requirement for a national population policy are all contributions made to the discussion by the neo-Malthusians. The condemnation aimed at neo-Malthusians of being racist, xenophobic, or misanthropic can only be addressed by a narrative that includes not only what needs to be restricted but also what needs to be expanded. Restraining the size of the human population is essential, not because we hate humans but because we care about and love the rest of Nature. The accusation of xenophobic misanthropy is as inappropriate regarding neo-Malthusians as the charge that advocates for civil rights for African Americans hate Caucasians or the false claim that fair and equal treatment for women could only be achieved by a diminution of opportunities for men. Caring more for Nature does not mean we care less about humans.

It is this extension of values to what Aldo Leopold called a "land ethic" that represents the contribution made by bioregionalism. The idea that people need to be rooted in a place and linked with a culture of simple consumption, close historical communities, and active participatory democracy is indispensable to producing both a healthy society and a balanced environment. The will to protect our environment must begin with the determination to defend those areas of the Earth where we live and conduct the daily affairs of our lives. Bioregionalism also must face up to its shortcomings, however. The failure of many writers in this genre to address the population issue or appreciate the role undomesticated animals and plants are glaring omissions. In this regard, respect for the pastoral needs to be joined with a love for the wild and that requires a bioregionalism that takes cognizance of deep ecology and the call for environmentalism to turn inside the human psyche.

Both faces of deep ecology have elements that can be part of our healing vision. Intellectually, humans need to be reminded that we are a perfectly good species that coevolved over hundreds of thousands of years within untamed surroundings. We do not need to destroy our environment in order to survive. We do not need perpetual consumption or a workaholic rat race to prove our worth. Perhaps what we most require is simply to psychologically relax to understand that the wild adventurous spirit within us is not some "beast" that needs to be tamed but a source of curiosity, wonder, and awe that is searching for expression. Although British playwright George Bernard Shaw once quipped that America is the only nation on earth to go from barbarism to decadence without passing through civilization, the truth may be that we are instead a country that has moved from infancy to adolescents and is finding difficulty transitioning into maturity. The culture of the juvenile, fealty to the idea that we can do anything, be anything, have everything, and that the entire world exists only for our exclusive entertainment and use needs to be supplanted with the culture of the mature adult; the appreciation that there are some things we are competent at and some that are beyond our scope, the humility that we share out surroundings with others who deserve consideration and respect, and the sure knowledge that we are limited individuals living on a limited planet whose purpose is to find meaning and love within the limited time we have to live. There were, no doubt, moments when it was fun being a reckless adolescent, but the time has arrived for America to end its Peter Pan complex and grow up.

This exposes the other face of deep ecology; the requirement to move from intellectual critique to direct action. Mature adults take charge of their lives and responsibility for their actions. If Americans permit the devastation of their environment, the resulting calamitous collapse will be the product of their failed political and economic institutions, but history will also point an accusing finger at all those who knew what was happening, were given abundant warning, and yet sat passively by. The positive message from deep ecology is that not only are we beings whose ontogeny draws us to develop and mature we are also creatures with an intensely imprinted urge to survive. We take action when threatened. We fight back and resist when confronted with danger. Since threats to our environment, the places we love, and the other beings with whom we share the Earth are everywhere, the most natural—the most human—thing to do is to fight back with each and every means available to us. The calls for active resistance from town hall meetings, to alternative life styles, to legislative and legal action, to demonstrations and protests, to civil disobedience, to subversive ecotage all have a place in the struggle.

If a picture of a healthy, harmonious future can be drawn, the question becomes whether such a vision can fit within the legacies of America's past, and, here, the positive message is that our history has bequeathed to us values that can be adopted and adapted to such an image. Dating back as far as the creation of the United States, components within the thought of each of our major Founding Fathers exist that can become part of this better, more pleasant America. James Madison may have left us with a Constitution premised on the conflictual model of gridlocked interest-group liberalism, but he also gave us a forewarning that his system could not continue forever. Madison's greatest contributions were his recognition and warning that environmental limits lie in our future, and the understanding that the system he helped create in Philadelphia in 1787 was time-limited. The American angst, combining a social optimism with anxiety that the good times cannot last forever, is part of Madison's environmental legacy. Even Alexander Hamilton, whose financial schemes, technological capitalism, and powerful central government can now be seen as the cause of much of our current plight has provided standards that will be needed in the future. His undeniable personal courage and unquestionable genius were able to take a political economy emerging in Europe and apply them to the setting of our country. The task ahead is to show the same courage and ingenuity in installing the replacement for Hamilton's legacy.

It is the legacy of Thomas Jefferson, however, that provides the most lasting and most hopeful values upon which to build our next transformation – the strong interest in America's natural environment; the faith in an active citizenry with a passionate, populist dedication to participating directly in their local democracies; and, most importantly, the belief that virtuous citizenship is built upon the conscience and moral sense of individuals who understand their conduct is observed and judged by others. Posterity will not only inherit the legacies we leave them, but also they will appraise what we have done. Whether we permitted our moral sense to be exercised and strengthened or weakened and atrophied, whether we struggled to have our sense of community expanded or contracted will be our final testament and the meaning of our lives. Thomas Jefferson's legacy, as we now know, is of a man with a towering intellect who articulated the very best of America's principles, but whose political and private life failed to put those values into effect; a man whose Declaration of Independence pronounced it was our right—it was our duty—upon occasion to abolish the forms to which we are accustomed but could not gather the will to abolish the form of slave aristocracy to which he had been accustomed. The future of the United States will be determined by our capacity to take Jeffersonian core values—a love of Nature, participatory Populism, the moral sense, and a duty to resist and revolt—and use them to alter or abolish the forms to which we are accustomed—growth capitalism, representative government, obscene consumption, constricted ethics, and ecological injustice. The bad news is that the values that might lead us to authoritarian capitalism, environmental devastation, and eventual collapse are deeply embedded in the legacies of American thought. The good news is the way out of this is also part of the rich heritage of America's past. We have a clear picture of where we are and how we got here. We know the alternative futures that present themselves, and we know that one of those futures is abysmal and one is hopeful. The only question is how, and if, we decide which path to follow.

Capacity and Will

Once we have an image of where we are, how we got here, and where we need to go, the final question is how to transition to that next stage in human and ecological development. This may, or may not, be the true "end of history," but it is the transformation currently underway. The capacity of our institutions and values to either retard or promote that transition must be the gauge by which they are judged and presents the opportunity to consider each of our major social organizations in order to determine which need to be supported, the changes that need to be made, and the elements of each that need to be resisted and fought. Near the top of this list are America's financial institutions.

The major equities exchanges, the brokerage houses and hedge funds that thrive on speculative stock trading, and the financial news media that both report on and support these institutions owe their existence to the promotion of the of the values of greed and fear, along with the devastation of our environment through discounting the worth of ecological preservation. In their essence, the price of stocks go up whenever greed and the hope for an unearned return dominate the beliefs of investors, and prices decline whenever fear and anxiety over the potential loss of wealth becomes the dominant passion. Coupled with these values is the view that money invested has the potential to do what natural resources can never do, that is, increase exponentially and perpetually. Therefore, according to the internal logic of financial capitalism, the economically rational thing to do is to convert Nature into products as quickly and completely as possible, sell the products, and invest the profits in equities whose value may increase (if greed prevails) or temporarily decline (if fear controls). Gambling on which companies will do best in producing products and convincing consumers to purchase them, which stocks are headed higher and which lower, and drawing in the discretionary incomes from more investors is the business of Wall Street. No real commodities are produced—no actual production of capital is undertaken (that was achieved by the initial issuance of the stock). All that is accomplished is the shuffling of money from the moderately wealthy to the super-rich, the intensification of greed and fear, and the perpetuation of Alexander Hamilton's delusional, crackpot, but very real world of finance, consumption, and ecological ransacking. The good news is that Hamilton's apparition is not the only image within America's economic legacies. Physiocracy's distrust and repugnance for gamblers and swindlers, Populism's creative attempts to build a financial system on something more permanent and real (such as the agricultural subtreasury plan), and the recent uprising of discontent as expressed by the Occupy Wall Street movement are signals that finance capitalism's grip on this nation is tenuous and failing.[24] A transition to a better world will require the collapse of Wall Street either through external pressure or internal contradiction. As the stock markets move with increased frequency from bubble to bubble—as the 1990s dot com bubble gets replaced with the 2000s housing mortgage bubble which gets replaced with the Federal Reserve's quantitative easing bubble—as Market volatility grows more and more extreme, and as world economic growth fails to expand at a rate necessary to justify escalating stock prices, those genuinely concerned about the future of the Earth await this painful but necessary and inevitable step. Greed and fear are not very fulfilling values upon which to build one's life, and, as greed becomes less rewarding, fear may draw people to search elsewhere for symbols of their worth. As this happens, hope for the future and attachment to one's community will become recognized as far better passions worthy of pursuit.

More real than the crackpot, fictitious operations of financial markets are the actual operations of economic institutions, and here the concern becomes the transition from operations whose scale and growth are too big and too cancerous to be sustainable to economies that are within the carrying capacity and replenishing capabilities of their surrounding ecosystem. As numerous writers and actual working models demonstrate, the technology, principles, and practices necessary to put a sustainable economy into practice exist and are being actualized all over America and the rest of the world every day. Transitioning here is a joint process of encouragement and resistance. It means strengthening local farmers and fighting agribusiness: promoting small, local businesses and combating big-box, resource guzzling conglomerates. It means recognizing that so-called economies of scale that are purchased by ignoring the externalizing of costs, shifting from human labor to fossil-fuel consumption, and downplaying the transaction costs of massive transportation systems are illusory gains that are far outweighed by the benefits of less consumption, closer connections with neighbors, and a stationary economy held within the scope of its bioregion. The movement to small-scale, no-growth economies must also be built on the containment of the size of the human population. Part of this transition, such as a stiffening of immigration requirements, will mollify conservatives and antagonize Progressive, while other required practices such as birth control, women's rights, and abortion will fit within liberal Progressive ideology and cause reaction on the right, but this simply points out that neither conservatives nor Progressive liberals have a monopoly on needed policies. Transforming our economy will involve states, counties, and local communities gaining the authority to declare themselves "built out" and no longer subject to immigration, and it will entail citizen action to stop harmful businesses while protecting the wild places still within their boundaries.[25] Local economies have all the capability necessary to provide us with the products we genuinely need, and there is a growing appreciation that more humans do not mean a better or happier life for those living or for future occupants of this nation. What is needed are government policies designed to encourage no-growth localism and a population with a specific, targeted, optimal size.

This turns our attention to the institutions of government, and here, the truth must be acknowledged that the written Constitution of the United States is a document that has probably outlived its practical usefulness. James Madison's Constitution may have provided a considerable amount of individual liberty, but his system of separation of powers, checks and balances, enumerated powers, and bicameralism has become what it was initially designed to be: a process of societal gridlock in the absence of well-intentioned, tolerant, and compromising bipartisanship. The significant difference is that Madison's liberal culture has degenerated, dwindled, and been replaced by elected politicians whose combined levels of incompetence, venality, and corruption serve only the purpose of alienating and angering growing numbers of Americans. Too many of these angry citizens believe that replacing the politicians might rectify the crisis when the actual source of our governmental incapacity lies not in those who staff the institutions, but the institutions themselves. Democracy requires a system where those in power are accountable—they can be punished or reward for the decisions they make—and that requires a system where responsibility for making decisions can be assigned. America's written Constitution so diffuses power and fragments authority (i.e., what separation of powers and bicameralism mean) that responsibility is virtually impossible to determine. There are reasons why virtually every new democracy established since 1787, when writing its constitution, has preferred a parliamentary system with strong parties, legislative supremacy, and majority rule; and why even modest reforms such as abolishing the practice of Senatorial "holds" or switching to a system of proportional representation in states with more than two Congressional seats (which would stop the practice of political gerrymandering) or viewing impeachment as a legislative vote of no-confidence in the Executive are practically impossible to contemplate in America. The Constitution of the United States, along with the uncritical worship with which we are all indoctrinated, makes transforming that eighteenth-century document nearly unimaginable. Article V of our Constitution allows two-thirds of the states to call a convention to propose amendments, and, although it is unclear how far those amendments could go (the closest we came was in 1980 when 30 states petitioned for a balanced budget amendment), a movement to redraft our Founding Charter would generate such trepidation and excitement that it might provide the agitating of the national soul we need. Of course, such a political and cultural revolution is highly unlikely. More likely is that Madison's gridlocked liberalism will transition into Hamilton's authoritarian capitalism with presidential power becoming more open-ended and more unaccountable. Hope lies not in Madison's representative government but in Jefferson's ward republics; not in Madison's "democracy" without citizens but in Jefferson's participatory communities who resist corruption and sacrifice for the common good.

In the past decade, the Tea Party movement has generated much enthusiasm at the grassroots level of American politics. Their arguments for a transition in our politics, their distrust of the national government, and their calls for a balanced Federal budget contain aspects of earlier Populism, but their unwillingness to level the same anger at Wall Street, their failure to see the national debt as a tool of Hamiltonian finance to force economic growth, and their rejection of sensible measures to address the deficit and debt, such as the Bowles-Simpson proposal for cuts in government spending, elimination of loopholes in the tax code and a return to a progressive tax rate on personal incomes demonstrate Tea Party supporters are more the (perhaps unintended) dupes of the capitalist elites than genuine heirs of the Boston Tea Party or the Gilded Age Populists.[26]

Another notable recent trend in American culture is the widespread practice of heaping extravagant praise and honor on members of the military armed services to pronounce them heroes who sacrifice their security and personal interests on behalf of the freedom and security of the rest of us. This recognition of sacrifice for the common good and the praise of civic virtue is testimony to our classical republican heritage and reveals the potential of the American collective will. But why should the men and women of the military be the only ones called upon to exhibit civic virtue? Why can't Wall Street brokers be asked, and required, to settle for yearly incomes of half a million dollars, rather than the hundreds of millions they currently pocket? Genuine republican patriotism in an age of limited resources and threatened ecosystems would call upon citizens to consume less, have fewer children, care about their neighbors a bit more, and demonstrate their attachment to the mountains, rivers, and wetlands they claim to love? Mustering the determination to transform America will be based, at least in part, on passionate, patriotic love and the willingness to forego some of our narrow self-interests. Such a moment presented itself after the attacks on September 11, 2001, but although then President George Bush proclaimed all Americans would have to sacrifice in the future, the opportunity for real cultural change was lost when the declaration was quickly dismissed by the unpatriotic advice that we should go about our normal business, travel, and go shopping—advice that our fashionable news media and their corporate sponsors quickly adopted.

Transitioning to a balanced and ecologically just society must also be at least partially premised on citizens having information and analysis that accurately reflects the world in which they live, the feasible alternatives open to them, and the consequences that follow from their choices. With the laudable exceptions of formats such as National Public Radio and the Public Broadcasting System, most mainstream media have been captured by a mentality of postmodernism that portrays every dispute as merely a difference of opinion and a relativism that confuses salacious voyeurism with a concern for challenges that genuinely affect the public. Moving to a better world requires mass media that are interested in and tell the truth; that remind us there is an objective difference between a healthy and a dying planet; and that point out the choices we make carry with them both intended and unintended consequences. Seeking out and detecting those truths requires a population that is educated, equipped with the skills of intellectual discernment, and ethically sensitive.

In addition to our town halls, our civic organizations, our churches, and our meetings with friends, our educational institutions must see themselves, and be agents of change. Far too often, our schools, colleges, and universities portray themselves as training grounds and job placement services. C.A. Bowers, The Culture of Denial: Why the Environmental Movement Needs a Strategy for Reforming Universities and Public Schools provides a timely, insightful, and convincing critique of liberal education and its tendency to view all change as beneficial, all values as relative, all perspectives as subject to individual validation, and all scientific/technological innovation as advantageously furthering the agenda that Francis Bacon called "enlarging the bounds of human empire." Bowers explains why this modern liberal pedagogy and its even more relativistic postmodernist version (what Murray Bookchin labels "yuppie nihilism") are inadequate for an ecological future that must be built on ecocentric ethics, strong communities, and a culture capable of distinguishing the elements of tradition that need to be altered from those that need to be revered. His attack is not aimed at environmentalism's enemies as much as those who think they are partisans of more ecologically responsible practices but who limit their support to offering environmental studies courses within an overall educational setting that supports anti-environmental institutions and fosters an anti-environmental culture.[27]

American colleges and universities have an important role in transitioning this society, demonstrating the cultural values that deserve support, and making the institutional decisions that further that transformation. Regents and trustees can make purchasing choices that support environmentally safe products and divest their endowments from stock holdings in fossil-fuel and nuclear power industries. Presidents and deans can make ecological literacy a basic requirement for all students during their first year of study and support the work of environmental scientists. If for no other reason than promoting a discussion of alternatives, every economics department in every college or university that claims to be an institution of higher learning must include at least one course in ecological economics with a clear distinction made between this perspective and the superficial approach of environmental economics. Faculty in every department must develop the interests and skills to bridge the narrow disciplines in which they were trained. Students must resist and rebel against an education that seeks only to train them for good, high-paying jobs in the twentieth century, for the fact is that the twentieth century is over. The twenty-first century will be quite different and will require both knowledge and wisdom that is interdisciplinary, historical, and capable of making ethical determinations regarding what is best for themselves, their nation, and the environmental community of which they are members. A small, but important step might be for some enterprising student to raise her or his hand in their macroeconomics class and ask the professor to explain how it is possible to have exponential economic growth on a finite planet.[28]

In whatever institutional or social setting we find ourselves, Americans must keep up the determined fight to transform this country in an economically, politically, and ecologically just direction. Perhaps for a motto, we might select the words of Edward Abbey, "If opposition is not enough, we must resist. And if resistance is not enough, then subvert." We can gain strength for this struggle by recognizing we are the heirs of a diverse and rich culture with legacies that date back to Thomas Jefferson's call for revolutionary action, the Populists who responded with rage against the robber barons and Wall Street bankers, and all of those patriots who fought throughout the decades against class, racial, gender, or environmental injustice. Our most powerful weapons are the selected cultural values contained in our environmental legacies, and perhaps the strongest of those is our consciences and the sure awareness that we are doing what is right and, in the long run, the only thing worth doing. Jefferson's advice to his daughter and nephew, given with love over 200 years ago, can be applied with equal truth today: exercise your moral sense, consider how you would act if all the world was watching, and remember you are a social creature who was created to be a member of a community. Today, that means we keep alive our passion and love for the Earth, the land, and the ecological communities that surround us. Most critically, it means we do not allow our moral sense to become atrophied and dull. Hike, hunt, camp, or fish. Stay in touch. Plant a garden and remember where your food comes from. Prepare your meals with patience, care, and respect. Say grace and thank Mother Nature for giving you sustenance. Above all else, do not allow yourself to go numb. If optimism is the faith that things will inevitably get better in the future, then there is little reason today to be optimistic. But if hope means the possibility exists that we can reach into our past, understand our choices, and summon the ethical courage to care, then Americans have plenty of reason to hope. A better future lies ahead.

Notes

1. Karl Polyanyi, The Great Transformation: The Political and Economic Origins of Our Time (Boston: Beacon Press, 2001).
2. Robyn Eckersley, Environmentalism and Political Theory: Toward an Ecocentric Approach, (Albany: State University of New York Press, 1992); Kirkpatrick Sale, Dwellers in the Land: The Bioregional Vision (Philadelphia, New Society Publishers, 1991); C.A. Bowers, The Culture of Denial: Why the Environmental Movement Needs a Strategy for Reforming Universities and Public Schools (Albany: State University of New York Press, 1997); James Gustave Speth, The Bridge at the Edge of the World (New Haven: Yale University Press, 2008).
3. David W. Orr, Ecological Literacy: Education and Transition to a Postmodern World (Albany: State University of New York Press, 1992), 112.
4. Herman E. Daly and John B. Cobb, Jr., For the Common Good: Redirecting the Economy Toward Community, the Environment and a Sustainable Future (Boston: Beacon, 1989), see especially 60–61.
5. Karl Polyanyi, The Great Transformation, Chapter 6, especially 75–76.
6. Ibid., 189 and 75.
7. Ibid., Chapter 7 is particularly important.
8. Murray Bookchin (published under pseudonym "Lewis Huber"), Our Synthetic Environment (New York: Alfred A. Knopf, 1962): Rachel Carson, Silent Spring (Boston: Houghton Mifflin Company, 1962).
9. Federalist 51, 322.
10. Roderick Nash, "Do Rocks Have Rights?" The Center Magazine, November/December 1977.
11. On environmental justice, see Robert D. Bullard, Dumping in Dixie (Boulder, CO.: Westview, 1990): Andrew Szasz, EcoPopulism: Toxic Waste and the Movement for Environmental Justice (Minneapolis: University of Minnesota Press, 1994): and Evan J. Ringquist. "Environmental Justice: Normative Concerns, Empirical Evidence," in Norman J. Vig and Michael E. Kraft, eds., Environmental Policy: New Directions for the Twenty-first Century, 5th ed. (Washington, D.C.: CQ Press, 2003) 249–273.
12. Plato, The Republic of Plato, translated with Introduction and Notes by Francis Macdonald Cornford (New York and London: Oxford University Press, 1945).

13. Aldo Leopold, <u>A Sand County Almanac: With Essays From Round River</u> (New York: Ballantine Books, 1966), 262.
14. Louis Hartz, <u>The Founding of New Societies</u> (New York: Harcourt, Brace, and World, Inc., 1964).
15. Andy Wood, <u>Riot, Rebellion, and Popular Politics in Early Modern England</u> (London and New York: Palgrave Macmillan Books, 2002). Carl J. Griffin, <u>Protest, Politics, and Work in Rural England: 1700–1850</u> (London and New York: Palgrave Macmillan Books, 2013).
16. Jared Diamond, <u>Collapse: How Societies Choose to Fail or Succeed</u> (New York: Viking, 2005).
17. <u>Ibid.</u>, 516.
18. <u>Ibid.</u>, 522.
19. <u>Ibid.</u>, 438.
20. <u>Ibid.</u>, and 433 and 440.
21. <u>Ibid.</u>, 498, emphasis in original.
22. Francis Fukuyama, <u>The End of History and the Last Man</u> (New York and London: The Free Press, 2006).
23. Derrick Jensen, "Democracy of Destruction," <u>Orion</u>, Vol. 31, No. 3, May–June 2012, 12–13. Jensen's writings deserve close attention. In particular, see Derrick Jensen, <u>Endgame: Vol. I, The Problem of Civilization</u> (New York: Seven Stories Press, 2006); and Derrick Jensen, <u>Endgame: Vol. II, Resistance</u> (New York: Seven Stories Press, 2006)
24. See, Todd Gitlin, <u>Occupy Nation: The Roots, The Spirit, and the Promise of Occupy Wall Street</u> (New York: HarperCollins Publishers, 2012).
25. For a consideration of Populist environmentalism, see Daniel Kemmis, <u>This Sovereign Land: A New Vision for Governing the West</u> (Covelo, CA: Island Press, 2001). For examples of successful local efforts, see Jay Erskine Leutze, <u>Stand Up That Mountain: The Battle to Save One Small Mountain Community in the Wilderness Along the Appalachian Trail</u> (New York: Scribner, 2012); and Judith M. Francis "Soaring With Eagles: The Triumph of Citizenship at Lake James," <u>Heartstone</u>, Vol. 6, No 1, Spring 2005, 111–121.
26. For an analysis of the Tea Party movement, see Kate Zernike, <u>Boiling Mad: Behind the Lines in Tea Party America</u> (New York: St. Martin's Griffin, 2010).
27. C.A. Bowers, <u>The Culture of Denial: Why the Environmental Movement Needs a Strategy for Reforming Universities and Public Schools</u> (Albany: State University of New York Press, 1997).
28. For a more thorough discussion of this issue, see David W. Orr, <u>Earth in Mind: On Education, Environment, and the Human Prospect</u> (Covelo, CA: Island Press, 1994).

Index

A
Abbey, Edward, 95, 353
Adair, Douglass, 264
administrative rationalism, 41
agrarian Populism, 308, 309, 316
  and bioregionalism, 308
  Gilded Age Populism, 309;
    differences from Progressives, 316; and Jeffersonian democracy, 309; as revolt of the underdog, 309
  subtreasury, 313
America's unwritten constitution, 18
  American political thought, 19
  political culture, 19
Anderson, Terry L. and Donald R. Leal, 35
anti-environmental cornucopians, 29–35
Aristotelian meaning of "natural", 295
authoritarian capitalism, 336
  methods of wealth distribution, 339

B
Bachrach, Peter and Morton, S. Baratz, 117
  "The Two Faces of Power", 117
Bailyn, Bernard, 142
Bartlett, R.V., 125
Berry, Wendell, 82
Bookchin, Murray, 70, 352
  Hegelian dialectics, 71
Bowers, C.A., 323, 352
Brundtland Commission, 48
  Our Common Future, 48
Bryan, William Jennings, 314
Bush, President George, 351

C
Carruthers, David, 52
carrying capacity, 32
Cato Institute, 35
Catton, William, 114
civic humanism, 69, 143

classical republican, 69
Harrington, James, 144
politics of nostalgia, 146
zoon politikon, 145
Clean Air Acts of 1990 and 1995, 45
climate change, 5, 328
Clinton, President Bill, 48
command-and-control. *See* direct regulation
Commoner, Barry, 107
comparative risk assessment, 49
Comprehensive Environmental Response, Compensation, and Liability Act (CERCLA), 40
cost-benefit analysis, 45
crackpot realism, 63, 117, 202, 316
Croly, Herbert, 297
 attacks Jefferson, Thomas, 298
 and civic humanism, 298
 as follower of Hamilton, Alexander, 298
 The Promise of American Life, 297
 and teachings of Comte, Auguste, 298

D
Dahl, Robert, 16
Daly, Herman, 64, 317
deep ecology, 88
 Earth First!, 95
 ecocentrism, 89
de Mirabeau, Marquis, 166
democratic pragmatism, 41
demographic transition, 31
de Tocqueville, Alexis, 17, 242
Diamond, Jared, 335
direct regulation, 42
Donnelly, Ignatius, 309
Dryzek, John, 29, 32, 41, 130, 341

E
ecological economics, 64–70, 342
 carrying capacity, 66
 free trade, 67
 no-growth economy, 66
 stationary-state economy, 68
ecological services, 5
economic growth, 9
economic theory, 164
 commercial society, 164
 corporatism, 165
 mercantilist economics, 164
 physiocracy, 166
Ehrlich, Paul, 54, 79
the Enlightenment, 89
environmental economics. *See* mainstream environmentalism
externalities. *See* market failure

F
Federalist Papers, The, 216
 no. 84, 163
 no. 83, 224
 no. 11, 216
 no. 45, 260
 no. 49, 270
 no. 55, 261
 no. 51, 260
 no. 70, 223
 no. 62, 265
 no. 10, 260
 no.37, 256
 no. 36, 164
 no. 23, 163
Federalist Society, 35
Foreman, Dave, 95
Franklin, Benjamin, 158
Freeman, A. Myrick, 46
Friedman, Thomas, 48
Fukuyama, Francis, 74, 338
 The End of History and the Last Man, 74

G
Gilded Age conservatism, 288
Goodwyn, Lawrence, 312
Gore, Al, 16

H

Hamilton, Alexander, 19, 215, 242, 245, 254, 346
- and broad construction of governmental authority, 230
- at Constitutional Convention, 222
- his contempt for the American people, 219
- his economic plan, 231
- his environmental legacy, 241–245; affluent society, 242; conspicuous consumption, 242; suicide, 245
- on human vanity, 240
- on "illusion" and "delusion", 230
- his intellectual links with David Hume, 220
- and interest-group liberalism, 221
- letter to Morris, Robert, 230
- on political theory, 218
- rejection of laissez-faire, 237
- Report on a National Bank, 235
- Report on the Subject of Manufactures, 235
- Report to Congress on Public Credit, 234

Hardin, Garrett, 80, 107
- Living Within Limits, 107
- tragedy of the commons, 111

Hartz, Louis, 334
Hayek, Friedrich, 35, 323
Hays, Samuel, 287
- Conservation and the Gospel of Efficiency, 304

Heinzerling, Lisa and Ackerman, Frank, 47
Heritage Foundation, 30
Hofstadter, Richard, 147
homo civitas, 40
homo economicus. *See* neo-classical economics
Hughes, J.D., 6
human exemptionalism, 114
Hume, David, 117
- "Of Commerce", 154, 239
- empiricist epistemology, 154
- History of England, 153
- Humean liberalism, 153, 158
- interest group liberal thought, 157
- "Of The Original Contract", 153
- "Of Public Credit", 232
- and theory of moral sentiments, 192

I

Inhofe, Senator James, 30
$I = P \times A \times T$, 79

J

Jackson, Andrew, 290
Jefferson, Thomas, 19, 159, 177, 218, 254, 346
- to Carr, Peter, 187
- as classical republican, 178
- on climate change, 179
- to his daughter Martha, 190
- Declaration of Independence, 200
- defining "republican government", 193
- "Dialogue between my Head and my Heart", 189
- "the Earth belongs to the living", 195
- and moral equality of blacks, 204
- moral sense, 180
- on native peoples, 180
- on natural rights, 160
- Notes on the State of Virginia, 179
- as physiocrat, 184
- and the primacy of ethics, 187–197
- on slavery, 200
- A Summary View of the Rights of British America, 159
- on tobacco growing, 185
- ward republics, 194

Jensen, Derrick, 95, 339
justice, varying concepts of, 330
- ecological justice, 334
- legal justice and economic justice, 331

K

Kahn, Herman, 31
Kemmis, Daniel, 81
   and Jefferson's legacy, 198
Kraft, Michael, 123
Kropotkin, Peter, 73
Kunstler, James Howard, 8

L

Laurens, John, 219
Lease, Mary Elizabeth, 309
Leibig's Law, 76
Leopold, Aldo, 84, 333, 344
   ecocentric ethics, 88
   moral sentiments, 84
   A Sand County Almanac, 84
libertarian free marketeers. *See also* mainstream environmentalism
Lienesch, Michael, 142
Lincoln, Abraham, 296
   legacy in Grand Army of the Republic, 297
   and origins of Progressive movement, 296
Lineberry, Robert, 128
Lloyd, Henry Demarest, 315
Locke, John, 32, 301
   rights-based liberalism, 142
   Second Treatise of Government, 149
   social contract, 142
Lockean liberalism, 152
Lomborg, Bjorn, 30

M

Machiavelli, Niccolo, 144
Macune, Charles, 313
Madison, James, 19, 161, 253, 329, 346
   address to the Agricultural Society of Albemarle, 274
   and economic opportunity, 272
   definition of "a republic", 193
   as Humean liberal, 262
   and individualism, 271
   on individual rights, 161
   Lewis, Meriwether and Clark, William, 275
   as Lockean liberal, 260
   and Malthus' Essay on Principle of Population, 273
   "Notes on Nullification", 270
   "Property", 260
   "Property" essay, 161
   and sustainability, 328
   toleration and compromise in liberal culture, 269
   "Vices of the Political System of the United States", 259
mainstream environmentalism, 29, 133
mainstream sustainable development, 48–53
Malthus, Thomas Robert, 76, 199
managed growth, 52
market failure, 44
McDonald, Forrest, 164, 231
McGee, W.J, 305
McKibben, Bill, 6
Merchant, Carolyn, 114
Mill, John Stuart, 66
Mills, C. Wright, 61
   The Causes of World War Three, 61
   crackpot realism, 63

N

Naess, Arne, 88
National Environmental Policy Act of 1969 (NEPA), 39–40
neo-classical economics. *See also* mainstream environmentalism
neo-Malthusians, 344
   the demographic transition, 78

immigration policy, 80
limiting factor, 76
positive checks and preventive
 checks, 76

O
Obama, President Barack, 132
Occupy Wall Street, 348
Ophuls, William and Boyan,
 Stephen, 127
Orr, David, 4, 87, 323

P
Paehlke, Robert, 41
paradigm shift, 20
Jevons Paradox, 33
Paul Ehrlich, 75
Paul, Congressman Ron, 38
Paul, Senator Rand, 39
peak oil, 5
 King Hubbert, M., 5
 natural gas, 13
 unconventional energy
  resources, 5
Pinchot, Gifford, 41, 112
 Breaking New Ground, 305
 epiphany in Rock Creek Park, 302
 and Resource Conservation, 302
Pocock, J.G.A., 142
 the "Machiavellian Moment", 259
Polanyi, Karl, 323
 fictitious commodities, 325
 The Great Transformation, 323
Pollution Prevention
  Act of 1990, 48
 collaborative planning, 49
 source reduction, 49
precautionary principle, 109
private property rights, 37
Prometheans. *See also* anti-
 environmental cornucopians

Q
Quesnay, Francois, 166

R
radical environmentalism, 64
 bioregionalism, 81
 ecological economics, 64
 deep ecology, 88
 neo-Malthusians, 75
 social ecology, 70
Reagan, Ronald, 37
reform environmentalism. *See*
  mainstream environmentalism
Resource Conservation and Recovery
  Act (RCRA), 40
Rodman, John, 92, 306
 liberation of Nature, 94
Rolston, Holmes, 82
Roosevelt, Theodore, 41, 300
 1908 Conference of Governors, 305
 stewardship theory, 305
Roszak, Theodore, 89
 green romanticism, 88
 Where The Wasteland Ends, 89
Rousseauian meaning of "Nature",
 295

S
Sagoff, Mark, 40
Sale, Kirkpatrick, 83
Schattschneider, E.E., 115
 "Laws of Ecology", 107
Shepard, Paul, 90
 Nature and Madness, 91
 The Tender Carnivore and the
  Sacred Game, 91
Simon, Julian, 30
Simpson, " Sockless" Jerry, 309
smart growth. *See* managed growth
Smith, Adam, 111
 invisible hand, 111

Theory of Moral Sentiments, 237
  on vanity as human motivation, 238
Smith, Zachary, 123
Social Darwinism, 73
  left-wing Social Darwinism, 344
  Spencer, Herbert and Huxley, T.H., 73
Sumner, William Graham, 288
  on inequality and justice, 292
  and the invisible hand, 293
  on private property, 291
  and Social Darwinism, 293

T
Tea Party, 350
technological innovation, 33
theories of power, 116
  bureaucracy, 116
  elitism, 116
  pluralism, 116
Tietenberg, Tom, 44
Toxic Substance Control Act (TSCA), 40

U
United States Constitution, 3, 118–120, 122, 123, 126, 128, 349
  Constitutional Convention, 141
  functions, 14
  policy-making process, 118; divided government, 119; fragmentation, 119; incrementalism, 126; iron triangles, 128; oversight function, 120; short-term bias, 123; unfunded mandates, 122

V
Veblen, Thorstein, 242
Vitek, William, 197

W
Walpole, Robert, 166
Ward, Lester Frank, 306
Warren, Senator Elizabeth, 10
Watson, Tom, 309
Weber, Max, 117
Wilson, E.O., 244
Witherspoon, John, 158
Wolin, Sheldon, 264

Z
zoon politikon. *See* homo civitas

The manufacturer's authorised representative in the EU is Springer Nature Customer Service Centre GmbH, Europaplatz 3, 69115 Heidelberg, Germany. If you have any concerns regarding our products, please contact ProductSafety@springernature.com

Printed and bound by CPI Group (UK) Ltd, Croydon, CR0 4YY
23/03/2026
02076734-0006